With the classification of finite groups, an era of research in the subject ended. Some of the key figures in the classification program organized a research year at Rutgers University to analyze future directions of research in finite group theory.

This volume is a record of the research year and will be vital reading for all group theorists.

Proceedings of the Rutgers Group Theory Year, 1983–1984

Proceedings of the Rutgers Group Theory Year, 1983–1984

Edited by

Michael Aschbacher
California Institute of Technology
Pasadena, California

Daniel Gorenstein
Rutgers University
New Brunswick, New Jersey

Richard Lyons
Rutgers University
New Brunswick, New Jersey

Michael O'Nan
Rutgers University
New Brunswick, New Jersey

Charles Sims
Rutgers University
New Brunswick, New Jersey

Walter Feit
Yale University
New Haven, Connecticut

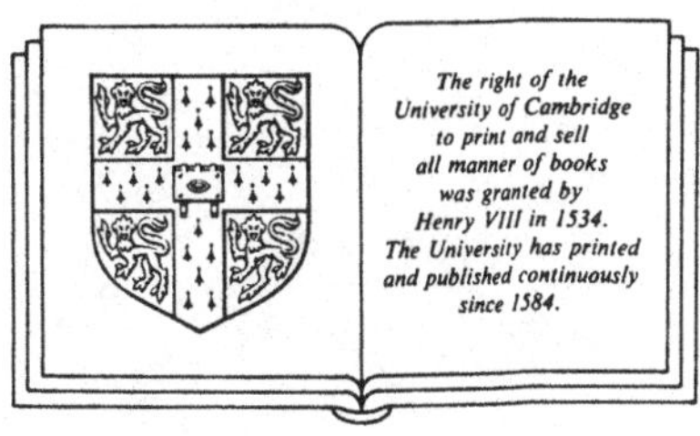
The right of the
University of Cambridge
to print and sell
all manner of books
was granted by
Henry VIII in 1534.
The University has printed
and published continuously
since 1584.

CAMBRIDGE UNIVERSITY PRESS

Cambridge
London • New York • New Rochelle
Melbourne • Sydney

CAMBRIDGE UNIVERSITY PRESS
Cambridge, New York, Melbourne, Madrid, Cape Town, Singapore, São Paulo, Delhi

Cambridge University Press
The Edinburgh Building, Cambridge CB2 8RU, UK

Published in the United States of America by Cambridge University Press, New York

www.cambridge.org
Information on this title: www.cambridge.org/9780521264938

First published 1984
This digitally printed version 2008

A catalogue record for this publication is available from the British Library

Library of Congress Cataloguing in Publication data

Main entry under title:

Proceedings of the Rutgers group theory year, 1983-1984.

1. Finite groups--Addresses, essays, lectures.
I. Aschbacher, Michael, 1944- . II. Title: Rutgers
group theory year, 1983-1984.
QA171.P917 1985 512'.55 84-7834

ISBN 978-0-521-26493-8 hardback
ISBN 978-0-521-09091-9 paperback

TABLE OF CONTENTS

OTHER DIRECTIONS

PREFACE

These Proceedings are an outgrowth of the Rutgers University Group Theory Year, (January, 1983 - June, 1984), supported in part by the National Science Foundation. The focus during the year was primarily on the following topics:

1) Revision of the classification of the finite simple groups.

2) Development of the properties of the known simple groups, including properties needed for the classification, the structure of maximal subgroups, and representations.

3) Applications of the classification to finite group theory, number theory, and geometry.

4) Chamber systems and amalgams.

5) Computational algorithms for groups.

A variety of additional topics were touched upon. Our sorting of the contributed papers by topic is admittedly somewhat crude, but at any rate it reflects the main themes of the year.

The participants were asked to include in their articles, wherever possible, open questions and suggested problem areas as a means of stimulating further research in finite group theory, and most of the papers of the Proceedings reflect this point of view.

A word about the "known" simple groups. Now that the simple groups have been classified, all finite groups are "known", so the term is clearly superfluous. However, as the proofs of many results in group theory are completely independent of the classification theorem, while others require the simple group G under investigation to be a "known" simple group, while still others require not only G, but each of its proper simple sections to be "known", continued use of the term and the related notion of "K-group" is justified as a way of conveying the nature of the argument to be undertaken.

It is a pleasure to acknowledge the efforts of a number of people who helped make the Group Theory Year successful. The Rutgers Mathematics Department staff - particularly Lynn Braun, Maryanne Jablonski, Dorothy Kozu, Judith Lige, and Arlene Tarbart - have contributed in many ways. The splendid typing job was done by Annette Roselli.

Finally, the editors extend their special gratitude to Professor Ronald Solomon of Ohio State University, who edited a number of the articles. His name rightfully belongs on the title page with the rest of ours; it is not there only because of our oversight.

New Brunswick, N.J.
July, 1984

A COMPUTATIONAL TOOLKIT FOR FINITE PERMUTATION GROUPS

J. Cannon*
Rutgers University and University of Sydney

1 INTRODUCTION

Over the past two decades a range of powerful algorithms has been designed and implemented for computing with permutation groups. For example, if G is a permutation group acting faithfully on the finite set Ω, and X is a set of generators for G, then it is possible to compute the order of G even in situations where the cardinality of Ω is 10,000 or more.

In this paper we outline most of the algorithms that are currently available giving, where possible, an indication as to their efficiency. Implemented in a group theory package such as Cayley (Cannon [15]) these techniques have found wide application to problems arising from both inside and outside group theory.

The availability of these techniques together with the classification of finite simple groups (Gorenstein [23]) mean that it is now possible to devise computer programs which, given as input a set of generators for a permutation group G, will output a description of the structure of G. The exact form of such a description will depend upon G and can range from the names of the composition factors of G to the name of G itself (in cases where G is a member of some class of groups that has an explicit name). Algorithms for performing such an analysis have been developed for 2-transitive groups by Cameron, Cannon and Neumann [11] and for arbitrary permutation groups by Cannon and Neumann [20]. Since these algorithms utilize most of the known permutation group algorithms, this paper may be regarded as setting the stage for the work described in those papers.

2 TRANSITIVITY, REGULARITY AND PRIMITIVITY

Permutation group algorithms divide fairly naturally into

*This research was supported by the Australian Research Grants Scheme and the National Science Foundation.

those that require a representation of the set of elements of the group and those that do not. In this section the latter class of algorithms will be discussed. Throughout the paper we let G denote a permutation group which acts faithfully on the finite set Ω of cardinality n. The symmetric group on Ω will be denoted by S_Ω. We suppose that G is generated by the set $X = \{g_1,\dots,g_m\}$.

The following simple algorithm can be used to compute the orbit of an element of Ω.

Algorithm ORBIT: Given a group G that acts faithfully on a set Ω, and a generating set $X = \{g_1,\dots,g_m\}$ for G, construct the orbit Δ of $\alpha \in \Omega$.

i) [Initialize] $\Delta \leftarrow \{\alpha\}$; $Q \leftarrow \{\alpha\}$.

ii) [Take the next unprocessed point from the queue Q]. If Q is the null set, stop. Otherwise, let β be some element of Q; $Q \leftarrow Q - \{\beta\}$, $i \leftarrow \emptyset$.

iii) [Form the image of β under the next generator]. $i \leftarrow i + 1$; If $i > m$, go to (ii); $\gamma \leftarrow \beta g_i$. If $\gamma \notin \Delta$, $\Delta \leftarrow \Delta \cup \{\gamma\}$, $Q \leftarrow Q \cup \{\gamma\}$. Go to (iii).

Algorithm ORBIT can easily be modified so as to construct all the orbits of G on Ω. Further, if $\Delta = \{\alpha_1,\dots,\alpha_r\}$, ORBIT can be modified so as to save a set of elements $\{x_1,\dots,x_r\}$ from G such that $\alpha_1 x_i = \alpha_i$. Provided that the algorithm is carefully implemented, it may be applied to groups having degrees well in excess of 100,000. The time complexity of the algorithm is $O(mn)$.

Although it is a straightforward task to generalize the algorithm so as to obtain orbits of G on Ω^k, $k \geq 2$, this is rarely practical because of the cardinality of Ω^k. If G is transitive and generators for G_α, $\alpha \in \Omega$, are known, the orbits of G on $\Omega\times\Omega$ may be obtained from the orbits of G_α on Ω by means of the following correspondence: If Γ is an orbit of G on $\Omega\times\Omega$, the correspondence $\Gamma \leftrightarrow \Gamma(\alpha) = \{\beta \mid (\alpha,\beta) \in \Gamma\}$ is a bijection between the orbits of G on $\Omega\times\Omega$ and the orbits of G_α on Ω.

Consider the problem of determining whether G acts regularly on Ω. A well-known theorem says that G is semiregular if and only if the centralizer of G in S_Ω is transitive (see Wielandt [37], pg.9). The following lemma is a consequence:

Lemma 2.1. Let $G = \langle X \rangle$ be a transitive group acting on the set Ω and suppose $\alpha \in \Omega$. If, for each x in X there is an element z in S_Ω such that

a) z centralizes X, and

b) $\alpha x = \alpha z$,

then G is regular.

The following algorithm of Sims is based on this lemma:

Algorithm REGULAR: Given a group G acting faithfully on the set $\Omega = \{\alpha_1,\ldots,\alpha_n\}$, and a generating set $X = \{g_1,\ldots,g_m\}$ for G, test if G acts regularly on Ω.

i) [Choose a point α] Let α be some point in Ω; $i \leftarrow 0$.

ii) [Take the next generator] $i \leftarrow i + 1$; If $i > m$, then G is regular, so stop. Otherwise, $\gamma \leftarrow \alpha g_i$; $j \leftarrow 0$.

iii) [Construct a mapping $z:\Omega\to\Omega$] $j \leftarrow j + 1$. If $j > n$ go to (iv). Otherwise, $\beta \leftarrow \alpha_j$; Let w be an element of G mapping α to β and define the image of β under z to be γw; Go to (iii).

iv) [Test whether z is a bijection] If z is a permutation of Ω which commutes with each element of X, go to (ii). Otherwise, G does not act regularly, so stop.

The time complexity of REGULAR is $O(m^2n)$. It is not difficult to devise a generalization of this algorithm which will test an intransitive group for being semiregular. Algorithm REGULAR is applicable to groups having degree up to 100,000 provided that the generating set X is small.

We now consider the problem of determining whether a transitive group G acts primitively on Ω. This corresponds to showing that there are no proper G-invariant partitions of Ω. Our approach is to determine the finest G-invariant partition such that the pair $\{\alpha,\beta\}$, $\alpha,\beta \in \Omega$, is contained in the same subset (block). The group is primitive if for a fixed α and for each choice of β from the set $\Omega - \{\alpha\}$, only the trivial partition is obtained.

Algorithm PRIMITIVE: Given a transitive group G acting faithfully on the set $\Omega = \{\alpha_1,\ldots,\alpha_n\}$, and a generating set $X = \{g_1,\ldots,g_m\}$ for G, test whether G is primitive.

i) [Choose α] $\alpha \leftarrow \alpha_1$; $i \leftarrow 1$.

ii) [Choose β] $i \leftarrow i + 1$; If $i > n$, then G is primitive, so stop. Otherwise, $\beta \leftarrow \alpha_i$, and create the partition $\alpha,\beta|\gamma_1|\ldots|\gamma_{n-2}$, where $\gamma_1,\ldots,\gamma_{n-2}$ are the distinct elements of $\Omega - \{\alpha,\beta\}$.

iii) [Initialize loop over the generators] $j \leftarrow 0$.

iv) [Apply a generator to the current partition] $j \leftarrow j + 1$; If $j > m$ go to (v). $g \leftarrow g_j$. Suppose the current partition is $\Delta_1|\Delta_2|\ldots|\Delta_t$; form the partition $\Delta_1 g|\Delta_2 g|\ldots|\Delta_t g$; If for each k we have $\Delta_k g \subseteq \Delta_\ell$ for some ℓ, go to (iv). Otherwise, form a new partition by merging Δ_p and Δ_q, $p \neq q$, where both Δ_p and Δ_q intersect some $\Delta_k g$ nontrivially; go to (iii).

v) [Proper partition?] If the partition consists of two or more sets, then it is a system of imprimitivity for G, so stop. Otherwise, go to (ii).

A particularly efficient version of this algorithm is described in Atkinson [1]. Various means of speeding up the algorithm are described by Atkinson. Suppose a subgroup $H \leq G_\alpha$ can be found. If $\Sigma = \{\beta_1,\ldots,\beta_p\}$ is a set of representatives for the orbits of H, then it suffices to let β run through the elements of $\Sigma-\{\alpha\}$ rather than the elements of $\Omega-\{\alpha\}$. A large subgroup of G_α can be computed cheaply using the random Schreier algorithm (section 3).

The time complexity of the primitivity test is $O(mn^2)$. Provided that all the speed-up tricks are used, algorithm PRIMITIVE may be applied to groups having degree up to 100,000.

It is important to be able to rapidly determine if G contains the alternating group A_Ω. The most useful result for this purpose is the following theorem of Jordan:

<u>Theorem 2.2</u>. Let G be a primitive group of degree n. If G contains a p-cycle, where $p \leq n-3$ is a prime, then G is the alternating or symmetric group of degree n.

This result can be used as the basis of a rather efficient test for A_n or S_n. Firstly, Algorithm PRIMITIVE is used to test G for primitivity. Assuming that it is, a number of random elements are examined in the hope of finding one that satisfies the conditions of the above theorem. Rather than looking directly for an element that is a p-cycle, it is much more effective to look for an element x containing a p-cycle whose length is coprime with the lengths of all the other cycles of x. If, after examining a predetermined number of random elements, no p-cycle of the appropriate length has been found, some version of the Schreier algorithm (section 3) is invoked to determine the order of G. It is possible to design an algorithm along these lines which can recognize that a group is the alternating or symmetric group for degrees up to 100,000 (see Cannon [16]).

3 BASES AND STRONG GENERATING SETS

In this section we introduce some ideas which allow us to represent the set of elements of a finite permutation group G in a particularly compact manner. This representation enables us to test membership in G and to run through the elements of G without repetition. Further, the order of G can be obtained immediately from this representation.

A sequence $B = \{\beta_1,\ldots,\beta_k\}$ of distinct points from Ω is called a base for G if the only element of G that fixes B pointwise is the identity. If $B = \{\beta_1,\ldots,\beta_k\}$ is a base for G, it is convenient to define the following symbols:

$$G^{(i)} := G_{\beta_1,\ldots,\beta_{i-1}}, \text{ so that } G^{(1)} = G \text{ and } G^{(k+1)} = 1;$$

$$\Delta_i := \text{orbit of } G^{(i)} \text{ containing } \beta_i \text{ (the } i^{th} \text{ basic orbit);}$$

$$U_i := \text{a right transversal for } G^{(i+1)} \text{ in } G^{(i)}.$$

A subset S of G is said to be a strong generating set for G relative to B if $G^{(i)} = \langle S \cap G^{(i)} \rangle$ $(i = 1,\ldots,k)$. Thus, S contains generators for each subgroup in the chain of stabilizers.

$$G = G^{(1)} > G^{(2)} > \ldots G^{(k)} > G^{(k+1)} = 1.$$

Given a base B and strong generating set S for G, the pair of sets (Δ_i,U_i) $(i = 1,\ldots,k)$ can be constructed simultaneously using a variant of Algorithm ORBIT. The notions of base and strong generating set were introduced by Sims in 1970 (see [30] and [31]).

The cardinality of the set B is called the length of the base. A regular group must have a base of length one, while the symmetric group of degree n must have a base of length n-1. However, the length of a base is not a group invariant. One reason for the importance of the notion of a permutation group base is the observation that important classes of groups will more often than not have bases whose lengths are small compared to the degree. In particular, if the alternating groups are excluded, this is true for simple groups. For example, the length of a typical base for Held's group (order 4,030,387,200) in its degree 2058 permutation representation lies between 8 and 12. Babai has shown that a primitive but not 2-transitive group has a base of length at most $4n^{\frac{1}{2}}\log n$. It is an easy exercise to show that the number of strong generators is bounded by n^2.

Every element g of G has a unique expression as a product $u_k u_{k-1} \ldots u_1$ with $u_i \in U_i$ $(i = 1,\ldots,k)$. The following algorithm may be used to determine whether the permutation g is an element of G.

<u>Algorithm STRIP</u>: Given a permutation group G acting on a set Ω, and an element g of S_Ω, determine whether g is an element of G. It is assumed that the sets S, B, Δ_i $(i = 1,\ldots,k)$ and U_i $(i = 1,\ldots,k)$ are defined for G.

i) [Initialize] $i \leftarrow 0$; $h \leftarrow g$.

ii) [Find the next factor u_i in the product $u_k \ldots u_1$] $i \leftarrow i + 1$; If $\beta_i h \notin \Delta_i$, then g is not in G, so stop. Otherwise, $h \leftarrow hu_i^{-1}$, where u_i is the unique element of U_i such that $\beta_i h = \beta_i u_i$.

iii) [Reached end of base] If $i < k$, go to (ii). Otherwise, if $h = 1$ then g is in G, while if $h \neq 1$, then g is not in G.

This process is often referred to as <u>stripping</u> the element g. If on termination $h \neq 1$, then g is not an element of G, and h is called the <u>residue</u> of G relative to B and S.

In 1967, Sims [29] described a method for constructing a base and strong generating set for G which is based on the following lemma of Schreier (Hall [24], p. 96):

<u>Lemma 3.1</u>. Let G be a group with generating set X, and let H be a subgroup of G. If U is a right transversal for H in G such that U contains the identity element then H is generated by the set

$$\{ux\phi(ux)^{-1} \mid u \in U,\ x \in X\}$$

*

where $\phi(ux)$ is the unique element of U such that $Hux = H\phi(ux)$.

In theoretical terms, the method may be described as follows: For β_1 choose any point not fixed by every element of X. Use Algorithm ORBITS, to construct Δ_1 and U_1. Now apply Schreier's lemma to construct a set of generators for $G_{\beta_1} = G^{(2)}$. By induction we successively construct the base points $\beta_2,\ldots,\beta_k$ and generators for the stabilizers $G^{(3)},\ldots,G^{(k+1)}$. The algorithm terminates when we reach the trivial subgroup.

This method is impractical because of the large number of Schreier generators that will be constructed $(1 + [G:G_{\beta_1}](|X|-1)$ for $G_{\beta_1})$. In fact, however, a tiny fraction of these generators will suffice. Sims' idea was the following: As soon as a non-trivial generator is obtained for $G^{(2)}$, it is used to generate part of the orbit Δ_2 (and hence part of U_2). Sufficient information is now available to deduce a generator of $G^{(3)}$. If this generator is non-trivial, it is immediately used in the same way to deduce a generator for $G^{(4)}$ etc. Whenever a new Schreier generator t is formed, Algorithm STRIP is used to determine whether or not it lies in the subgroup of G generated by the currently known generators of $G^{(2)},\ldots,G^{(k+1)}$. If t has a non-trivial residue after the application of STRIP, then that residue is added to S. By this means, the number of generators actually saved for a subgroup $G^{(i)}$ $(i = 2,\ldots,k)$ is typically five or six.

This algorithm, labelled the Schreier algorithm by Sims, is sufficiently powerful for it to be applicable to groups having degrees up to 500. It has been implemented by a number of people and it is perhaps the most important existing algorithm for permutation groups. A number of variants of this algorithm will be briefly described. First, however we make the following observation: If g is an element of G and $B = \{\beta_1,\ldots,\beta_k\}$ is a base for G, then g is uniquely determined by the sequence, $\{\beta_1 g,\ldots,\beta_k g\}$. This sequence will be referred to as the _base image representation_ of G.

Variant 1: _Known base Schreier algorithm_. Suppose a base B is known for the group G and it is necessary to compute strong generators for G relative to B. Most formation of products of permutations x and y which arise in the Sims algorithm may be replaced by computation of the base images corresponding to these products: $\{(\beta_1 x)y,\ldots,(\beta_k x)y\}$.

If k is significantly smaller than n, this can result in a considerably faster algorithm. This variant is particularly important when it is necessary to construct bases and strong generating sets for a number of distinct subgroups of a fixed group G, for a base for G is also a base for any subgroup of G. See Butler and Cannon [8] for details.

Variant 2: Schreier Todd-Coxeter algorithm. The standard form of the Sims algorithm is rather inefficient when one considers the fact that only a tiny fraction of the Schreier generators constructed are actually needed: the vast majority are redundant and are simply discarded. The problem is recognizing when a sequence of points B and a set of elements S constitutes a base and strong generating set for the group G. In 1974, Sims [32] described an approach to this problem which is based on the use of the Todd-Coxeter algorithm (see Cannon et al [17]) for enumerating cosets in a finitely presented group. This form of the Schreier algorithm was subsequently refined by Leon [26,27]. Given a sequence of points B and a set of elements S belonging to the group G this algorithm may be regarded as a means of verifying whether B and S constitute a base and strong generating set, respectively, for G. Commencing with $G^{(k)}$, the algorithm inductively constructs a presentation for each stabilizer $G^{(i)}$ $(i = k,\ldots,i = 1)$. The Todd-Coxeter algorithm is used to define $|\Delta_i|+1$ cosets of $G^{(i+1)}$ in the group defined by those relations currently known to hold for $G^{(i)}$. Letting representatives for these cosets act on β_i and comparing the resulting sequence of points with the orbit Δ_i, we are led either to a new relation for $G^{(i)}$, or a new strong generator for $G^{(i)}$. The process terminates for $G^{(i)}$ when the Todd-Coxeter algorithm terminates with exactly $|\Delta_i|$ cosets defined.

Because of overheads, this algorithm is often slower than the standard Schreier algorithm for groups having degree less than 100. As the degree increases the Schreier Todd-Coxeter algorithm becomes more competitive. For degrees greater than 1,000 it is typically several times faster than the standard Schreier algorithm. In particular, it enables us to construct bases and strong generating sets for groups of degree 10,000.

Variant 3: The random Schreier algorithm. In many situations it is useful to be able to rapidly construct an "approximation" to a base and strong generating set for a group G. We say that the sets B and S approximate a base and strong generating set for G if either B and

S already constitute a base and strong generating set, or they do so upon the addition of a small number of elements. In his work on the Schreier Todd-Coxeter algorithm Leon devised the following algorithm:

Algorithm APPROX: Given a permutation group G, construct an approximation for a base and strong generating set for G. The algorithm is to terminate after m consecutive unsuccessful attempts to find a new strong generator.

i) [Initialize]. Create a base B and strong generating set for the trivial group acting on Ω, $i \leftarrow 0$.

ii) [Consider a new random element]. $i \leftarrow i + 1$. If $i > m$ then stop. Generate a random element g of G and apply Algorithm STRIP so as to obtain the residue h. If $h = 1$, $i \leftarrow i + 1$ and go to (ii). Otherwise, add h to S and, if necessary, add a new point to B; $i \leftarrow 0$; Go to (ii).

Random elements may be obtained by generating random words in the given generators X of G. It is desirable that the lengths of these random words vary with the lower bound on $|G|$ given by the current values of B and S. It has been found experimentally that if the stopping parameter m is set to 20, then almost always this algorithm will return a complete base and strong generating set for G.

There are a number of important applications of this algorithm:

a) It can be used to quickly obtain a large subgroup H of a one-point stabilizer G_α in a transitive group G. Knowledge of the orbits of H may greatly reduce the number of pairs $[\alpha,\beta]$ that must be considered by the primitivity algorithm (algorithm PRIMITIVE).

b) If the order of G is known in advance, the random Schreier algorithm gives a very fast method for computing a base and strong generating set for G.

c) The efficiency of the Schreier Todd-Coxeter algorithm may be enhanced by first using the random Schreier algorithm to obtain a close approximation to a base and strong generating set.

For many applications it is desirable to have a particular sequence of points from Ω appear as an initial segment of a base for G. Given a base $B = \{\beta_1,\ldots,\beta_k\}$ and a sequence of points $C = \{\gamma_1,\ldots,\gamma_\ell\}$, there exists an efficient algorithm due to Sims [31] which constructs a strong generating set for G relative to the base $\{\gamma_1,\ldots,\gamma_\ell,\delta_1,\ldots,\delta_m\}$, where $\delta_1,\ldots,\delta_m$ are points from Ω chosen to complete the sequence

$\{\gamma_1,\ldots,\gamma_\ell\}$ to a base. At the heart of the Sims algorithm is a procedure which modifies the current strong generating set so as to correspond to the base obtained by interchanging two adjacent points in the current base. The algorithm involves mainly the computation of orbits. In [6], Butler describes an improvement to the algorithm whereby he precedes the Sims algorithm with the conjugation of G by an element g in S_Ω which maps as large an initial segment of B as possible into the corresponding initial segment of G.

The existence of a fast algorithm for changing base means that it is possible to obtain the pointwise stabilizer $G_{\gamma_1,\ldots,\gamma_\ell}$ of any sequence of distinct points from Ω very cheaply.

4 CONSTRUCTING SUBGROUPS I: NORMAL CLOSURE

Let H be a subgroup of G. The normal closure $ncl_G(H)$ may be constructed using the following algorithm which forms the closure of H under conjugation by the generators of G:

Algorithm NCL: Given a subgroup $H = \langle y_1,\ldots,y_s\rangle$ of the permutation group $G = \langle x_1,\ldots,x_r\rangle$, determine the normal closure of H in G. At the outset we assume that a base and strong generating set are known for H.

i) [Initialize] $K \leftarrow H$; $i \leftarrow 0$; $t \leftarrow s$.

ii) [Choose the next generator of K] $i \leftarrow i + 1$; If $i > t$, then $ncl_G(H) \leftarrow K$, and stop. Otherwise, $j \leftarrow 0$.

iii) [Choose the next generator of G] $j \leftarrow j + 1$. If $j > r$, go to (ii). Otherwise, $g \leftarrow y_i^{x_j}$. If $g \in K$, go to (iii).

iv) [Extend K] $t \leftarrow t + 1$; $y_t \leftarrow g$; $K \leftarrow \langle K,g\rangle$. Use the Schreier algorithm to construct a base and strong generating set for K. Go to (iii).

Algorithm STRIP is used to perform the element membership test in step (iii). The most time consuming part of this algorithm is the construction of new strong generating set each time K is extended. Usually a base will be known for K (since a base for G is also a base for K), so that for groups of moderate degree Variant 1 of the Schreier algorithm is appropriate. For groups of large degree it may be necessary to employ the Schreier Todd-Coxeter algorithm. In cases where the order of $ncl_G(H)$ is known in advance the random Schreier algorithm is the method of choice.

If $H = \langle y_1,\ldots,y_s\rangle$ and $K = \langle z_1,\ldots,z_t\rangle$ are normal subgroups of G, the commutator subgroup $[H,K]$ can be constructed as the normal closure of the subgroup

$$<[y_i,z_j] \mid 1 \leq i \leq s,\ 1 \leq j \leq t>.$$

This observation provides us with an effective means of computing the derived series and lower central series. For details of these and related algorithms see Butler and Cannon [8] and Cannon [12].

As a final application we note that the ability to construct the normal closure of a subgroup is at the heart of an algorithm which computes the normal subgroup lattice for a group of moderate order (see Cannon and Neubüser [18]).

5 CONSTRUCTING SUBGROUPS II: BACKTRACK ALGORITHMS

Consider the problem of constructing a subgroup K of the group G, where K is defined in terms of some property P: For example, P might be the property of centralizing some subgroup H. If $B = \{\beta_1,\ldots,\beta_k\}$ is a base for G, then each element x of K is uniquely represented by the sequence $\{\beta_1 x,\ldots,\beta_k x\}$. One possible method of constructing K is to use a backtrack search on the set of base images of B for those that correspond to elements having property P.

A sequence $C = \{\gamma_1,\ldots,\gamma_\ell\}$, $0 \leq \ell \leq k$, of distinct points from Ω is called a partial (base) image. If $g \in G$, define C_g to be the sequence $\{\gamma_1 g,\ldots,\gamma_\ell g\}$. Let $E_G(C) = \{g \in G \mid \{\beta_1,\ldots,\beta_\ell\}g = C\}$. A backtrack search through the images of B is based on knowing some necessary condition Q for P such that if the elements of the set $E_G(C)$ have property Q, then so do the elements of the set $E_G(D)$, for any initial segment D of C.

Let L be some known subgroup of K. Then it is only necessary to examine one element from each double coset of L in G. Suppose that X is totally ordered so that $\beta_1,\ldots,\beta_k$ are the first k elements. This induces an ordering both on the elements of G and on the set of partial images of B. Since we lack an efficient means of recognizing whether or not an element x is first in the double coset LxL, we might instead insist that x be first both in the right coset Lx and in the left coset xL.

Lemma 5.1. Let Ω_i $(i = 1,\ldots,k)$ denote the ith basic orbit of L.

i) An element x is first in its right coset Lx if and only if, for each i, $1 \le i \le k$, $\beta_i x$ is first in the set $\Omega_i x$.

ii) An element x is first in its left coset xL if and only if, for each i, $1 \le i \le k$, $\gamma_i = \beta_i x$ is first in the $L_{\gamma_1,\ldots,\gamma_{i-1}}$-orbit of γ_i.

Part (i) appears in Sims [30], while part (ii) appears in Butler [3]. In practice, the test for being first in a left coset is seldom used since its application usually involves a change of base.

The backtrack method is perhaps best illustrated by its application to the construction of the intersection K of a pair of subgroups G and H of the symmetric group S_Ω. We assume that G and H both have strong generating sets relative to the base $\beta_1,\ldots,\beta_k$. Let Δ_i and Γ_i $(i = 1,\ldots,k)$ denote the ith basic orbits for G and H, respectively. Let $C = \{\gamma_1,\ldots,\gamma_\ell\}$. A necessary condition for the set $E_{G\cap H}(C)$ to be non-empty is that each of the sets $E_G(C)$ and $E_H(C)$ be non-empty. Let $\gamma \in \Omega$ be such that $\gamma \neq \gamma_i$ $(i = 1,\ldots,\ell)$ and write C' for the sequence $\{\gamma_1,\ldots,\gamma_\ell,\gamma\}$. Then $E_{G\cap H}(C')$ can be nonempty only if $\gamma \in \Delta^g_{\ell+1} \cap \Gamma^h_{\ell+1}$, where $g \in E_G(C)$ and $h \in E_H(C)$. For further details see Sims [31] or Butler [3].

To summarize, an efficient backtrack search of a permutation group G depends upon being able to choose the base of G in such a way that very strong restrictions can be placed on the choice of possible images for the ith base point β_i. The search is further enhanced by only looking at those base images which correspond to elements x which are first in their right coset Lx, where L is some subgroup of the subgroup K that is being constructed. Backtrack algorithms are organized so as to produce a base and strong generating set for K.

Backtrack searches have been designed and implemented in the following situations:

i) Centralizer of an element (Sims [30], Butler [2,3]);

ii) Centralizer of a subgroup (Butler [2,3]);

iii) Testing conjugacy of elements (Butler [2,3]);

iv) Intersection of subgroups (Sims [31], Butler [2,3]);

v) Normalizer (Butler [4]);

vi) Testing conjugacy of subgroups (Butler [4]);

vii) Set stabilizer (Butler [2,3]);

viii) Sylow subgroup (see section 7);

ix) Canonical representative of a double coset (Butler [5]);

x) Automorphism group (Robertz [28]).

The backtrack algorithms for computing centralizers and testing elements for conjugacy are quite fast. For example, the construction of the centralizer of an involution in ${}^2F_4(2)$ (degree 1755, order $2^{10}.3^3.5^2.13$) takes about 100 seconds of central processor time on a CYBER 72. The average running time for the intersection algorithm is also good. The algorithms for computing normalizers and testing subgroups for conjugacy, as described in Butler [4], perform satisfactorily whenever G is not the full symmetric group. Even though Butler developed a special algorithm for computing normalizers in the full symmetric group, its application is restricted to groups having degree less than 20. The performance of the set stabilizer algorithm is heavily dependent upon the cardinality of the set and, in practice, its application is limited to sets having cardinality less than 6 or 7.

6 HOMOMORPHISMS

Suppose G has a natural action on a set Δ. The following choices of Δ give rise to three homomorphisms that are central to the study of permutation groups:

i) Δ is a G-invariant subset of Ω (constituent homomorphism);

ii) Δ is a G-invariant partition of Ω (blocks homomorphism);

iii) Δ is the coset space of some subgroup H of G (coset homomorphism).

For each of these homomorphisms ϕ it is desirable to have efficient algorithms for computing the image of ϕ (im ϕ), the kernel of ϕ (ker ϕ), the image of an element or subgroup, and the preimage of an element or subset. More precisely, we wish to be able to obtain a base and strong generating set for im ϕ and ker ϕ directly from that of G.

Suppose that $\Delta = \{\alpha_1,\dots,\alpha_r\}$ is a G-invariant subset of Ω and that ϕ is the homomorphism of G into S_Δ obtained by restricting the action of elements of G to Δ. Using the base change algorithm if necessary, we may assume that G has base $\{\alpha_1,\dots,\alpha_r,\beta_1,\dots,\beta_s\}$. If S denotes the union of the generators for the stabilizers $G^{(1)},\dots,G^{(r)}$, then

$\phi(S)$ is a strong generating set of $\text{im}\ \phi$ relative to the base $\{\alpha_1,\ldots,\alpha_r\}$. The union of the generating sets for the stabilizers $G^{(r+1)},\ldots,G^{(r+s)}$ form a strong generating for $\ker \phi$ relative to the base $\{\beta_1,\ldots,\beta_s\}$. Finally, we note that if the element $y \in \text{im}\ \phi$ has base image $\{\gamma_1,\ldots,\gamma_r\}$ (relative to the base $\{\alpha_1,\ldots,\alpha_r\}$), then any element of $E_G(\{\gamma_1,\ldots,\gamma_r\})$ will lie in the preimage of y.

Suppose that $\Delta = \{\Delta_1,\ldots,\Delta_t\}$ is a G-invariant partition of Ω and that ϕ is the homomorphism corresponding to the induced action of G on Δ. The elements of Δ are often referred to as <u>blocks</u>. Suppose $\{\Delta_{i_1},\ldots,\Delta_{i_r}\}$ is a base for $\text{im}\ \phi$. This corresponds to the following chain of stabilizers in G:

$$G_{\Delta_{i_1}} \geq G_{\Delta_{i_1},\Delta_{i_2}} \geq \cdots \geq G_{\Delta_{i_1},\ldots\Delta_{i_r}} \tag{1}$$

If S denotes the union of the generating sets for these subgroups, then $\phi(S)$ is a strong generating set of $\text{im}\ \phi$ relative to the base $\{\Delta_{i_1},\ldots,\Delta_{i_r}\}$. The kernel of ϕ is given by

$$\ker \phi = \bigcap_{i=1}^{t} G_{\Delta_i} = G_{\Delta_{i_1},\ldots,\Delta_{i_r}}$$

Our problem therefore reduces to the problem of computing the stabilizer of a block. This may be found using the following observation of Butler [6]:

<u>Lemma 6.1</u>. Let Γ be a block for G in its action on Ω and suppose $\gamma \in \Gamma$. If A is a subset of G such that

a) A contains a generating set of G_γ, and

b) $\gamma^{\langle A\rangle} = \gamma^G$,

then $G_\Gamma = \langle A\rangle$.

This result forms the basis of an algorithm for constructing block stabilizers which is much faster than the backtrack algorithm for stabilizers of arbitrary sets.

A detailed account of techniques for computing with the three standard permutation group homomorphisms may be found in Butler [6]. It is worth noting that computations with the constituent and blocks homomorphisms are very efficient and are readily applicable to groups of degree up to 10,000.

7 SYLOW p-SUBGROUPS

Algorithms designed to analyze the structure of a permutation group G sometimes need to be able to find a Sylow p-subgroup of G. There are two methods currently in use: a backtrack algorithm and a recursive algorithm. We outline the backtrack algorithm:

Algorithm SYLOW: Given a group G and a prime p such that $p \| G|$, construct a Sylow p-subgroup P of G.

i) [Find a p-element] By examining a selection of random elements and their powers, find an element x whose order is a power of p; $Q \leftarrow \langle x \rangle$. If $|Q| = |P|$, $P \leftarrow Q$ and stop.

ii) [Look for a subgroup of G whose Sylow p-subgroup is a proper extension of Q]. By examinating the elements of $Z(Q)$ that have order p, locate one, y say, such that the Sylow p-subgroup of $C_G(y)$ has order greater than the order of Q; $C \leftarrow C_G(y)$.

iii) [Construct the Sylow p-subgroup of C]. Using a backtrack search over the elements of C, extend Q to the Sylow p-subgroup of C. The search is used to find an element z such that

(a) $z \notin Q$;

(b) $z \in N(Q)$; and

(c) $z^{p^a} \in Q$, some positive integer a.

Having found such a z, Q is replaced by the subgroup $\langle Q,z \rangle$. The search is repeated, if necessary, until Q is a Sylow p-subgroup of C.

iv) [Finished?] If $|Q| = |P|$, $P \leftarrow Q$ and stop. Otherwise, go to (ii).

The complete algorithm is very complicated (see Butler and Cannon [9] for details). As might be expected the performance of the algorithm depends upon the size of the exponent of p. For exponents greater than 15 or 16, the algorithm can be very slow. Recently, Butler and the author have experimented with a new recursive algorithm which is much less sensitive to the size of the exponent of p and which shows considerably promise. (See Butler and Cannon [10]).

Given a Sylow p-subgroup P of G, one can use the intersection algorithm (section 5) to intersect conjugates of P and thereby obtain $O_p(G)$. By constructing $O_p(G)$ for each prime p dividing $|G|$, it is a trivial matter to construct the Fitting subgroup of G.

8 OTHER ALGORITHMS

In this section we make brief mention of a number of other permutation group algorithms.

i) In [30], Sims describes an approach to the problem of computing representatives for the conjugacy classes of a permutation group. In unpublished work, Butler and Cannon have devised an effective method for computing the conjugacy classes of a near-simple group. However, no satisfactory method currently exists for computing the classes of an arbitrary permutation group.

ii) Cannon and Neumann [19] have designed and implemented a very fast algorithm for computing an elementary abelian regular normal subgroup of a primitive permutation group.

iii) Sims [34,36] has devised special techniques for constructing certain simple groups as permutation groups of very large degree. For example Leon and Sims used these techniques to construct the Baby Monster as a permutation group having degree about 14×10^9.

iv) Algorithms which name the composition factors of a permutation group are described in Cameron, Cannon and Neumann [11], Cannon [14], and in Cannon and Neumann [20].

v) An elegant algorithm for computing the Schur multiplier of a permutation group has been developed by Holt [25].

vi) As part of his work on the Schur multiplier algorithm, Holt devised a very fast algorithm for constructing a power-commutator presentation of a p-group P, given a base and strong generating set for P. (See Holt [25] and Butler [7]).

vii) If G is a finitely presented group and H is a subgroup of G having finite index, then the Todd-Coxeter algorithm may be used to construct a permutation representation of G on the cosets of H, provided that the index [G:H] is not too large. (See Cannon, Dimino, Havas and Watson [17] for details.)

viii) The Schreier Todd-Coxeter algorithm (section 3) constructs a presentation for G on a set of strong generators. Cannon [13] describes an algorithm which will construct a presentation for a small group on a given generating set.

REFERENCES

1. Atkinson, M.D. (1975). An algorithm for finding the blocks of a permutation group. Math. Comp. 29, 911-913.
2. Butler, G. (1979). Computational Approaches to Certain Problems in the Theory of Finite Groups. Ph.D. Thesis, University of Sydney.
3. Butler, G. (1982). Computing in permutation and matrix groups II: Backtrack algorithm. Math. Comp. 39, 671-680.
4. Butler, G. (1983). Computing normalizers in permutation groups. J. Algorithms, 4, 163-175.
5. Butler, G. (1982). On computing double coset representatives in permutation groups. Procs. of the LMS Symposium on Computational Group Theory (Dunham, August, 1982). Edited by M.D. Atkindon, Academic Press, New York, 1984.
6. Butler, G. (1982). Effective computation with group homomorphisms, TR184, Basser Department of Computer Science, University of Sydney, 41 pages.
7. Butler, G. (1984). Proof and implementation of Holt's algorithm TR210. Basser Department of Computer Science, University of Sydney.
8. Butler, G. & Cannon, J.J. (1982). Computing in permutation and matrix groups I: Normal closure, commutator subgroups, series. Math. Comp. 39, 663-670.
9. Butler, G. & Cannon, J.J. (1979). Computing in permutation and matrix groups III: Sylow subgroups. Manuscript, 15 pages.
10. Butler, G. & Cannon, J.J. (1983). Using homomorphisms to compute Sylow subgroups of permutation groups. Manuscript, 25 pages.
11. Cameron, P.J., Cannon, J.J. & Neumann, P.M. An algorithm for the recognition of 2-transitive groups. Manuscript, 17 pages.
12. Cannon, J.J. (1971). Computing local structure of large finite groups, SIAM-AMS Proceedings, Vol. 4, Amer. Math. Soc., Providence, R.I., 161-176.
13. Cannon, J.J. (1973). Construction of defining relations for finite groups. Discrete Math. 5, 105-129.
14. Cannon, J.J. (1980). Effective procedures for the recognition of primitive groups. Proceedings of Symposia in Pure Mathematics, Vol. 37, Amer. Math. Soc., Providence, R.I., 487-493.
15. Cannon, J.J. (1982). An introduction to the group theory language Cayley, Proceedings of the LMS Symposium on Computational Group Theory (Durham, August, 1982). Edited by M.D. Atkinson, Academic Press, New York, 1984, 143-182.
16. Cannon, J.J. A fast algorithm for recognizing the alternating and symmetric groups. In preparation.
17. Cannon, J.J., Dimino, L.A., Havas, G. & Watson, J.M. (1973). Implementation and analysis of the Todd-Coxeter algorithm. Math. Comp. 27, 463-490.
18. Cannon, J.J. & Neubüser, J. A normal subgroup algorithm. In preparation.
19. Cannon, J.J. & Neumann, P.M. An algorithm for constructing an elementary abelian regular normal subgroup in a primitive group. In preparation.
20. Cannon, J.J. & Neumann, P.M. An algorithm for determining the composition factors of a finite permutation group. In preparation.
21. Felsch, V. (1972). Programs for permutation groups. Permutations (Actes Colloq., Univ. Rene-Descartes, Paris, 1972), Gauthier-Villars, Paris, (1974), 241-250.
22. Ferber, K. (1967). Ein Programm zur Bestimmung der Ordnung grosser Permutationsgruppe. Manuscript, Kiel, 8 pages.

23. Gorenstein, D. (1983). Finite Simple Groups. Plenum, New York.
24. Hall, M. Jr. (1959). The Theory of Groups. Macmillan, New York.
25. Holt, D.F. (1982). A computer programme for the calculation of the Schur multiplier of a finite permutation group. Proceedings of the LMS Symposium of Computational Group Theory (Durham, August, 1982), Edited by M.D. Atkinson, Academic Press, 1984.
26. Leon, J.S. (1980). On an algorithm for finding a base and strong generating set for a group given by generating permutations. Math. Comp. 35, 941-974.
27. Leon, J.S. (1980). Finding the order of a permutation group. Proceedings of Symposia in Pure Mathematics, Vol. 37. Amer. Math. Soc., Providence, R.I., 511-517.
28. Robertz, H. (1976). Eine Methode zur Berechnung der Automorphismengruppe einer endlichen Gruppen. Diplomarbeit, RWTH Aachen.
29. Sims, C.C.(1967). Computational methods in the study of permutation groups. Computational Problems in Abstract Algebra (Oxford, 1967). Edited by John Leech, Pergamon, Oxford, 1970, 169-183.
30. Sims, C.C. (1971). Determining the conjugacy classes of a permutation group. SIAM-AMS Proceedings, Vol. 4. Amer. Math. Soc., Providence, R.I., 191-195.
31. Sims, C.C. (1971). Computation with permutation groups. Proceedings of the Second Symposium on Symbolic and Algebraic Manipulation (Los Angeles, 1971). Edited by S.R. Petrick, Assoc. Comp. Mach., New York, 1971, 23-28.
32. Sims, C.C. (1974). Some algorithms based on coset enumeration. Manuscript. Rutgers University, 52 pages.
33. Sims, C.C. (1978). Some group theoretic algorithms. Topics in Algebra (Canberra, 1978). Edited by M.F. Newman. Lecture Notes in Math., Vol. 697, Springer, Verlin, 1978, 108-124.
34. Sims, C.C. (1978). A method for constructing a group from a subgroup. Topics in Algebra (Canberra, 1978). Edited by M.F. Neuman. Lecture Notes in Math., Vol. 697, Springer, Verlin, 1978, 125-136.
35. Sims, C.C. (1978). Group-theoretic algorithms, A survey. Proc. Internal. Congr. Math.,(Helsinki, 1978), Vol. 2. Academia Scientiarum Fennica, Helsinki, 1980, 979-985.
36. Sims, C.C. (1978). How to construct a baby monster. Proceedings of a Research Symposium on Finite Simple Groups (Durham, 1978), Edited by M.J. Collins, Academic Press, London, 1980, 339-345.
37. Wielandt, H. (1964). Finite Permutation Groups. Academic Press. New York-London.

NOTE ON POLYNOMIAL-TIME GROUP THEORY

W.M. Kantor*
University of Oregon

Given a "small" subset Γ of S_n, what properties of $G = \langle\Gamma\rangle$ can be found efficiently? This question has been important for the construction of sporadic groups, and for Cannon's Cayley [2], where "efficient" meant "reasonably cheap to implement on a computer". There is another meaning of the word "efficient" within the context of theoretical Computer Science: "requiring polynomial time". This note is a brief survey of some recent work on polynomial-time group theoretic algorithms.

Consider $G \leq S_n = \mathrm{Sym}(X)$, where $|X| = n$. We are concerned with algorithms requiring at most $p(n)$ steps, where $p(n)$ is a polynomial in n. (Here, we are assuming that $|\Gamma|$ is small: of polynomial size.) For example, in $n(n-1)/2$ steps you can examine every 2-element subset of X. On the other hand, it would take $n!$ steps to examine each element of S_n, and many subgroups of S_n also fail to have polynomial orders. Thus, polynomial time (i.e., a polynomial number of steps) imposes restrictions not normally encountered in ordinary group theory.

The most basic algorithm is that of Sims [9] (cf. [3]), which finds $|G|$ and generators for G_x, $x \in X$, in polynomial time. More generally, given $Y \subseteq X$, this algorithm finds the pointwise stabilizer of Y in G in polynomial time. Moreover, Sims' algorithm produces a generating set of size $\leq n^2$ for G. (N.B. - We are dealing with generating permutations, not with generators and relations: no relations are available.)

The following can also be found in polynomial time, and the proofs are all easy ([1], [3], [8]): all orbits of G; if G is transitive, a complete system of imprimitivity Σ such that $|\Sigma| > 1$ and G^Σ is primitive; $\langle S^G\rangle$ for any given subset S of G; the derived series of G; and the center of G.

Centralizers present an enormous stumbling block. It is not

*Supported in part by NSF Grant MCS 7903130-82.

difficult to show that, if a polynomial-time algorithm for finding (generators for) $C_G(t)$, $t \in G$, $t^2 = 1$, were available, then there would be a polynomial-time algorithm for the Graph Isomorphism problem ("Given two n-vertex graphs, are they isomorphic?"). This is an important open problem in Computer Science (cf. [7]). Therefore, centralizers cannot be used, so that many of the familiar techniques of group theory must be avoided. On the other hand, this prohibition also suggests that group theoretic algorithms can lead to new results within Computer Science [7]. In any event, the restriction on time forces a rethinking of many group theoretic methods and problems.

The classification of finite simple groups has entered into this area. There is a polynomial-time algorithm for finding a composition series of G [8]. The validity of this algorithm depends upon the truth of Schreier's conjecture. There is also a polynomial-time algorithm for finding an element of order p, given a prime p [5]. Here, one easily reduces to the case in which G is simple and primitive on X, at which point tedious use of the results of [4] and linear algebra produce the desired element. Once again, the validity of the algorithm depends upon the classification. (N.B. - The algorithm used by Cayley involves the random selection of a few elements of G, each of which is checked to see if p divides its order. This process requires exponential time, as does Sims' centralizer algorithm used by Cayley.

Polynomial-time versions of Sylow's theorem have yet to be found. (One cannot use the familiar approach involving centralizers.) However, special cases have been obtained, when G is simple [5], solvable [6], or has all its noncyclic composition factors suitably restricted [6]. Specifically, assume that a bound b is given, and that each noncyclic composition factor either has order $\leq b$, is an exceptional Chevalley group, or is a classical group of dimension $\leq b$. Then the following can be done in polynomial time (the polynomial depending on b) [6]: any given p-subgroup can be embedded in a Sylow p-subgroup; and given (generators for) two Sylow p-subgroups of G, an element of G can be found conjugating the first to the second. In [6] there are also polynomial-time versions of the standard results concerning Hall subgroups of solvable groups, as well as versions of both parts of the Schur-Zassenhaus theorem when G is restricted as above.

It should be clear that the results just described primarily concern algorithmic versions of undergraduate-level group theory (even though some proofs require the classification). But then, the subject of this note is just in its infancy. In order to make this fact even clearer, we conclude with some problems that are presently OPEN. (1) Prove Sylow's theorem for arbitrary G (both finding Sylow subgroups and conjugating them). (2) Given m, determine whether G has an element of order m. (This is even open if m is a prime power.) (3) Find a minimal normal subgroup of G. (4) If $G \triangleright N$ and the extension splits, find a complement. (5) If G is solvable, find a system normalizer and a Carter subgroup. (6) If $G, H \leq S_n$, find (generators for) $G \cap H$. (A polynomial-time algorithm for this problem would, however, settle the Graph Isomorphism problem.)

REFERENCES

1. Atkinson, M.D. (1975). An algorithm for finding the blocks of a permutation group. Math. of Comp. 29, 911-913.
2. Cannon, J.J. (1980). Effective procedures for the recognition of primitive groups. Proc. Symp. Pure Math. 37, 487-493.
3. Furst, M., Hopcroft, J. & Luks, E. (1980). Polynomial-time algorithms for permutation groups. Proc. 21st IEEE Symp. Found. Comp. Sci., 36-41.
4. Kantor, W.M. (1979). Permutation representations of the finite classical groups of small degree or rank. J. Algebra. 60, 158-168.
5. Kantor, W.M. Polynomial-time algorithms for finding elements of prime order and Sylow subgroups. (submitted)
6. Kantor, W.M. & Taylor, D.E. Polynomial-time versions of Sylow's theorem. (in preparation)
7. Luks, E. (1982). Isomorphism of graphs of bounded valence can be tested in polynomial time. J. Comp. Syst. Sci. 25, 42-65.
8. Luks, E. (unpublished)
9. Sims, C.C. (1978). Some group-theoretic algorithms. Springer Lecture Notes in Math. 697, 108-124.

COSET TABLE METHODS

C.C. Sims
Rutgers University, New Brunswick, NJ 08903

The effort to classify the finite simple groups had a profound effect on computational group theory. In order to construct certain sporadic simple groups, such as described in [9], techniques were developed for computing in permutation groups of fairly large degree, sometimes exceeding 100,000. Algorithms for determining orders of groups, for computing centralizers, and for performing many other basic constructions were devised. A survey of group-theoretic algorithms, including algorithms dealing with permutation groups, may be found in [10]. A more complete description of permutation group algorithms is given in [3]. These algorithms have recently been used in proofs that various graph-theoretic problems have polynomial-time solutions. Examples may be found in [8].

Although work on permutation group algorithms continues, there is a growing effort to develop algorithms dealing with other classes of groups, finitely generated nilpotent groups, finite solvable groups, and finitely generated subgroups of free groups. This paper provides a summary, largely without proofs, of some recent developments concerning algorithms related to finitely generated subgroups of free groups.

Throughout this paper we shall assume that X is a finite set and that we are studying finitely generated subgroups of the free group F on X. The set X^{-1} will be a set of formal inverses for the elements of X and $X^{\pm}$ will denote $X \cup X^{-1}$. A word will be a finite sequence of elements from $X^{\pm}$ and the free monoid generated by $X^{\pm}$ is the set M of all words. The elements of F are the equivalence classes of free equivalence on M. If U is in M, then the element of F containing U will be written $[U]$. If $U = a_1a_2\ldots a_r$ with each a_i in $X^{\pm}$, then U^{-1} will denote the word $a_r^{-1}\ldots a_2^{-1}a_1^{-1}$. In F we have $[U][U^{-1}] = 1$. Moreover, $(UV)^{-1} = V^{-1}U^{-1}$ for all U and V in M. We shall assume that $X^{\pm}$ has been given a fixed linear order $<$. In examples, X will be a two-element set $\{x,y\}$ and the order will be $x < x^{-1} < y < y^{-1}$.

The basic objects used here to study finitely generated subgroups of F are coset tables. The definition of a coset table given here is a minor variant of that used in [7]. We consider maps $t: \Omega \times X^{\pm} \to \Omega \cup \{0\}$, where Ω is a finite set of positive integers. Normally we shall omit explicit references to t and write $t(\alpha,a)$ as α^a for α in Ω and a in $X^{\pm}$. We define 0^a to be 0 for all a in $X^{\pm}$. If U is the empty word, then α^U is defined to be α for all α in Ω. If $U = Va$ for some word V and some a in $X^{\pm}$, then α^U is defined recursively to be $(\alpha^V)^a$.

We may define one of these maps t by a table. For example, the table

	x	x^{-1}	y	y^{-1}
1	4	0	0	3
3	0	4	7	0
4	1	3	0	1
7	7	1	3	3

defines a map $t: \Omega \times X^{\pm} \to \Omega \cup \{0\}$, where $\Omega = \{1,3,4,7\}$. Here $3^y = 7$ and $3^U = 4$, where $U = yx^{-1}y^{-1}x^{-1}$.

A <u>coset table</u> relative to X is a triple $T = (\Omega, t, \alpha)$, where Ω is a finite set of positive integers, α is an element of Ω, and t is a map from $\Omega \times X^{\pm}$ to $\Omega \cup \{0\}$ satisfying the following conditions:

i) If β is in Ω, a is in $X^{\pm}$ and $\beta^a = \gamma \neq 0$, then $\gamma^{a^{-1}} = \beta$.

ii) For all β in Ω there is a word U such that $\alpha^U = \beta$.

The set Ω is often referred to as the set of <u>active cosets</u> of T. The point α will be called the <u>base point</u> of T. Normally 1 is in Ω and 1 is the base point, but this will not always be the case.

The example above with $\Omega = \{1,3,4,7\}$ is not a coset table since $1^x = 4$ and $4^{x^{-1}} = 3 \neq 1$. The table

	x	x^{-1}	y	y^{-1}
1	4	0	2	2
2	0	3	1	1
3	2	0	0	0
4	0	1	0	0

does define a coset table. The usual data structure for a coset table is an array. However, for sparse tables, ones in which most entries are 0, other data structures may be appropriate.

It is possible to replace a coset table $T = (\Omega, t, \alpha)$ by a directed graph whose edges are labeled with elements of $X^{\pm}$. If β and γ are in Ω and a is in $X^{\pm}$, then there is an edge from β to γ labeled by a if and only if $t(\beta,a) = \gamma$. However, since the most common operation is determining for some given β and a whether there is an edge from β with label a, the formulation in terms of tables or arrays seems the more natural to me.

<u>Lemma 1</u>. Let $T = (\Omega, t, \alpha)$ be a coset table, let β be in Ω and let U be in M. Suppose $\beta^U = \gamma \neq 0$. Then $\gamma^{U^{-1}} = \beta$. Moreover, if V is in $[U]$, then β^V is γ or 0, and if V is freely reduced, then $\beta^V = \gamma$.

Let $T = (\Omega, t, \alpha)$ be a coset table. We define $H(T)$ to be $\{[U] \in F | \alpha^U = \alpha\}$.

<u>Lemma 2</u>. The set $H(T)$ is a finitely generated subgroup of F and every finitely generated subgroup of F occurs as $H(T)$ for some T.

The coset table T contains all the information needed to decide membership in $H(T)$ and, more generally, to decide equality of cosets.

<u>Lemma 3</u>. Let U and V be reduced words and write $U = AB$ and $V = CD$, where A and B are the longest prefixes of U and V, respectively, such that α^A and α^C are not 0. Then $H(T)[U] = H(T)[V]$ if and only if $\alpha^A = \alpha^C$ and $B = D$.

It is quite easy to construct a set of free generators for $H(T)$ from T. The procedure will be sketched later.

Different coset tables may define the same subgroup of F. Let T_1 be

	x	x^{-1}	y	y^{-1}
1	0	0	0	0

In T_1 we have $1^U = 1$ only for U the empty word. Thus $H(T_1)$ is the trivial subgroup. Now let T_2 be

	x	x^{-1}	y	y^{-1}
1	2	0	0	0
2	0	1	0	0

In T_2 we have $1^U = 1$ if and only if $U = (xx^{-1})^k$ for some $k \geq 0$. But any such word is freely equivalent to the empty word and so $H(T_2)$ is trivial also.

In order to get a 1-1 correspondence between coset tables and finitely generated subgroups of F, we must add further conditions on the coset tables. A coset table $T = (\Omega,t,\alpha)$ is said to be <u>reduced</u> if for all β in $\Omega-\{\alpha\}$ there are at least two elements a in $X^{\pm}$ such that $\beta^a \neq 0$. The coset table T_2 is not reduced since $2^a \neq 0$ only for $a = x^{-1}$. Note that the definition of a coset table requires that if β is in $\Omega-\{\alpha\}$, then $\beta^a \neq 0$ for at least one a in $X^{\pm}$.

<u>Lemma 4</u>. For any coset table T there is a reduced coset table T' such that $H(T) = H(T')$.

There is a simple algorithm for computing a table T' as described in Lemma 4. Suppose β is in $\Omega-\{\alpha\}$ and $\beta^a \neq 0$ for only one element a of $X^{\pm}$. Then $\beta^a = \gamma \neq \beta$. Set $\Omega' = \Omega-\{\beta\}$ and define t' to be the restriction of t to $\Omega'\times X^{\pm}$ with one change, $t'(\gamma,a^{-1})$ is set to 0, not β. Then $T' = (\Omega',t',\alpha)$ is a coset table with $H(T') = H(T)$. If T' is still not reduced, repeat this process.

The basic step in the above reduction algorithm is an example of a coset table operation, a procedure for converting one coset table into another. We shall refer to this operation as <u>deletion</u> of the coset β. It is useful to think of coset table operations as analogues of elementary row and column operations on matrices.

If $T = (\Omega,t,\alpha)$ is a coset table, then we shall order the elements (β,a) of $\Omega\times X^{\pm}$ first by β and then by a. An <u>occurrence</u> of an element γ of Ω in T is an element (β,a) of $\Omega\times X^{\pm}$ such that $\beta^a = \gamma$. We say that T is <u>standard</u> if $\Omega = \{1,\ldots,|\Omega|\}$, $\alpha = 1$, and for $1 < \beta < \gamma \leq |\Omega|$ the first occurrence of β in T is earlier than the first occurrence of γ.

<u>Lemma 5</u>. For any coset table T there is a standard coset table T' such that H(T) = H(T').

The standardization algorithm uses another coset table operation, switching. Let $T = (\Omega, t, \alpha)$ be a coset table and let i and j be distinct positive integers. Define f to be the permutation of the set of nonnegative integers which interchanges i and j and leaves all other points fixed. Switching i and j in T means forming a new coset table $T' = (\Omega', t', \alpha')$, where $\Omega' = f(\Omega)$, $\alpha' = f(\alpha)$ and for β in Ω' and a in $X^{\pm}$ we have $t'(\beta,a) = f(t(f^{-1}(\beta),a))$. Of course $f^{-1} = f$, but the definition given makes sense when f is any permutation of the set of nonnegative integers which fixes 0. If neither i nor j is in Ω, then $T' = T$. Normally we convert T to T' in place, rather than creating a new table. The time to do this is proportional to $|X|$.

The following algorithm standardizes a coset table $T = (\Omega, t, \alpha)$.

```
1.    If α ≠ 1 then
          Begin
2.            Switch 1 and α in T;
3.            α ← 1
          End
4.    c ← s ← 1;
5.    For a in X± (taken in order) do
          Begin
6.            u = c^a;
7.            If u > s then
                  Begin
8.                    s ← s + 1;
9.                    If u ≠ s then switch u and s
                  End
          End
10.   c ← c + 1;
11.   If c ≤ s then go to 5
```

The main result connecting coset tables with finitely generated subgroups of F is given in the following theorem.

Theorem 6. There is a 1-1 correspondence between the finitely generated subgroups of F and the reduced, standard coset tables relative to X.

To decide whether two coset tables describe the same subgroup of F we reduce and standardize each of them and check whether they are now the same.

Suppose the finitely generated subgroup H of F corresponds to the reduced, standard coset table T. Although T depends on the linear order which has been fixed for $X^{\pm}$, the number of active cosets in T does not. Let us call a right coset of H *important* if there are words U and V in M such that

a) The coset is $H[U]$.

b) UV is reduced.

c) $[UV]$ is in H.

We allow U and V to be empty, so H is an important coset of itself.

Lemma 7. A subgroup H of F is finitely generated if and only if H has finitely many important cosets. A finitely generated subgroup K of F has finite index in F if and only if every right coset of K is important.

It should be noted that Lemma 7 does not hold if X is infinite.

Lemma 8. Let T be a reduced coset table and let $H = H(T)$. There is a 1-1 correspondence between the important cosets of H and the active cosets of T given by $H[U] \longleftrightarrow \alpha^U$.

Let us order the words in M first by length and then lexicographically according to the order on $X^{\pm}$. If T is a standard coset table, then for each active coset β we can find the first word $W = W(\beta)$ such that $\alpha^W = \beta$ by the following recursive procedure:

```
Procedure W(β):
Begin
    If β = α then return the empty word;
    Let γ be the smallest active coset of the form β^a;
    U ← W(γ);
    Let b be the first element in X^± such that γ^b = β;
    Return Ub
End
```

If T is standard, then we can get a set of free generators for $H(T)$ by forming the set S of elements $[W(\beta)aW(\gamma)^{-1}]$, where β and γ are active cosets, a is in $X^{\pm}$, $\gamma = \beta^a$, (β,a) is not the first occurrence of γ and either $\beta < \gamma$ or $\beta = \gamma$ and $a < a^{-1}$. The elements of S are Nielsen reduced.

In 1954, Howson [5] proved that the intersection of two finitely generated subgroups of a free group is finitely generated. Let T_1 and T_2 be two coset tables. We can construct a coset table T_3 such that $H(T_3) = H(T_1) \cap H(T_2)$. This construction is sketched in [6]. For $i = 1$ and 2 let Ω_i be the set of active cosets of T_i and let α_i be its base point. Number the elements of $\Omega_1 \times \Omega_2$ from 1 up to $N = |\Omega_1| \cdot |\Omega_2|$ with (α_1,α_2) numbered 1. Set $\Delta = \{1,\ldots,N\}$ and define a map $t: \Delta \times X^{\pm} \to \Delta \cup \{0\}$ as follows: If j is in Δ, let (β,γ) be the j-th element of $\Omega_1 \times \Omega_2$. For a in $X^{\pm}$, define $t(j,a)$ to be 0 if either $\beta^a = 0$ in T_1 or $\gamma^a = 0$ in T_2. Otherwise, set $t(j,a) = k$, where (β^a,γ^a) is the k-th element of $\Omega_1 \times \Omega_2$.

The triple $(\Delta,t,1)$ may not be a coset table because the transitivity property may not hold. That is, for j in Δ there need not be a word U such that $1^U = j$. However, if we let Ω_3 be the set of j in Δ_3 such that there is a word U with $1^U = j$ and we restrict t to $\Omega_3 \times X^{\pm}$, then $T_3 = (\Omega_3,t,1)$ is a coset table. Let U be a reduced word. Then $[U]$ is in $H(T_i)$ if and only if $\alpha_i^U = \alpha_i$ in T_i. But this means $[U]$ is in $H(T_1) \cap H(T_2)$ if and only if $1^U = 1$ in T_3. Thus $H(T_3) = H(T_1) \cap H(T_2)$.

From the proceeding construction, we see that if H_1 and H_2 are two finitely generated subgroups of F, then the number of important cosets of $H_1 \cap H_2$ is at most the product of the numbers for H_1 and H_2.

Suppose T is a coset table, $H = H(T)$, and W is a word in M. It is easy to construct a table defining the conjugate $[W]^{-1}H[W] = H^{[W]}$. To do so, we use a coset table operation called <u>defining a new coset</u>. Suppose β is in Ω, a is in $X^{\pm}$, and $\beta^a = 0$. To define a new coset in T at (β,a), we choose a positive integer γ not in Ω and form the table $T' = (\Omega',t',\alpha)$, where $\Omega' = \Omega \cup \{\gamma\}$, $t'(\gamma,b) = 0$ for all b in $X - \{a^{-1}\}$, $t'(\gamma,a^{-1}) = \beta$, and the restriction of t' to $\Omega \times X^{\pm}$ is t, except that $t'(\beta,a) = \gamma$. It is easy to see that $H(T') = H(T)$ and deleting γ in T' yields T. Normally we replace T by T' when making a definition.

Let T be a coset, let β be an active coset in T, and let

$W = a_1 \ldots a_r$ be a word. The following algorithm defines what it means to trace W from the left at β with definitions.

1. $\gamma \leftarrow \beta$;
2. For $i = 1$ to r do
   begin
3. $\delta = \gamma^{a_i}$;
4. If $\delta = 0$ then define a new coset at (γ, a_i) and call it δ;
5. $\gamma \leftarrow \delta$
   end

After this algorithm has been executed, $H(T)$ is unchanged and β^W is not 0.

To describe the conjugate of $H(T)$ by $[W]$, we trace W from the left at α with definitions and change the base point to α^W. If T' is the resulting table, then $H(T') = H(T)^{[W]}$. Since we can decide whether $H(T') = H(T)$, we can decide whether $[W]$ normalizes $H(T)$.

If H is a nontrivial finitely generated subgroup of F, then H has finite index in its normalizer N. We can compute N easily using coset tables. If u is in N and v is in F, then $u(Hv) = uHu^{-1}uv = Huv$. Thus multiplication of a right coset of H on the left by u yields another right coset. In general, this map does not take important cosets to important cosets. However, if we replace H by an appropriate conjugate of H, then multiplication on the left by elements of the normalizer of H will map the set of important cosets of H into itself.

Let $T = (\Omega, t, \alpha)$ be a coset table. We say that T is <u>trivial</u> if $\Omega = \{\alpha\}$ and $\alpha^a = 0$ for all a in $X^{\pm}$. We call T <u>cyclically reduced</u> if either T is trivial or for all β in Ω there are at least two elements a of $X^{\pm}$ such that $\beta^a \neq 0$.

<u>Lemma 9</u>. If T is a coset table, then there is a cyclically reduced coset table T' such that $H(T)$ and $H(T')$ are conjugate.

To construct the coset table T' of Lemma 9, we first reduce T. If T is nontrivial, it can fail to be cyclically reduced only if there is exactly one a in $X^{\pm}$ such that $\alpha^a \neq 0$. Let $\beta = \alpha^a$. Then $\beta \neq \alpha$. Change the base point to β and delete α. Repeating the process if necessary, we arrive at a cyclically reduced table T'. Since deletions do not change the group of a table and changing the base point replaces the group by a conjugate, H(T) is conjugate to H(T').

Lemma 10. Let T be a nontrivial cyclically reduced coset table and let H = H(T). If u is in the normalizer N of H in F, then Hu is an important coset of H.

By Lemma 10, if $T = (\Omega,t,\alpha)$ is nontrivial and cyclically reduced, then the normalizer N of H = H(T) consists of the elements [U] of F such that $\beta = \alpha^U \neq 0$ and H = H(T'), where $T' = (\Omega,t,\beta)$. We can determine the relevant β's and determine a set of generators for N.

The basic point of this paper is to demonstrate that for most purposes coset tables are a useful representation for finitely generated subgroups of free groups. If this is really the case, then there should be a good algorithm for solving the following problem: Given a finite subset $\mathcal{U}$ of M, construct a coset table T such that $H(T) = \langle [U] \mid U \in \mathcal{U} \rangle$. There is such an algorithm and it is called coset enumeration. Actually all we shall need is a special case of the general coset enumeration algorithm. The general case is described in [4] and its references.

Coset enumeration involves two more coset table operations. Let $T = (\Omega,t,\alpha)$ be a coset table. Suppose β and γ in Ω and a in $X^{\pm}$ satisfy the conditions $\beta^a = \gamma^{a^{-1}} = 0$. Let $t' : \Omega \times X^{\pm} \to \Omega \cup \{0\}$ agree with t, except that $t'(\beta,a) = \gamma$ and $t'(\gamma,a^{-1}) = \beta$. Then $T' = (\Omega,t',\alpha)$ is a coset table and is said to be formed from T by entering the deduction $\beta^a = \gamma$.

Lemma 11. If T' is obtained from $T = (\Omega,t,\alpha)$ by entering the deduction $\beta^a = \gamma$, then H(T') is generated by H(T) and $[UaV^{-1}]$, where U and V are any words satisfying $\alpha^U = \beta$ and $\alpha^V = \gamma$ in T.

The last coset table operation is called forcing a coincidence. Let $T = (\Omega,t,\alpha)$ and suppose β and γ are distinct points of Ω. Let $\sim$ be the finest equivalence relation on Ω such that

1. $\beta \sim \gamma$.
2. If $\delta \sim \varepsilon$, a is in $X^{\pm}$, and $\delta^a \neq 0 \neq \varepsilon^a$, then $\delta^a \sim \varepsilon^a$.

Select a set Ω' of representatives for the equivalence classes of $\sim$. Let α' be the element of Ω' equivalent to α. Define

$t':\Omega'\times X^{\pm} \to \Omega' \cup \{0\}$ as follows: If δ is in Ω', a is in $X^{\pm}$, and for some ε equivalent to δ we have $\varepsilon^a \neq 0$ in T, then set $t'(\delta,a)$ equal to the element of Ω' equivalent to ε^a. If no such ε exists, set $t'(\delta,a) = 0$. Then $T' = (\Omega',t',\alpha')$ is a coset table, the table obtained from T by forcing the coincidence of β and γ.

<u>Lemma 12</u>. Suppose T' is obtained by forcing the coincidence of β and γ in T. Then $H(T')$ is generated by $H(T)$ and $[UV^{-1}]$, where U and V are any words satisfying $\alpha^U = \beta$ and $\alpha^V = \gamma$ in T.

For fixed X, the time to construct T' in Lemma 12 can be made nearly linear in $|\Omega|-|\Omega'|$ using the fast union-find algorithm in [1].

Earlier we defined what it means to trace a word from the left with definitions. Now we need to trace words from both ends. Let $T = (\Omega,t,\alpha)$ and suppose β is in Ω. To trace W from both ends at β in T we do the following:

1. Let A be the longest prefix of W such that $\beta^A \neq 0$;
2. $\gamma \leftarrow \beta^A$;
3. Write $W = AD$;
4. Let C be the longest suffix of D such that $\beta^{C^{-1}} \neq 0$;
5. $\delta \leftarrow \beta^{C^{-1}}$;
6. Write $D = BC$;
7. If B is empty then
8.   If $\gamma \neq \delta$ then force the coincidence of γ and δ

    Else

        Begin

9.   Let b be the first term of B;
10.   If B has length 1 then note the deduction $\gamma^b = \delta$

            Else

                Begin

11.   Define a new coset at (γ,b);
12.   Go to 1

                End

        End

Lemma 13. If T' is the coset table resulting from tracing W from both ends at β in T, then $H(T')$ is generated by $H(T)$ and $[UWU^{-1}]$, where U is any word such that $\alpha^U = \beta$ in T.

The transfer to step 1 in step 12 above is only to make the description short. The actual time to perform the trace is essentially proportional to the length of W plus the number of cosets removed if the force in line 8 is actually done. As usual, the changes are made in T.

Given a set $\mathcal{U}$ of words, we can construct a coset table T such that $H(T) = \langle [U] \mid U \in \mathcal{U} \rangle$ with the following algorithm:

1. Initialize T to the trivial table with $\Omega = \{1\}$;
2. For each U in $\mathcal{U}$ trace U from both ends at 1.

For fixed X, the time to construct T is essentially linear in the sum L of the lengths of the elements of $\mathcal{U}$. If we construct the free generators for $H(T)$ as described earlier, we have in effect performed a Nielsen reduction of the set $\mathcal{U}$, again in a time which for practical purposes is linear in L. A Nielsen reduction algorithm is sketched in [2]. The approach described here is faster, even allowing for the differences in computational model.

REFERENCES

1. Aho, A.V., Hopcroft, J.E. & Ullman, J.D. (1974). The Design and Analysis of Computer Algorithms. Reading, Mass.: Addison-Wesley.
2. Avenhaus, J. & Madlener, K. (1981). P-complete problems in free groups. Lecture Notes in Computer Science 104. Springer-Verlag, pp. 42-51.
3. Cannon, J.J. A computational toolkit for finite permutation groups. These Proceedings.
4. Cannon, J.J., Dimino, L.A., Havas, G. & Watson, J.M. (1973). Implementation and analysis of the Todd-Coxeter algorithm. Math. Comp. 27, 463-490.
5. Howson, A.G. (1954). On the intersection of finitely generated free groups. J. London Math. Soc. 29, 428-434.
6. Imrich, W. (1977). Subgroup theorems and graphs. Lecture Notes in Mathematics 622. Springer-Verlag, pp. 1-27.
7. Leon, J.S. (1980). On an algorithm for finding a base and strong generating set for a group given by generating permutations. Math. Comp. 35, 941-974.
8. Luks, E.M. (1982). Isomorphism of graphs of bounded valence can be tested in polynomial time. J. of Computer and System Sciences 25, 42-65.

9. Sims, C.C. (1973). The existence and uniqueness of Lyons' group. Finite Groups '72. North-Holland Mathematics Studies 7. North-Holland/American Elsevier.
10. Sims, C.C. (1978). Group-theoretic algorithms, a survey. Proceedings of the International Congress of Mathematicians. Helsinki.

SUBGROUP STRUCTURE OF FINITE GROUPS

M. Aschbacher
California Institute of Technology
Pasadena, CA

During the seventies there was a tendency to view the Classification Theorem as an end unto itself; even now with the Classification complete that attitude lingers on. My own viewpoint is somewhat different; I regard the Classification Theorem as the cornerstone of a powerful theory of finite groups which proceeds by reducing problems about general finite groups to questions about simple groups and then invoking the Classification and our knowledge of the known simple groups to answer these questions. From this point of view the Classification is more than just an important theorem; it is also a means for proving other important theorems.

Actually this point of view is just a special case of a somewhat more general philosophy. To explain this philosophy I need the concept of a representation. I'll use the term in a more general sense than usual. Namely if C is a category then a C-<u>representation</u> of a group G is a group homomorphism of G into the group of automorphisms of an object of C. Among such representations, the permutation representations and linear representations seem to be most basic. There is a notion of "indecomposible representation" induced by the product or coproduct in C and often there is a notion of "irreducible representation" corresponding to the absence of suitable G-invariant equivalence relations on the object. For example the simple groups act irreducibly on themselves by conjugation in the category of groups and homomorphisms.

Most of the finite group theoretical questions which interest me can be regarded as questions about representations of groups, where the term representation is used in the very general sense defined above. Frequently these questions are best answered via a reduction to a question about simple (or almost simple) groups, followed by appeals to the

Classification Theorem and knowledge of the irreducible permutation and linear representations of the known simple groups.

Recently a number of problems of this sort have come up in other areas of mathematics. I regard this as an encouraging development and look forward to more applications of finite group theory now that the Classification Theorem makes it possible to answer so many more group theoretic questions.

This point of view also has implications for problem selection among finite group theorists. It suggests (at least to me) that the most fundamental questions about finite groups come from one of the following three areas:

a) Classification of the finite simple groups.

b) Description of the irreducible permutation representations and linear representations of the almost simple groups.

c) Generation of techniques to reduce questions about general finite groups to questions about almost simple groups, or more generally generation of techniques to reduce questions about general representations to questions about irreducible representations.

In the remainder of this article I'll pursue this philosophy as it applies to the study of permutation representations of finite groups or equivalently to the study of the subgroup structure of finite groups.

The indecomposible permutation representations of a group G are its transitive representations, and the irreducible representations are the primitive representations. The study of the first class of representations is equivalent to the study of the subgroup structure of G, while the study of primitive representations amounts to a study of the conjugacy classes of maximal subgroups of G. To reduce questions about transitive representations to questions about primitive representations, one needs information about maximal overgroups of subgroups of G. These remarks suggest we should study maximal subgroups of finite groups and in particular study maximal overgroups of suitable subgroups of G. That is to say we can hope that in most interesting transitive permutation representations the stabilizer of a point is reasonably large in the sense that it contains certain easily recognizable subgroups. Hence the maximal overgroups of such subgroups become of particular interest.

Following the philosophy advanced above, two goals suggest themselves: First, to generate techniques to reduce the study of maximal subgroups of the general finite group to the study of maximal subgroups of almost simple groups. (For purposes of permutation representations a group G is almost simple if F*(G) is a nonabelian simple group.) A result of this sort, obtained in collaboration with L. Scott, is discussed in section 2.

The second goal is to describe the maximal subgroups of almost simple groups, particularly the maximal overgroups of suitable subgroups. It's unclear to me what form this description should take. As a matter of fact such descriptions no doubt depend on the type of subgroup to be covered. Still one would like the description to be reasonably elegant, to have a nice proof, and perhaps most important, to be applicable simultaneously to most of the almost simple groups. Such descriptions are extremely useful in establishing a variety of properties of finite groups. Their utility was quite apparent, for example, in the last stages of the Classification, where it was necessary to prove various facts about K-groups.

Sections 3 and 4 discuss two results of this second type. Section 3 provides a (limited) description of subgroups of the classical groups in terms of the representation of the group on its natural module. The result should be regarded as the analogue of a theorem of O'Nan and Scott for the alternating groups. (See the appendix of [3] for a statement of the O'Nan-Scott Theorem.) Since one can hope to enumerate the maximal subgroups of the sporadic groups and the exceptional groups of Lie type by ad hoc methods, there is some reason to treat the alternating groups and classical groups from the point of view of their natural representations, even though such representations do not appear to be available (or at least as useful) for all the simple groups.

Section 4 provides a description of the overgroups of Sylow p-groups in sporadic groups G such that p^2 divides the order of G. The description is part geometric and part group theoretic, and extends a similar description of the overgroups of Sylow p-groups of groups of Lie type and characteristic p. The description can be used to establish an analogue of the Borel-Tits Theorem.

Section 5 contains a list of some open problems on subgroup structure.

2 PRIMITIVE PERMUTATION REPRESENTATIONS OF THE GENERAL FINITE GROUP

Let G be a finite group and $\mathcal{N} = \mathcal{N}_G$ the set of maximal subgroups of G. I'm going to discuss a description of $\mathcal{N}$ obtained in collaboration with L. Scott in [3]. There is significant intersection of this work with that of F. Gross and L. Kovacs in [4].

Let $M \in \mathcal{N}$ and denote by $\ker_M(G)$ the kernel of the permutation representation of G on the cosets of M. With little loss of generality we can take $\ker_M(G) = 1$ by passing to $G/\ker_M(G)$. That is we assume M is in $\mathcal{N}^* = \mathcal{N}^*_G$, the set of N in $\mathcal{N}$ with $\ker_N(G) = 1$. Let $\mathcal{C}^*$ denote the set of orbits of G on $\mathcal{N}^*$ via conjugation.

Theorem 1. Let $D = F^*(G)$ and $M \in \mathcal{N}^*$. Then one of the following holds:

A) D is an elementary abelian p-group and M is a complement to D in G. $\mathcal{C}^* \cong H^1(M,D)$.

B) D is the direct product of the G-conjugates of LK, where L and K are isomorphic nonconjugate simple components of G, $G = MD$, $M = N_G(M \cap D)$, and $M \cap D$ is the direct product of the M-conjugates of a full diagonal subgroup of LK. $\mathcal{C}^* \cong m\cdot(C_{\mathrm{Out}(L)}(\mathrm{Out}_G(L)))$, where m is the number of conjugates U of K such that $N_G(U) = N_G(L)$ and there is a $N_G(L)$-invariant full diagonal subgroup of UL.

C) D is the direct product of the G-conjugates of some simple component L of G. $\mathcal{N}^*$ is the disjoint union of G-stable subsets $\mathcal{N}^*_i$, $1 \leq i \leq 3$, where:

1) $\mathcal{N}^*_1$ consists of complements M to D with $\mathrm{Inn}(L) \leq \mathrm{Aut}_M(L)$.

2) $\mathcal{N}^*_2$ consists of those subgroups M of G such that $G = MD$, $M = N_G(M \cap D)$, and $M \cap D$ is the direct product of the M-conjugates of a full diagonal subgroup of $\prod_{X \in \Gamma} X$ for some $\Gamma^G \in P^*(G)$.

3) $\mathcal{N}^*_3$ consists of those subgroups M of G such that $G = MD$, $M = N_G(M \cap D)$, $\mathrm{Aut}_M(L) \in \mathcal{N}^*_{\mathrm{Aut}_G(L)}$, and $M \cap D$ is the direct product of the M-conjugates of $M \cap L$. $\mathcal{C}^*_3 \cong \mathcal{C}^*_{\mathrm{Aut}_G(L)}$.

A <u>full diagonal subgroup</u> of a direct product $L_1 \times \ldots \times L_n$ is a subgroup X such that the projection maps from X to L_i are isomorphisms for each i. In Case C2, $P^*(G)$ denotes the set of G-invariant partitions Γ^G of L^G, such that $|\Gamma| > 1$ and Γ^G possesses no proper refinement satisfying these constraints. $H^1(M,D)$ is of course the 1-cohomology group of M on D. There are parameterizations of $\mathcal{C}^*_1$ and $\mathcal{C}^*_2$ given in [3], but I omit them here as they are rather technical.

In some sense Theorem 1 reduces the study of primitive permutation representations of finite groups to two difficult but well defined problems:

<u>Problem 1</u>. Determine C_G^* for G an almost simple group.

<u>Problem 2</u>. Determine $H^1(G,V)$ for G an almost simple group and V an irreducible $GF(p)G$-module.

This type of reduction is in the spirit of the philosophy of the introduction. Problem 1 arises out of Case C3, while Problem 2 arises out of Case A, given the following theorem from [2]:

<u>Theorem 2</u>. Let G be a finite group, K a field of prime characteristic p, and V a faithful irreducible KG-module with $H^1(G,V) \neq 0$. Then:

1) $F^*(G)$ is the direct product of the G-conjugates of a simple component L with $p \in \pi(L)$.

2) V is the direct sum of the G-conjugates of $U = [L,V]$.

3) $N_G(L)$ is irreducible on U with $C_G(L) = C_G(U)$.

4) $H^1(G,V) \cong H^1(\mathrm{Aut}_G(L),U)$.

5) $\dim(H^1(G,V)) \leq \dim(H^1(L,E))$ for each nontrivial irreducible KL-submodule E of V.

3 <u>SUBGROUPS OF CLASSICAL GROUPS</u>

In this section I wish to discuss the following result from [2]:

<u>Theorem 3</u>. Let G be an almost simple group such that $F^*(G) = G_0$ is a classical group. If $G_0 \cong P\Omega_8^+(q)$ assume no member of G induces a triality automorphism on G_0. Let $H \leq G$ with $G = G_0H$ and V the natural module for the covering group of G_0. Then either H is contained in a member of the collection S_G of subgroups of G defined below, or the following hold:

1) $F^*(H)$ is a nonabelian simple group.

2) Let L be the covering group of $F^*(H)$ and F the field of definition of V. Then V is an absolutely irreducible FL-module.

3) The representation of L on V is defined over no proper subfield of F.

4) If $a \in \mathrm{Aut}(F)$, V^* is the dual of V, and V is FL-isomorphic to V^{*a} then either:

i) $a = 1$ and G_0 is orthogonal or symplectic, or

ii) a is an involution and G_0 is unitary.

There exists a covering Ω of G_0 and a form f on V such that Ω is the derived group of the isometry group of (V,f). For example if $G_0 \simeq L_n(q)$ then f is trivial. Let Γ be the group of nonsingular semilinear transformations of V preserving f up to a scalar multiple and a twist by a field automorphism. Usually $\mathrm{Aut}(G_0) = P\Gamma$ is the image of Γ under the projective map P, so that G is the image of some subgroup $\tilde{G}$ of Γ under P. For purposes of exposition I'll assume G is such an image. Then the members of S_G are the images of the stabilizers in $\tilde{G}$ of certain natural structures on V described in the next paragraph. Again for purposes of exposition, the definitions will be somewhat simplified and hence not entirely accurate.

The first class of structures are the nondegenerate or totally singular subspaces of V. Second are the sets S of isometric subspaces such that V is either the orthogonal direct sum of the members of S, or S consists of a pair of complementary maximal totally singular subspaces. Third are vector space structures over some extension of F of prime degree. Fourth are subspaces U of V over a subfield K of prime degree in F such that $V = F\otimes_K U$. Fifth are certain tensor product structures on V. The sixth type of stabilizer is the collection of normalizers of r-subgroups of Γ of symplectic type acting irreducibly on V. Finally we have the stabilizers of certain forms on V.

In addition the isomorphism type of the stabilizers is determined as are the orbits of G on the structures.

The following problem is still open as far as I know:

Problem 3. Determine which members of S_G are maximal in G.

Problem 3 can certainly be solved with available techniques and the Classification, but the next problem seems much more difficult:

Problem 4. Let L be a quasisimple group and $\pi: L \to O(V,f)$ an absolutely irreducible FL-representation defined over no proper subfield of F. When is $N_{\tilde{G}}(L\pi)$ maximal in $\tilde{G}$?

Gary Seitz has made some progress on Problem 4; see his article elsewhere in this book.

4 OVERGROUPS AND LOCAL STRUCTURE IN SPORADIC GROUPS

In this section I'll discuss descriptions of the maximal overgroups of Sylow p-groups in sporadic groups and the p-local structure of sporadic groups. The theory is analogous to that for groups of Lie type in characteristic p. This work is still in progress so the results are tentative.

The theory is in part geometric. The relevant notion of geometry is due to J. Tits. According to Tits, a geometry over an index set I is a triple $(\Gamma,\tau,*)$, where Γ is a set of objects, $\tau:\Gamma \to I$ is a type function, and $*$ is a symmetric incidence relation on Γ such that objects u and v of the same type are incident if and only if $u = v$. Γ is said to be of rank $|I|$. A flag of Γ is a set T of objects each pair of which is incident. $\tau(T)$ is the type of T and $|\tau(T)|$ is the rank of T. The residue of T is the geometry Γ_T on $I - \tau(T)$ whose objects are those not in T but incident with each member of T, and whose incidence and type are inherited from Γ. A morphism of geometries is a map preserving type and incidence. A group G of automorphisms of Γ is flag transitive on Γ if G is transitive on flags of type J for each $J \subseteq I$. Γ is connected if the graph $(\Gamma,*)$ is connected. Γ is residually connected if each residue of rank at least 2 is connected and each residue of rank 1 is nonempty.

If Γ admits a flag transitive group of automorphisms then associated to Γ is a "diagram" $D(\Gamma)$ analogous to the Dynkin diagram that describes the building of a group of Lie type. (Incidentally that building is a geometry in the sense of Tits.) The "nodes" of $D(\Gamma)$ are the elements of I and the "edge" between nodes i and j is the isomorphism type of the residue of a flag of cotype (i,j). This diagram provides a concise notation for describing the structure of Γ.

Here's a construction which produces geometries. Let G be a group and $F = (G_i : i \in I)$ a family of subgroups of G. $\Gamma = \Gamma(G,F)$ is the geometry whose set of objects of type i is the coset space G/G_i and with two cosets incident when their intersection is nontrivial. Evidently G is represented as a group of automorphisms of Γ by right multiplication.

See [1] and [8] for more about geometries, diagrams, and the construction of the last paragraph.

Let G be a sporadic group, p a prime such that p^2 divides the order of G, and T a Sylow p-group of G. I determine the maximal overgroups of T and use them to define a geometry Γ preserved by G. $\Gamma = \Gamma(G,F)$, where usually F is a set of representatives for the orbits of $N_G(T)$ on the maximal overgroups of T. Γ is residually connected and G acts flag transitively on Γ. The diagram $D(\Gamma)$ is well behaved. In these and in other respects, Γ is analogous to the building of a group of Lie type. Γ provides a useful description of the overgroups of T and their intersections. For example the residual connectivity of Γ and the flag transitive action of G say these intersections are well behaved (cf. [1]). The diagram serves as an excellent tool for recording this description.

Define a p-<u>superlocal</u> of G to be a p-local H such that $H = N_G(O_p(H))$. The Borel-Tits Theorem says that in a group of Lie type and characteristic p, the parabolics are the p-superlocals. It is an elementary observation that if K is a p-local then there exists a p-superlocal H with $K \leq H$ and $O_p(K) \leq O_p(H)$. Thus to describe the p-locals it suffices for most purposes to describe the p-superlocals, and indeed this approach seems to provide the optimal description of the local structure of a group.

A p-local H is of <u>characteristic</u> p-<u>type</u> if $F^*(H) = O_p(H)$. The correct analogue of the Borel-Tits Theorem for sporadic groups seems to be the statement that almost all p-superlocals of characteristic p-type are superlocals of members of $\mathcal{F}$. As the p-locals which are not of characteristic p-type are easily determined using standard local group theoretic methods, we get a nice description of the local structure of G in terms of Γ via this approach.

The determination of the maximal overgroups of a Sylow p-group of a group G of Lie type and characteristic p and the proof of the Borel-Tits Theorem depend upon the existence of the Bruhat decomposition for G. Geometrically, the Bruhat decomposition corresponds to the existence of apartments in the building of G (cf. [8]). On the other hand my analysis of the sporadic groups proceeds from a different point of view. This suggests Problems 5 and 6 stated in the next section. The Bruhat decomposition can be used to define root groups in groups of Lie type, so perhaps a solution to Problem 5 would suggest a solution to Problem 6.

5 SOME OPEN PROBLEMS

Here's a list of some problems on subgroup structure which to my knowledge are open and seem to me to be important and interesting. The first six problems have already been mentioned in earlier sections.

Problem 1. Determine the conjugacy classes of maximal subgroups of the almost simple groups.

Problem 2. Determine $H^1(G,V)$ for the almost simple groups G and the irreducible modules V for G.

Problem 3. Let G be an almost simple group such that $F^*(G)$ is classical. Determine which members of S_G are maximal in G, where S_G is the set of subgroups of G described in section 3.

Problem 4. Let L be a quasisimple group, $G = O(V,f)$ a classical group over a finite field F, and $\pi:L \to G$ an absolutely irreducible FL-representation defined over no proper subfield of F. When is $N_G(L\pi)$ maximal in G?

Problem 5. Is there a useful analogue of the Bruhat decomposition for the sporadic groups? Do the geometries of the sporadic groups possess useful analogues of apartments?

Problem 6. Do the sporadic groups possess some reasonable analogue of root groups?

Problem 7. Answer the analogues of Problems 3 and 4 when G is an alternating group (cf. [5] and the appendix to [3]).

Problem 8. Extend Theorem 3 in some way to the exceptional groups of Lie type and the sporadic groups. For example replace the module in Theorem 3 by the geometry of the group.

Problem 9. Describe the superlocals in the groups of Lie type and the alternating groups.

Problem 1 amounts to a determination of the primitive permutation representations of the almost simple groups. Perhaps this goal is too ambitious. However it does seem reasonable to hope that Problem 1 can be reduced to Problems 4 and 7, and that at least some qualitative statement can be made about the maximality of the subgroups discussed in those problems: e.g. "usually" $N_G(L\pi)$ is maximal in G.

Even if Problem 1 is solved our information about the primitive permutation representations of the almost simple groups is far from complete. For example what is the permutation rank of any given representation? For this question and others one needs information about the intersection of maximal subgroups and the action of one maximal subgroup on conjugates of another. Such information and other important facts are perhaps best retrieved from nice representations of the group. One candidate for such a representation is that of the group on one of its overgroup geometries. Solutions to Problems 5, 6, and 7 would assist in the investigation of sporadic groups and exceptional groups of Lie type using this representation.

The Lie theory gives a good hold on the centralizers of semisimple elements in groups of Lie type. Recent work of Seitz [6] on overgroups of maximal tori gives a means for determining more general locals. Of course Theorem 3 supplies some maximal overgroups of locals in the classical groups. Seitz relates this representation to the Lie theory in [7]. These remarks suggest a possible approach to Problem 9.

REFERENCES

1. Aschbacher, M. (1983). Flag structures on Tits geometries. Geom. Ded. 14, 21-32.
2. Aschbacher, M. On the maximal subgroups of the finite classical groups. Preprint.
3. Aschbacher, M. & Scott, L. Maximal subgroups of finite groups. To appear in J. Alg.
4. Gross, F. & Kovacs, L. Maximal subgroups in composite finite groups. Preprint.
5. Scott, L. (1980). Representations in characteristic p. Proc. Sym. Pure Math. 37, 319-332.
6. Seitz, G. The root subgroups for maximal tori in finite groups of Lie type. (1983). Pac. J. Math. (106), 153-244.
7. Seitz, G. (1982). On the subgroup structure of classical groups. Comm. Alg. 10, 875-885.
8. Tits, J. (1981). A local approach to buildings. In The Geometric Vein, Springer, 517-547.

PROBLEMS REDUCIBLE TO THOSE ON FINITE SIMPLE GROUPS

M. Hall Jr.*
California Institute of Technology
Pasadena, CA

1 INTRODUCTION

Many problems on finite groups can be reduced to problems on finite simple groups. Now that the classification of the finite simple groups is complete this makes such a problem the examination of the known list of the simple groups.

Some of these problems have been completely resolved. Using a classic result of W.G. Burnside [1], Peter Cameron [3] has given a complete list of the multiply transitive finite permutation groups.

Using this result and the 1972 Oxford dissertation for the Master of Science degree of J.I. Hall, the writer [7] has now confirmed the correctness of his 1960 conjecture [5] "A Steiner triple system with a group of automorphisms doubly transitive on its points is either an affine geometry over GF(3) or a projective geometry over GF(2)."

The writer [6] has shown that for each prime p there is a family of numbers, including prime powers $q_p = r^s$, r is a prime, with $q_p \equiv 1 (\mathrm{mod}\ p)$ and $\{n_p\}$ where n_p is the number of Sylow p-subgroups in a finite simple group. The number of Sylow p-subgroups in an arbitrary group is a product of numbers q_p and n_p, and any such product can arise in some group. This last can be done simply by taking the direct product of groups with respectively the q_p Sylow p-subgroups and the n_p Sylow p-subgroups. It would be interesting to find the numbers n_p. Note that any odd number can be the number of Sylow 2-subgroups in a group.

In his 1954 Ohio State Ph.D. dissertation, Richard A. Zemlin [8] considered a conjecture of Frobenius. "If G is a finite group of order mn and if the number of solutions of $x^n = 1$, $x \in G$ is exactly n, then these n elements form a characteristic subgroup of G." He reduced the truth of this to the case in which G is simple. Here this amounts

*This research was supported in part by the National Science Foundation under Grant Number MCS 82-17596.

to showing that if $n \neq 1$, then $x^n = 1$, $x \in G$ always has more than n solutions.

There are doubtless many other problems of this kind which remain to be settled.

2 THE NUMBER OF SYLOW p-SUBGROUPS IN A GROUP

By the Sylow theorems, the number n of Sylow p-subgroups in a group G, is a divisor of $|G|$ and $n \equiv 1 (\mathrm{mod}\ p)$.

Theorem 2.1 [6]. Let $q_p = r^s \equiv 1 (\mathrm{mod}\ p)$ r a prime and let n_p be the number of Sylow p-subgroups in a simple group S. Then in an arbitrary group G, the number n of Sylow p-subgroups is a product of numbers q_p and n_p.

Here any odd number can be the number of Sylow 2-subgroups in a group. For $p \neq 2$, it appears that there are some n's with $n \equiv 1 (\mathrm{mod}\ p)$ which are not the number of Sylow p-subgroups in any group. Of course $n = 1$ means that a Sylow p-subgroup is normal. For $p = 5$, $n = 6$ in A_5, $n = 11$ and 16 are q_5's and $n = 21$ is the first value in question for Sylow 5-groups $S(5)$'s.

Lemma 2.1. There is no group G with exactly 21 $S(5)$'s.

Proof. Since $21 = 3.7$ and $3 \not\equiv 1(5)$, $7 \not\equiv 1(5)$ the Theorem says that we may suppose G to be a simple group. Thus G has a transitive faithful representation on 21 points and the stabilizer of a point is an $N(5)$, the normalizer of an $S(5)$. Since $25 > 21$, an $S(5)$ has orbits of length 1 and length 5. Thus $S(5)$ is a subdirect product of cyclic groups of order 5 and so is Abelian. By a result of J.S. Brodkey [2] there will be two $S(5)$'s whose intersection is the intersection of all 21 $S(5)$'s and so is a normal subgroup and in this case necessarily the identity. Let $S(5)$ be of order 5^t, $1 \leq t \leq 4$. Let $S_i(5)$ fixing i and $S_j(5)$ fixing j be the two $S(5)$'s whose intersection is the identity. Now in $S_i(5)$ j is in a orbit of length 1 or 5 so that $S_i(5)$ has a subgroup H of index at most 5 which fixes j. As H fixes j and is a 5 group, H is a subgroup of $S_j(5)$. Here $|H| = 5^{t-1}$ and $H = 1$ so that an $S(5)$ is of order 5. If an $S(5)$ fixes more than one point, then $N(5)$ will be transitive on the points fixed by $S(5)$, contrary to the fact that $N(5)$ is the stabilizer of a point. Hence with appropriate numbering we may take $S_1(5) = \langle a\rangle$ where $a = (1)(2,3,4,5,6)(7,8,9,10,11)$ $(12,13,14,15,16)(17,18,19,20,21)$. Here if $C_1(5)$ is the centralizer of $S_1(5)$, then $[N_1(5): C_1(5)] = 1, 2,$ or 4 and $|C_1(5)|$ divides $4!\cdot 5 = 120$. Hence the order of G is a multiple of $21.10 = 210$ and a divisor of $21 \cdot 4 \cdot 5 \cdot 24 = 10{,}080$. The only simple group whose order satisfies this is A_7 of order 2520. But in A_7 $N(5)$ is of order 20 and A_7 has 126 $S(5)$'s. Hence no group has exactly 21 $S(5)$'s.

A similar argument shows that no group has 15 $S(7)$'s and a slightly more complicated argument that no group has 22 $S(3)$'s.

3 A CONJECTURE OF FROBENIUS

In a finite group G of order $g = mn$ let A_n be the set of elements x in G such that $x^n = 1$.

In 1895 Frobenius [4] proved the following interesting result:

$$|A_n| = kn \quad \text{for an integer } k \geq 1.$$

Clearly since the identity belongs to A_n the set is never empty.

He made the following conjecture:

Frobenius conjecture: *If $|A_n| = n$, then the n elements of A_n form a characteristic subgroup of* G.

Clearly if the elements of A_n form a subgroup it is necessarily a characteristic subgroup since every automorphism of G takes the elements of A_n into themselves.

In his 1954 Ph.D. dissertation at Ohio State, R.A. Zemlin [8] reduced the issue of the truth of the Frobenius conjecture to the issue when G is simple. He showed even more, as given by the following theorem.

Theorem 3.1. The Frobenius conjecture is valid for all groups G provided that it is valid in the following restricted case.

G is a group of order $q = mn$ and n is a proper divisor of g such that

i) G is a simple group

ii) G is generated by A_n

iii) $g \leq 2^{n-1}$

iv) Either $(2,m) = 1$ or $(2,n) = 1$

v) If p is a prime dividing both m and n then the Sylow p-groups of G are cyclic.

vi) The conjecture is valid for all groups of order less than g. For groups of order g the conjecture is valid for all divisors n* of g which are less than n.

Since a simple group has no proper characteristic subgroup, this amounts to saying for G simple that $|A_n| > n$ for every proper divisor n of g. The conditions (iv) and (v) are very strong. Here (vi) is merely a base for an inductive proof.

REFERENCES

1. Burnside, W.G. (1911). The Theory of Groups. Second edition. Cambridge University Press.
2. Brodkey, J.S. (1963). A note on finite groups with an Abelian Sylow group. Proc. Amer. Math. Soc. 14, 132-133.
3. Cameron, P.J. (1981). Finite permutation groups and finite simple groups. Bulletin L.M.S. 13, 1-22.
4. Frobenius, G. (1895). Verallgemeinerung des Sylowschen Satze. Berliner Sitzungsberict, 981-993.
5. Hall, M. (1962). Automorphisms of Steiner Triple Systems. A.M.S. Proceedings of Symposia in Pure Mathematics 6, 47-66.
6. _______. (1967). On the number of Sylow subgroups in a finite group. J. of Algebra 7, 363-371.
7. _______. Steiner triple systems with a doubly transitive automorphism group. To appear in J. Combinatorial Theory (Series A).
8. Zemlin, R.A. (1954). On a conjecture arising from a theorem of Frobenius. Ph.D. dissertation. The Ohio State University.

FINITE GROUPS WITH EXACTLY ONE p-BLOCK

M.E. Harris
School of Mathematics, University of Minnesota
Minneapolis, Minn. 55455

Let p denote a prime integer and let G be a finite group. As developed by R. Brauer, the irreducible complex representations of G are distributed into disjoint systems called the p-blocks of G, which are naturally related to the primitive central idempotents of the group algebra of G over the algebraic closure of the field of p elements (cf. [3]). In particular, every p-block of G always contains at least one irreducible complex representation of G and G always has at least one p-block: the p-block of G that contains the trivial complex representation of G - called the principal p-block of G and denoted by $B_o(G)$. An old problem in this realm (cf. [3, IV, 5, (V)]) is:

Can the number of p-blocks be characterized in terms of the structure of G?

Utilizing the recently completed classification of finite simple groups, properties of finite simple groups and the theory of p-blocks, we can demonstrate:

<u>Theorem 1</u>. Let G be an arbitrary finite group. The following two conditions hold:

a) if p is odd, then $B_o(G)$ is the only p-block of G if and only if $F^*(G) = O_p(G)$; and

b) $B_o(G)$ is the only 2-block of G if and only if (i) $O_{2'}(G) = 1$ and (ii) all components of G are of type M_{22} or M_{24}.

A complete proof of this result will appear in [5] and involves the following concept introduced by R. Brauer in [1, Section IV]:

<u>Definition 1</u>. If r is a non-negative integer, then G is said to be of p-deficiency class r if all non-principal p-blocks of G have defect less than r.

Thus G is of p-deficiency class 0 if and only if $B_o(G)$ is the only p-block of G and G is of p-deficiency class 1 if and only if all non-principal p-blocks of G (if any) have defect 0. Also if G is of p-deficiency class r, then certainly it is of p-deficiency class $r + 1$.

These observations suggest:

Definition 2. Let G be of p-deficiency class r. Then G is said to be of exact p-deficiency class r if $r = 0$ or if $r > 0$ and G is not of p-deficiency class $r - 1$.

As is well known, a finite p-group is of exact p-deficiency class 0 and a finite non-identity p'-group is of exact p-deficiency class 1 (cf. [3, Chapter III]).

In fact, [5] contains a proof of the following p-local characterization of the exact p-deficiency class of a finite group. To state this result, we require:

Definition 3. Suppose that $|G|_p = p^\alpha$. If p is odd, set $\delta(p,G) = \{n \in Z | 0 \leq n \leq \alpha$ and $F^*(N_G(Q)) = O_p(N_G(Q))$ for every subgroup $Q \leq G$ with $|Q| = p^n\}$. For $p = 2$, set $\delta(2,G) = \{n \in Z | 0 \leq n \leq \alpha$ and $N_G(Q)$ satisfies (i) $O_{2'}(N_G(Q)) = 1$ and (ii) all components of $N_G(Q)$ are of type M_{22} and M_{24} for every subgroup $Q \leq G$ with $|Q| = 2^n\}$.

In [5] , we prove:

Theorem 2. Let $P \in Syl_p(G)$ and suppose that $|G|_p = |P| = p^\alpha$. Then:

a) if $C_G(P) \neq Z(P)$, then $\delta(p,G) = \emptyset$ and G is of exact p-deficiency class $\alpha + 1$; and

b) if $C_G(P) = Z(P)$, then $\delta(p,G) \neq \emptyset$ and G is of exact p-deficiency class $\min \delta(p,G)$.

Our proofs of Theorems 1 and 2 depend on the following two results whose proofs utilize a survey of all finite simple groups (via the recently completed classification):

Theorem 3 (P. Brockhaus and G.O. Michler, [2, Theorem]). If $p \neq 2$ and G is a non-abelian finite simple group, then G has at least two p-blocks.

Theorem 4 ([5, Theorem 4]). The following three conditions hold:

a) M_{22} and M_{24} are the only finite simple groups of 2-deficiency class 0;

b) the following finite simple groups are of exact 2-deficiency class 1:

i) all finite simple groups of Lie type over a finite field of characteristic 2,

ii) $PSL(2,p^f)$ with p an odd prime, f a positive integer and $p^f = 9$ or p^f a Fermat or Mersenne prime,

iii) A_n with n = 5, 6 or 8,

iv) PSL(3,3), PSU(3,3), PSU(4,3), PSp(4,3) and $G_2(3)$, and

v) M_{11}, M_{23}, J_3, J_4, .2 and F_3; and

c) all other finite simple groups are of 2-deficiency class at least 2.

Clearly Theorem 1 is a direct generalization to all finite groups of the simple group case of Theorems 3 and 4(a).

In [5], we used the general theory of the p-deficiency class of a finite group to derive Theorem 1 from Theorems 3 and 4(a) and to derive Theorem 2 by induction from Theorem 1.

It is possible to utilize Brauer's First Main Theorem directly to give a different derivation of Theorem 1 from Theorems 3 and 4(a). To see this, first observe that if $H \trianglelefteq \trianglelefteq G$ and G is of p-deficiency class r (resp. G has a p-block of defect 0) then H is also of p-deficiency class r (resp. H also has a p-block of defect 0) (cf. [5, Lemmas 3.10(b) and 3.11]). Suppose that $B_o(G)$ is the only p-block of G (i.e. G is of p-deficiency class 0). Then $O_{p'}(G) = 1$, and [3, V, Lemma 4.5] and Theorems 3 and 4(a) imply that $F^*(G) = O_p(G)$ if $p \neq 2$ and that (i) and (ii) of Theorem 1(b) hold if p = 2. On the other hand, it is well known (cf. [3, V, Corollary 3.11]) that $F^*(G) = O_p(G)$ implies that $B_o(G)$ is the only p-block of G. Assume that p = 2 and that G satisfies conditions (i) and (ii) of Theorem 1(b). Since $F^*(G) \neq 1$, it follows that G does not possess a 2-block of defect 0. Utilizing induction, passage to $\bar{G} = G/Z(G)$, [3, V, Lemma 4.5] and [4, Lemmas 2.17 and 2.18], we may assume that $O_2(Z(G)) = Z(G) = 1$. Also, it follows from the theory of components that if Q is a 2-subgroup of G, then $N_G(Q)$, $QC_G(Q)$ and $C_G(Q)$ also satisfy conditions (i) and (ii) of Theorem 1(b) (cf. [5, Lemma 2.9]). Let B be a 2-block of G and let D be a defect group of B. Then $D \neq 1$ and $H = DC_G(D)$ also satisfies (i) and (ii) of Theorem 1(b). Here $1 \neq Z(D) \leq Z(H)$ and hence $H < G$. By induction, $B_0(H)$ is the unique 2-block of H. Applying Brauer's First Main Theorem (cf. [3, V, Lemma 5.2]), we conclude that $B_0(H)^G = B$. On the other hand, if K is any conjugacy class of G, then $|K| \equiv |K \cap C_G(D)| \pmod 2$. From this and [3, III, Sections 7 and 9] it follows that $(B_0(H))^G = B_0(G)$. Thus $B = B_0(G)$ and Theorem 1 follows.

These results suggest several questions and problems that seem worthy of pursuit.

Find additional connections between the p-blocks of a finite group G and the components, p-components, p-layer, etc. of G. In particular, can necessary and sufficient conditions for G to have a p-block of defect 0 be given in these terms?

Since the simplicity and number of p-blocks of a finite group G can be easily determined from the character table of G, can Theorems 3 and/or 4(a) be derived without invoking the entire classification of finite simple groups?

REFERENCES

1. Brauer, R. (1964). Some applications of the theory of blocks of characters of finite groups, I. J. Algebra 1, 152-167.
2. Brockhaus, P. & Michler, G.O. Finite simple groups of Lie type have non-principal p-blocks, $p \neq 2$. Preprint.
3. Feit, W. (1982). The Representation Theory of Finite Groups. North-Holland, New York.
4. Harris, M.E. (1982). Finite groups containing an intrinsic 2-component of Chevalley type over a field of odd order. Trans. Amer. Math. Soc. 272, 1-65.
5. Harris, M.E. On the p-deficiency class of a finite group. To appear in the Journal of Algebra.

SOME RECENT RESULTS ON FINITE PERMUTATION GROUPS

M.W. Liebeck
DPMMS, University of Cambridge

J. Saxl
DPMMS, University of Cambridge

Over the last few years much work has been done on applying the classification of finite simple groups to problems (both classical and new) concerning permutation groups and related combinatorial objects. In this paper we shall report on some results in this area. The classification of finite simple groups will be assumed throughout.

NOTATION AND TERMINOLOGY

In what follows, G is a primitive permutation group of degree n acting on a set Ω. The product of all minimal normal subgroups of G is the _socle_ of G. We say that G is _affine_ if its socle is an elementary abelian p-group for some prime number p (so that $G \leq AGL(m,p)$ for some integer m), and G is _almost simple_ if its socle is a non-abelian simple group T (here then $T \leq G \leq \mathrm{Aut}\, T$).

A Doubly transitive groups

While we have no new results to report on multiple transitive groups, we include this section as an excellent illustration of the sort of results one aims for.

Theorem 1. All finite 2-transitive groups are known.

For a complete list and a discussion see [5] and [13]. The starting point of the proof is the classical theorem of Burnside, reducing considerations to affine groups or almost simple groups. The former case has been solved by Huppert when the group G is soluble and by Hering [10] if G is insoluble. In the latter case, where G is almost simple, the analysis splits naturally into three subcases: the case where G is alternating or symmetric, the case where G is of Lie type and finally the case where the socle of G is sporadic. Here the alternating and symmetric groups were investigated by Maillet last century and the sporadic groups yield to ad hoc arguments. The largest part of the proof is concerned with groups of Lie type — these were analysed by Curtis, Kantor and Seitz in [9] (see also Howlett [11]).

Kantor [13] recently used Theorem 1 to classify all 2-transitive Steiner systems.

B Odd degrees

The classification of primitive permutation groups of odd degree is now complete.

Theorem 2 ([20]). Let G be a primitive permutation group of odd degree n.

a) If G is almost simple then the action of G is explicitly known.

b) If G is not almost simple then either

G is affine (with $n = p^m$ for some odd prime p and an integer m); or

$G \leq G_0$ wr S_m, where G_0 is an almost simple primitive group of an odd degree n_0 (as in a)) and the action of G is the product action of degree n_0^m.

For a complete list of groups in a) see [20]. The proof of part b) is not at all difficult, using the O'Nan-Scott reduction theorem. We outline the proof of part a). Let H be the stabilizer of a point. As $|G:H| = n$ is odd, H contains a Sylow 2-subgroup of G. Consider first the case where G is alternating or symmetric of degree c. If H is intransitive or imprimitive in the natural action on c points then we know what H is by its maximality in G. On the other hand, if H is primitive on the c points, then since it contains a Sylow 2-subgroup of A_c, we have $c \leq 8$. If G is of Lie type of even characteristic then the action is parabolic by a theorem of Tits. In the remaining cases, the situation is considerably more complicated. If G is sporadic, the actions are known by a recent theorem of Aschbacher [4].

Hence we may assume that G is of Lie type of odd characteristic. If G is classical, we use Aschbacher's theorem in [3]. Most cases there are easy to deal with, but more care is needed if the dimension is small. Finally, let G be exceptional or twisted of odd characteristic. Here the required technology is developed in Aschbacher's papers [1] and [2]: let $\Omega(G)$ be the set of fundamental subgroups of G, let R(G) be the set of the quaternion subgroups of G that are Sylow 2-subgroups of some fundamental subgroup. For S a Sylow 2-subgroup of G, one defines $\mathrm{Fun}_G S = \{X \in \Omega(G) : \exists R \in R(G) \text{ with } R \leq S \cap X\}$. Since S can be taken to be contained in H, one can investigate the relationship between $\mathrm{Fun}_G S$ and the corresponding object for H. This is not difficult (modulo [1]) if H is not local, but for H local (and in particular H 2-local), a fairly detailed analysis of the situation is required.

Theorem 2 was also obtained by W.M. Kantor. As a consequence Kantor obtained a classification theorem for projective planes with an automorphism group acting transitively on flags.

C More special degrees

Much of the early work on permutation groups was concerned with groups of special degrees. The classification of transitive groups of prime degree of course follows from Theorem 1. Kantor [14, Example 5] discusses primitive groups of prime power degree. We have a result concerning the case of degree kp:

<u>Theorem 3</u> ([19]). Let G be a primitive permutation group of degree kp where p is a prime number and k is an integer smaller than p. Then G is explicitly known.

In fact, we prove a more general result classifying all primitive groups containing an element of order p with fewer than p cycles of length p.

The proof of Theorem 3 involves the use of bounds discussed in the next section. After reduction to the case where G is a group of Lie type, the knowledge of $|G|$ implies that p is small compared to $|G|$, so that the order of the stabilizer H of a point is large. For example, if G is of type Sp(2m,q) then $p < q^{2m}$, so that $|H| > |G|/q^{4m}$ and Theorem 5 below applies.

One of the motivations in the investigation of groups of special degrees was to show that the set $F = \{n \in \mathbb{N}$: the only primitive groups of degree n are A_n and $S_n\}$ is infinite. In this direction, we now have the surprising theorem of Cameron, Neumann and Teague [6]: not only is F infinite, but it has density 1 in $\mathbb{N}$ in the sense that if $e(x) = |\{n \in \mathbb{N}\backslash F : n \leq x\}|$ then $e(x) \sim 2x/\log x$.

D Bounds

One of the classical problems in group theory was to obtain "good" upper bounds on the orders of maximal subgroups of alternating and symmetric groups. The question can be extended naturally to other families of almost simple groups G. Quite apart from very considerable independent interest, we have found such bounds an invaluable starting point in attacking a number of problems – see Theorem 3 above, section F below and also [22].

Cameron applied the classification of finite simple groups to obtain (quite easily) the following result.

Theorem 4 (Cameron [5, Theorem 6.1]). If H is a maximal subgroup of A_m or S_m then one of the following holds:

i) H is affine;

ii) H is explicitly known; or

iii) $|H| \leq \exp(10 \log m \log\log m)$.

Here iii) can be strengthened if one is willing to enlarge the list in ii). Also, i) can be transferred into ii) using Theorem 5 below.

For groups G other than the alternating or symmetric, the situation is more complicated. For classical groups, a satisfactory answer was supplied in [17]:

Theorem 5 (Liebeck [17]). Let T be a classical simple group with natural (projective) module V of dimension m over GF(q), let $T \leq G \leq \mathrm{Aut}\, T$ and let H be a maximal subgroup of G. Then either

i) H is explicitly known, or

ii) $|H| < q^{2m+4}$.

The proof here starts with an application of the Aschbacher theorem [3] to reduce the considerations to subgroups H that are almost simple with a covering group absolutely irreducible on V. If H is alternating or symmetric, the results in [12] can be applied. If H is sporadic one has to use ad hoc arguments together with the information in the Atlas [8]. For example, if H is the Conway group Co_1, then the existence of subgroups of type $3^6 2M_{12}$ and $2^{11}M_{24}$ leads to the conclusion that $m \geq 24$, and in fact Co_1 appears in i) as a subgroup of $O^+_{24}(2)$. The case where H is of Lie type of characteristic not dividing q is covered by the results of Landazuri and Seitz [16]. Hence it remains to consider the groups H of Lie type of the same characteristic as G. Here the methods are those used in the modular representation theory of these groups. We illustrate the two main steps involved by an example where H is $Sp(2\ell,q_0)$ as an absolutely irreducible subgroup of $GL(m,q)$. Firstly, if $q_0 = p^a$ and $q = p^b$ then it can be shown using the Steinberg tensor product theorem that $m \geq (2\ell)^c$, where $c = a/(a,b)$. Secondly, if $q_0 = q$ and $m < 2\ell^2$ then it is shown that one of the following holds: a) $m = 2\ell$ and V is the natural module for H; b) $m = \ell(2\ell-1)-\varepsilon$ with ε either 1 or 2; c) $m = 2^\ell$, q is even and V is a spin module for H; d) $m = 14$, $\ell = 3$ and q is odd.

Finally, to conclude this section, we report that work is in progress on bounds for groups of exceptional or twisted type. We give a sample theorem.

<u>Theorem 6</u> ([21]). Let H be a maximal subgroup of $E_8(q)$. Then one of:

i) H is parabolic;

ii) H is of type $D_8(q)$, $A_1(q)\circ E_7(q)$ or $E_8(q^{\frac{1}{2}})$; or

iii) $|H| < q^{118}$.

Here the bound in iii) is motivated by Proposition 10 in section F. We have similar results for the other exceptional and twisted groups.

The results of this section should be sufficient to complete the classification of primitive rank 3 permutation groups — see also section F.

E <u>The Sims conjecture</u>

It was suggested by Charles Sims that in a primitive permutation group, the length of any non-trivial suborbit should bound the size of the stabilizer of a point (and hence the length of any other suborbit). This has now been proved:

<u>Theorem 7</u> ([7, Theorem 1]). The Sims conjecture is true.

As a consequence, a result concerning distance transitive graphs was obtained:

Theorem 8 ([7, Theorem 2]). There are only finitely many finite distance transitive graphs of any given valency greater than 2.

Further discussion of such graphs appears in the next section.

F Distance transitive graphs

Recall that a connected, undirected graph with no loops and no multiple edges is said to be distance transitive if, whenever α, β, γ and δ are points with the distance between α and β being the same as the distance between γ and δ, there is an automorphism taking α to γ and β to δ. (Here the distance between two points is the length of the shortest path between them.)

Examples. 1. Any primitive rank 3 permutation group of even order gives rise to a pair of complementary distance transitive graphs of diameter 2.

2. For many families of groups G of Lie type there are parabolic examples of distance transitive graphs. These families include most classical groups as well as groups of type G_2, ${}^3D_4$, ${}^2F_4$, E_6 and E_7. Also, there are two families of graphs Γ with $\mathrm{Aut}\,\Gamma$ a symmetric group.

3. Various affine examples. For example, if the set Ω of vertices is taken to be the set of all $r\times s$ matrices with entries in $GF(q)$, joining two such if the rank of their difference is 1, we obtain a distance transitive graph admitting $\Omega\cdot(GL(r,q)\times GL(s,q))$.

4. Hamming graphs: If $\Sigma = \{1,\ldots,m\}$, we take Ω to be Σ^k, with two elements joined if they differ in exactly one position. This leads to a distance transitive graph Γ with $\mathrm{Aut}\,\Gamma \simeq S_m$ wr S_k, the wreath product of S_m with S_k in the product action on Ω.

So, while distance transitive graphs are highly symmetric structures, there are plenty of examples. Nevertheless, we now hope to be able to prove a classification theorem for these in the near future and we describe in this section what our hope is based on. We are working on this with Nick Inglis and Cheryl Praeger.

Firstly, there is a reduction theorem due to D.H. Smith [25]: if G is distance transitive on Γ but not primitive on the set Ω of vertices then Γ is either antipodal (that is the relation of being at maximal distance is an equivalence relation) or bipartite, and in either case there is a reduction procedure for obtaining a smaller distance transitive graph Γ' out of Γ with $\mathrm{Aut}\,\Gamma'$ primitive on the set of vertices of Γ'. Hence we shall assume that our distance transitive graph Γ admits an automorphism group G primitive on the set of vertices.

Next we apply the O'Nan-Scott reduction theorem for primitive groups. While this has turned out to be very much more difficult than in all the other applications so far, it has now been completed:

Theorem 9 (Praeger and Saxl [23]). Suppose that G is a group acting distance-transitively on the graph Γ and primitively on the set of vertices of Γ. Then one of the following holds:

i) G is almost simple;

ii) G is affine; or

iii) Γ is a Hamming graph or its complement.

We now consider cases i) and ii) separately. Let $H = G_\alpha$, the stabilizer of a vertex.

Case i). The starting point here is the well-known fact that the permutation character 1_H^G is multiplicity free (as all the suborbits are self-paired, due to the fact that distance is symmetric). If G is alternating or symmetric, the graphs have been classified in [24]. While the case where G is sporadic is wide open at present, we believe that it will yield to ad hoc arguments using the information in the Atlas [8]. We therefore concentrate on the case where G is a group of Lie type.

Let now B be a Borel subgroup of G, let W be the Weyl group of G. The following observation is essentially [24, 3.b]:

Proposition 10. If 1_H^G is multiplicity free then $|H| > |G:B|/|W|$.

This is the point where the bounds of section D come in. For example, if G is classical then either the dimension of G is small or H is known. It remains to analyze which of the "suspect" characters 1_H^G are actually multiplicity free (or, at least, which lead to distance transitive graphs) and Nick Inglis is working on this. Surprisingly, a non-trivial proportion of the candidates H do give rise to multiplicity free characters 1_H^G. For instance, if G is $GL(m,q)$ with $m \geq 8$ then the set of maximal subgroups H of G with 1_H^G multiplicity free consists precisely of the following: a) parabolic subgroups; b) m even and $H \trianglerighteq GL(\frac{1}{2}m, q^2)$ or $H \trianglerighteq Sp(m,q)$; c) q a square and $H \trianglerighteq GL(m,q^{\frac{1}{2}})$ or $H \trianglerighteq GU(m,q^{\frac{1}{2}})$. That 1_H^G is multiplicity free for cases b) and c) is due to Inglis, Gow and Kljačko.

Case ii). Here $G \leq AGL(m,p)$ for some prime number p and an integer m. The first observation is that the permutation rank is not too large:

Proposition 11. The permutation rank of G is at most $m+1$, so that $|H| \geq (p^m-1)/m$.

This is not difficult to prove. To contrast it, if the rank is 3, we can finish:

Theorem 12 (Liebeck [18]). The affine rank 3 permutation groups are known.

The ingredients of the proof of this are bounds and representation theory. We hope to be able to extend the techniques to cover all affine distance transitive graphs.

Acknowledgements

The authors thank Rutgers University for their hospitality. The second author also thanks SERC for their financial support.

REFERENCES

1. Aschbacher, M. (1977). A characterization of Chevalley groups over fields of odd order. Ann. Math. 106, 353-468.
2. Aschbacher, M. (1980). On finite groups of Lie type and odd characteristic. J. Alg. 66, 400-424.
3. Aschbacher. On the maximal subgroups of the finite classical groups. To appear in J. Alg.
4. Aschbacher, M. To appear.
5. Cameron, P.J. (1981). Finite permutation groups and finite simple groups. Bull. London Math. Soc. 13, 1-22.
6. Cameron, P.J., Neumann, P.M. & Teague, D.N. (1982). On the degrees of primitive permutation groups. Math. Z. 180, 141-149.
7. Cameron, P.J., Praeger, C.E., Saxl, J. & Seitz, G.M. (1983). On the Sims conjecture and distance transitive graphs. Bull. London Math. Soc. 15, 499-506.
8. Conway, J.H. et al. Atlas of finite simple groups. To appear.
9. Curtis, C.W., Kantor, W.M. & Seitz, G.M. (1976). The 2-transitive permutation representations of the finite Chevalley groups. Trans. Amer. Math. Soc. 218, 1-57.
10. Hering, C. To appear in J. Alg.
11. Howlett, R.B. (1974). On the degrees of Steinberg characters of Chevalley groups. Math. Z. 135, 125-135.
12. James, G.D. (1983). On the minimal dimensions of irreducible representations of symmetric groups. Proc. Cambridge Phil. Soc. 94, 417-424.
13. Kantor, W.M. Homogeneous designs and geometric lattices. To appear.
14. Kantor, W.M. Some consequences of the classification of finite simple groups. To appear.

15. Kantor, W.M. & Liebler, R.A. (1982). The rank 3 permutation representations of the finite classical groups. Trans. Amer. Math. Soc. 271, 1-71.
16. Landazuri, V. & Seitz, G.M. (1974). On the minimal degrees of projective representations of the finite Chevalley groups. J. Alg. 32, 418-443.
17. Liebeck, M.W. On the orders of maximal subgroups of finite classical groups. To appear in Proc. London Math. Soc.
18. Liebeck, M.W. Affine permutation groups of rank 3. In preparation.
19. Liebeck, M.W. & Saxl, J. Primitive permutation groups containing an element of large prime order. Submitted to London Math. Soc.
20. Liebeck, M.W. & Saxl, J. The primitive permutation groups of odd degree. Submitted to London Math. Soc.
21. Liebeck, M.W. & Saxl, J. On the orders of maximal subgroups of exceptional and twisted groups of Lie type. In preparation.
22. Liebeck, M.W., Praeger, C.E. & Saxl, J. On the maximality of primitive permutation groups with non-simple socles. In preparation.
23. Praeger, C.E. & Saxl, J. Distance transitive graphs and finite simple groups. To appear.
24. Saxl, J. (1981). On multiplicity free permutation representations. In Finite geometries and designs, London Math. Soc. Lecture Notes Series 49, 337-353.
25. Smith, D.H. (1971). Primitive and imprimitive graphs. Quart. J. Math. 22, 551-557.

SHARPLY 2-TRANSITIVE SETS OF PERMUTATIONS

M.E. O'Nan
Rutgers University, New Brunswick, NJ 08903

A set S of permutations on a set X is said to be sharply transitive if given $a,b \in X$, there is a unique $f \in S$ with $f(a) = b$. In what follows X will be finite, and, in this case, it follows that $|S| = |X|$.

Similarly, a set S of permutations of X is sharply 2-transitive if given (a,b), (c,d) with $a,b,c,d \in X$, and $a \neq b$, $c \neq d$, there is a unique $f \in S$, with $f(a) = c$ and $f(b) = d$. If S is sharply 2-transitive and $g,h \in S(X)$, the symmetric group on X, it is easily verified that gSh is still sharply 2-transitive. Thus, without loss, one may assume that a sharply 2-transitive set contains the identity. If $|X| = n$, simple counting shows:

1) $|S| = n(n-1)$.

2) If $a \in X$ and S_a denotes those $f \in S$ with $f(a) = a$, then $|S_a| = n-1$.

3) If $1 \in S$, the number of elements of S having no fixed point is $n-1$.

4) Define $f \sim g$ if either $f = g$ or $g^{-1}f$ has no fixed point. Then, $\sim$ is an equivalence relation, and the equivalence classes have size n.

By a result of Witt [4], a projective plane of order n exists if and only if there is a sharply 2-transitive set of permutations of degree n.

In fact, if Y is a projective plane of order n and ℓ_1, ℓ_2 lines of Y, let a be the common point of ℓ_1 and ℓ_2. Then, $|\ell_1 - a| = |\ell_2 - a| = n$, and we obtain functions from $\ell_1 - a$ to $\ell_2 - a$ as follows. If $t \in Y - (\ell_1 \cup \ell_2)$, we define $f_t\colon \ell_1 - a \to \ell_2 - a$ by the rule: if $x \in \ell_1 - a$, let m be the line determined by x and t. Let y be the intersection of m with $\ell_2 - a$. Then, $f_t(x) = y$. One verifies easily enough given $r,u \in \ell_1-a$, $r \neq u$, $v,w \in \ell_2-a$, $v \neq w$, there exists a unique t with $f_t(r) = v$, $f_t(u) = w$. Identifying ℓ_1-a and ℓ_2-a, we obtain the sharply 2-transitive set of permutations of degree n.

By a theorem of Curtis, Kantor, Seitz [1], and the classification, all doubly-transitive groups of degree not a prime power are known. We examine those that do not contain the alternating group to determine if they contain a sharply 2-transitive set of permutations. The study was begun by Lorimer [2] who showed that if $q \geq 5$, $m \geq 2$, $P\Gamma L(m,q)$ contains no sharply 2-transitive set of permutations (in the natural representation on projective space). Here we prove:

<u>Theorem</u>. If $m \geq 2$, $P\Gamma L(m,q)$ contains no sharply 2-transitive set of permutations unless $m = 2$, $q = 2,3,4$.

Our goal is to extend Lorimer's counting methods by the use of character theory.

Let G be a finite group with subgroup H. We begin by giving a criterion for determining whether the permutation representation of G on cosets of H contains a sharply transitive set of permutations.

<u>Lemma 1</u>. Let G be a group with subgroup H. Let S be a subset of G with $|S| = |G:H|$. Then the following are equivalent:

i) S is sharply transitive on cosets of H.

ii) If ψ is a non-principal irreducible constituent of 1_H^G, then $\sum_{f \in S} f$ (in $\mathbf{Z}[G]$) lies in $\mathrm{Ker}(\psi)$.

<u>Proof</u>. We regard G as a permutation group on cosets of H. If $f \in G$, let $P(f)$ be the corresponding permutation matrix and V the corresponding permutation module.

By complete reducibility over C, we obtain projections $\pi_1, \pi_2, \ldots, \pi_k$ on V with $\pi_i \pi_j = 0$ if $i \neq j$, and $\pi_1 + \pi_2 + \ldots + \pi_k = 1$. We assume π_1 is the projection onto the trivial module.

To say that S is sharply transitive on cosets of H is equivalent to saying:

$$\sum_{f \in S} P(f) = \begin{bmatrix} 1 & 1 & 1 & \cdots & 1 \\ 1 & 1 & 1 & \cdots & 1 \\ 1 & 1 & 1 & \cdots & 1 \\ & & \vdots & & \\ 1 & 1 & 1 & \cdots & 1 \end{bmatrix} = J.$$

Also observe

$$\pi_1(P(f)) = \frac{1}{|G:H|}\begin{bmatrix} 1 & 1 & 1 & \cdots & 1 \\ 1 & 1 & 1 & \cdots & 1 \\ 1 & 1 & 1 & \cdots & 1 \\ & & \vdots & & \\ 1 & 1 & 1 & \cdots & 1 \end{bmatrix}.$$

It follows that $\sum_{f\in S} \pi_1(P(f)) = J$, since $|S| = |G:H|$.

Thus, if S is sharply transitive, $\sum_{f\in S} P(f) = \pi_1(\sum_{f\in S} P(f))$.
Thus, if $i \neq 1$, $\pi_i(\sum_{f\in S} P(f)) = 0$. Thus, if ψ denotes the corresponding character, $\sum_{f\in S} f \in \mathrm{Ker}(\psi)$. Reversing the argument yields the converse.

Using this lemma we obtain:

Lemma 2. Suppose G admits a sharply transitive set of permutations on cosets of a subgroup H. Let K be a subgroup of G, such that all irreducible constituents of 1_K^G are constituents of 1_H^G. Then, $|G:K| \mid |G:H|$.

Proof. Now let $P(f)$, $f \in G$, be the permutation matrix on cosets of K, V the permutation module, $\pi_1,\pi_2,\ldots,\pi_\ell$ the projections onto the irreducible constituents, with π_1 being the projection onto the trivial module. As in the last lemma, let J be $|G:K|\times|G:K|$ matrix all of whose entries are 1. Then, $\pi_1(P(f)) = \frac{1}{|G:K|} J$

Also,

$$\sum_{f\in S} P(f) = (\pi_1+\ldots+\pi_\ell)(\sum_{f\in S} P(f)) = \pi_1(\sum_{f\in S} P(f)),$$

since $\sum_{f\in S} \pi_i(P(f)) = 0$, if $i > 1$, by the last lemma and the assumption that any irreducible constituent of 1_K^G is an irreducible constitutent of 1_H^G.

It follows that

$$\sum_{f\in S} P(f) = \pi_1(\sum_{f\in S} P(f)) = \frac{|G:H|}{|G:K|} J.$$

Since $\sum_{f\in S} P(f)$ is an integer matrix it follows that $|G:K|$ divides $|G:H|$.

If H is a subgroup of G, we say a subgroup K is a contradicting subgroup for H if

1) All irreducible constituents of 1_K^G are constituents of 1_H^G, and

2) $|G:K|$ does not divide $|G:H|$.

Thus, by the last lemma, if H has a contradicting subgroup, G does not have a sharply transitive set of permutations on cosets of H. For example, let $G = S_n$ with n even. Let H be the setwise stabilizer of a 2-element set. Then, $|G:H| = \frac{n(n-1)}{2}$. Let K be a one-point stabilizer. It is well-known that all constituents of 1_K^G are constituents of 1_H^G. Since n does not divide $\frac{n(n-1)}{2}$, K is a contradicting subgroup for H.

We now turn to the case in which G is a known 2-transitive group on a set X. To say that a subset S of G is sharply 2-transitive on X is equivalent to saying that S is sharply transitive on cosets of the two-point stabilizer G_{xy}, $x,y \in X$, $x \neq y$. We shall hunt for a contradicting subgroup for G_{xy}.

We suppose $G = P\Gamma L(n,q)$, U is an n-dimensional vector space over $GF(q)$ affording $G\Gamma L(n,q)$, X the one-dimensional subspaces of U and X' the hyperplanes of U. Then, G is 2-transitive on X and X'. Let V and V' be the corresponding permutation modules. Let ψ_1 be the trivial character and ψ_2 the non-trivial irreducible constituent of the permutation character on X. If $x,y \in X$, $x \neq y$, then, since G_{xy} has two (or more) orbits on X, ψ_2 is a constituent of $1_{G_{xy}}^G$.

First observe that $V \cong V'$, as G-modules, since a semi-linear transformation fixes the same number of points and hyperplanes. It follows that $V\otimes V \cong V\otimes V'$, as G-modules. Note that $V\otimes V$ is the permutation module for the action of G on $X\times X$, while $V\otimes V'$ is the permutation module for the action of G on $X\times X'$.

Next the orbits of G on $X\times X$ are $Y_1 = \{(x,x)|x \in X\}$ and $Y_2 = \{(x,y)|x,y \in X,\ x \neq y\}$. The permutation character on Y_1 is $\psi_1 + \psi_2$ and on Y_2 is $1_{G_{xy}}^G$. Since ψ_1 and ψ_2 are constituents of $1_{G_{xy}}^G$, it follows that all irreducible constituents of $V\otimes V$ occur in $1_{G_{xy}}^G$.

It follows that all irreducible constituents of the permutation character of G on $X\times X'$ occur in $1_{G_{xy}}^G$. Next the orbits of G

on $X{\times}X'$ are $Y_1' = \{(x,P) | x \in X, P \in X', x \notin P\}$ and $Y_2' = \{(x,P)|, x \in X, P \in X', x \in P\}$. Let K be the stabilizer of a point in Y_1'. Then, all irreducible constituents of 1_K^G occur in $1_{G_{xy}}^G$.

By conjugation we may take $K \cap PGL(n,q)$ to be the image of

$$\left\{ \left[\begin{array}{c|c} a & 0 \\ \hline 0 & A \end{array}\right] \;\middle|\; a \in GF(q)^{\times}, A \in GL(n-1,q) \right\}.$$

Thus, the index $|G:K|$ is $q^{n-1}(1+q+\ldots+q^{n-1})$, while $|G:G_{xy}| = (1+q+\ldots+q^{n-1})(q+\ldots+q^{n-1})$. If $n \geq 3$, q^{n-1} does not divide $q+\ldots+q^{n-1}$. Thus, K is a contradicting subgroup. Thus, the theorem stated in the beginning follows if $n \geq 3$. For $n = 2$, $q \geq 5$, Lorimer's result yields it.

The same methodology shows that $P\Gamma U(3,q)$, $\Gamma R(q)$, and $\Gamma Sz(q)$ do not possess sharply 2-transitive sets of permutations, as Lorimer has obtained elsewhere [3].

Work on the groups $Sp(2n,2)$ of degrees $2^{2n-1} \pm 2^{n-1}$ is in progress. Interestingly, $Sp(6,2)$ of degree 28 has no contradicting subgroup. However, using Lemma 1 directly, one can show that it does not have a sharply 2-transitive set of permutations.

REFERENCES

1. Curtis, C.W., Kantor, W.M., Seitz, G.M. (1976). The 2-transitive representations of the finite Chevalley groups. Trans. Amer. Math. Soc. 218, 1-59.
2. Lorimer, P. (1973). A property of the groups $P\Gamma L(m,q)$, $q \geq 5$. Proc. Amer. Math. Soc. 37, 393-396.
3. Lorimer, P. (1974). Finite projective planes and sharply 2-transitive subsets of finite groups. Proc. of Second International Conference on the Theory of Groups, pp. 432-436. Lecture Notes in Mathematics, Vol. 372, Berlin: Springer.
4. Witt, E. (1938). Über steinersche Systeme. Abh. Hamburg 12, 265-275.

SCHUR MULTIPLIERS OF THE KNOWN FINITE SIMPLE GROUPS, III

R.L. Griess, Jr.
Department of Mathematics, University of Michigan
Ann Arbor, Michigan 48109

1 INTRODUCTION

The purpose of this note, whose appearance was foretold in [Gr6], is to clear up a few points about Schur multipliers of the finite simple groups.

We prove that the Schur multiplier of the Rudvalis group has order 2 (an argument for this was given by Feit and Lyons, but not published), that the Schur multiplier of F_2 has order 2 and that F_1 (the Monster) has trivial Schur multiplier. Finally, we close with a few comments.

2 THE SCHUR MULTIPLIER OF RUDVALIS' GROUP

Let G be the simple group discovered by Rudvalis. $|G| = 2^{14}3^3 5^3 7.13.29$. We refer to [Rul] for the facts we need. Clearly the multiplier is a $\{2,3,5\}$-group.

Since G contains the perfect group $\widehat{3M_{10}}$, which has trivial multiplier, the multiplier is a $\{2,5\}$-group. The normalizer of a Sylow 5-subgroup $P \cong 5^{1+2}$ contains an involution j which is -1 on $P/Z(P)$ and is $+1$ on $Z(S)$. An elementary Lie ring argument shows that $<P,j>$ has trivial multiplier. So, the multiplier of G is a 2-group. The multiplier has even order, since G contains an involution t, with the property that t does not lie in any conjugate of a subgroup, isomorphic to ${}^2F_4(2)$, of index $4060 \equiv 4 \pmod 8$ [Rul] [Grl]. Therefore, it remains to show that the multiplier of G has order at most 2.

Let $1 \to A \to \hat{G} \xrightarrow{\pi} G \to 1$ be the covering sequence of G. Let $H \leq G$ have the form $2^{3+8}GL(3,2)$; see [Dem]. Since $O_2(H)/Z(O_2(H))$ is the Steinberg module for $H/O_2(H)$, the extension $1 \to O_2(H)/Z(O_2(H)) \to H/Z(O_2(H)) \to H/O_2(H) \to 1$ is split. Therefore H contains a subgroup L with the property that $H = O_2(H)L$, $O_2(H) \cap L = C_G(O_2(H))$. So, L is part of a short exact sequence $1 \to Z_2^3 \to L \to GL(3,2) \to 1$.

<u>Lemma 1</u>. Let $T \in Syl_2(G)$, $\hat{T} = T^{\pi^{-1}}$, $\langle z \rangle = Z(T) \cong Z_2$. Then $\langle z \rangle^{\pi^{-1}} = Z(\hat{T})$.

<u>Proof</u>. It suffices to take some $\hat{z} \in z^{\pi^{-1}}$ and show that $\hat{z} \in Z(\hat{T})$. Let $C = C_G(z)$. Then $|C| = 2^{14}3\cdot 5$, $C/O_2(C) \cong \Sigma_5$ and $C \cap C(\theta) \cong SD_{16} \times Z_3$, where $\theta \in C$, $|\theta| = 3$ [Rul]. By considering the irreducible $F_2\Sigma_5$-modules, one sees that the C-chief factors within $O_2(C)$ have dimensions 1, 4, 4, 1, 1, where θ acts fixed point freely on the 4-dimensional factors. Let D be the terminal member of the derived series of C. Then $D/O_2(D) \cong A_5$ and easily we see that $O_2(D) = D \cap O_2(C)$ has order 2^a, $8 \leq a \leq 11$. Also, $C = D \cdot C_C(\theta)$. By the three subgroups lemma, $C(\hat{z}) \geq D^{\pi^{-1}}$. Also $C(\theta) = 3\hat{M}_{10}$, i.e. $C(\theta)' = C(\theta)''$ is a perfect central extension of A_6 by Z_3, and $C(\theta)/\langle\theta\rangle \cong M_{10}$. Since $C(\theta)$ has SD_{16}-Sylow 2-subgroups, $C(\theta)$ has trivial Schur multiplier and so $C(\hat{z}) \geq C_C(\theta)^{\pi^{-1}}$. The lemma follows, since $C^{\pi^{-1}} = D^{\pi^{-1}} C_C(\theta)^{\pi^{-1}}$.

We take an involution $t \in G$ such that $C_G(t) \cong Z_2 \times Z_2 \times Sz(8)$ and $N_G(O_2(C_G(t)))$ contains an element of order 3 which acts nontrivially on $Z_2 \times Z_2$ and on $Sz(8)$. Without loss, H contains a Sylow 2-group, S, of $C(t)$ so that $H = N_G(Z(S \cap C(t)'))$.

<u>Lemma 2</u>. i) $L - O_2(L)$ contains involutions in the conjugacy class of z.

ii) $L^{\pi^{-1}}$ does not have $SL(2,7)$ as a homomorphic image.

<u>Proof</u>. Let $K \leq G$, $K \cong {}^2F_4(2)$, $Q \in Syl_2(K)$. Then $|Q| = 2^{12}$, $|\Omega_1(Q)| \geq 2^{10}$ and $\Omega_1(Q)$ has nilpotence class at least 3 (this is easily seen by looking at a subgroup of order $2^{11}\cdot 5$ in K'). Without loss, we may assume that $Q \leq H$. Since $\Omega_1(Q)$ has nilpotence class at least 3, $\Omega_1(Q) \not\leq O_2(H)$. Let $x \in \Omega_1(Q) - O_2(H)$, $|x| = 2$. Since $O_2(H)/Z(O_2(H))$ is a free $F_2\langle x\rangle$-module, if $y \in H$, $y \notin O_2(H)$ and $y^2 \in Z(O_2(H))$, then y is conjugate in H to an element of $xZ(O_2(H))$. Therefore, x is conjugate to an element of $L - O_2(L)$. Since every involution in K lies in the conjugacy class of z [Rul], (i) follows. The structure of H implies that $Z(O_2(H))^{\#}$ consists of conjugates of z. If $h \in L$, $|h| = 7$, then $[O_2(L)^{\pi^{-1}}, h^{\pi^{-1}}] \cong Z_2^3$. This, with (i), implies that $\langle z \rangle^{\pi^{-1}} \cong Z_2 \times Z_2$ and (ii) follows.

Lemma 3. Suppose that $1 \to A^* \to G^* \xrightarrow{\sigma} G \to 1$, $A^* \le Z(G^*)$, and G^* is perfect. Then $|A^*| \le 2$ and $A^* = 1$ if and only if $O_2(C(t))^{\sigma^{-1}}$ splits over A^*.

Proof. Let $S_1 = S^{\sigma^{-1}} \cap (C(t)^{\sigma^{-1}})'$, $U := O_2(C(t))$, $R := O_2(H)^{\sigma^{-1}}$. Since every involution in $Z(O_2(H))$ is central in a Sylow 2-group of G, Lemma 1 implies that $Z(O_2(H))^{\sigma^{-1}} = Z(R)$. Since $Z(R)$ is elementary abelian, $Z_1 := [Z(R),\langle h\rangle] \cong Z_2^3$, for an element $h \in (C(t)^{\sigma^{-1}})'$ of order 7. Since an indecomposable $F_2[GL(3,2)]$ module M with socle of dimension 1 and quotient of dimension 3 has the property that the 14 vectors of $M\text{-soc}(M)$ form an orbit, Lemma 1 tells us that Z_1 is normal in $H^{\sigma^{-1}}$. In particular, [A.G.1] implies that $(C(t)^{\sigma^{-1}})' \cong Sz(8)$ and we have that $C(t)^{\sigma^{-1}}$ splits over A^* if and only if $U^{\sigma^{-1}}$ does. Thus, R/Z_1 is either elementary of order 2^9 or an extraspecial group 2^{1+8}, since $H^{\sigma^{-1}}/R$ acts irreducibly on $R/Z(R)$. Since $N_G(U)$ induces Z_3 on U, $U^{\sigma^{-1}}$ is either elementary abelian or $Q_8 \times E$, where E is elementary abelian. Thus, $A_1 := (U^{\sigma^{-1}})'$ has order 1 or 2 and $C_G(t)^{\sigma^{-1}}/A_1$ splits over A^*/A_1.

Suppose momentarily that $A_1 = 1$ and A^* is cyclic of order 2. We claim that R/Z_1 is elementary abelian. Suppose $R/Z_1 \cong 2^{1+8}$. Since $C(t)^{\sigma^{-1}}$ splits over $\langle a\rangle$, R/Z_1 contains an elementary abelian subgroup S/Z_1 of order 2^6, a contradiction. Thus, R/Z_1 is elementary abelian. Since $R/Z(R)$ is the Steinberg module for $H\sigma^{-1}/R$, $R = [R,H^{\sigma^{-1}}] \times A^*$. Since G^* is perfect and H has odd index in G, $H^{\sigma^{-1}}/[R,H^{\sigma^{-1}}] \cong SL(2,7)$. Since $L^{\sigma^{-1}}$ covers this quotient group, we have a contradiction to Lemma 2.

The preceding paragraph implies that $A_1 = A^*$. Since $|A_1| \le 2$, the Lemma follows.

Proposition. The multiplier of Rudvalis' simple group has order 2.

Proof. Lemma 3 and the introductory remarks.

3 THE SCHUR MULTIPLIER OF F_2

Let G be the simple $\{3,4\}$- transposition group of Fischer, of order $2^{41}3^{13}5^6 7^2 11.13.19.23.31.47$. In [Gr4], it was shown that G has a 2-fold cover. To prove that $H^2(G, Q/\mathbf{Z}) \cong \mathbf{Z}_2$, it suffices to prove that the multiplier has order at most 2. The arguments require the basic results of Fischer on F_2, which were circulated in 1973 and summarized in Lemma 2.40 of [Gr4]. They are reproduced here:

Lemma 1. Let G be a finite simple group generated by a class D of $\{3,4\}$-transpositions such that if $d \in D$, $C_G(d) \cong 2.{}^2E_6(2).2$. Then

a) $|G| = 2^{41}3^{13}5^6 7^2 11.13.17.19.23.31.47$.

b) G has exactly four classes of involutions with centralizers of the forms

$$2.{}^2E_6(2).2, \quad (2^{1+22})(.2),$$
$$(2\times 2\times F_4(2))2, \quad (2\times 2^8)2^{16}.D_4(2).2.$$

c) If K is the last centralizer in (b), $N_G(Z(O_2(K))) \cong 2^8.2.2^{16}.Sp(8.2)$.

d) G has exactly two classes of elements of order 3; they have centralizer shapes $3\times F_{22}.2$ and $3^{1+8}.2_-^{1+6}.U_4(2)$.

e) G has exactly 2 classes of elements of order 5; they have centralizer shapes $5\times HiS.2$ and $5^{1+4}.2_-^{1+4}.A_5$.

f) G has exactly one class of elements of order 7; they have centralizer shape $7\times 2.L_3(4).2$.

g) The centralizers of the Sylow 11- and 13-groups have shape $11\times\Sigma_5$, $13\times\Sigma_4$.

Lemma 2. Suppose that H is a p-constrained group with $Q = O_p(H)$ extraspecial and Q/Q' an absolutely irreducible $F_p[H/Q]$-module. Set $d = \dim_{F_p} \mathrm{Ext}^1_{H/Q}(Q/Q', F_p) = \dim_{F_p} H^1(H/Q, Q/Q')$.

Suppose that $O_p(H^2(H/Q, \mathbb{Q}/\mathbb{Z})) = 0$. Then $O_p(H^2(H, \mathbb{Q}/\mathbb{Z})) \cong F_p^e$, for some $0 \le e \le d$.

Proof. A standard Lie ring argument. See [Gr2] or [Gr3] for examples.

Lemma 3. .2 has trivial Schur multiplier, A_5 and $U_4(2)$ have Schur multipliers of order 2 and $2\cdot L_3(4)\cdot 2$ has Schur multiplier of order dividing 24.

Proof. See [Gr6] for a summary of results on Schur multipliers.

Lemma 4. Let M be an irreducible 11-dimensional constituent in the permutation module of dimension 23 for F_2M_{23}. Then $\mathrm{Ext}^1_{M_{23}}(F_2,M) = \mathrm{Ext}^1_{M_{23}}(M,F_2) = 0$.

Proof. See [Gr3].

Lemma 5. Let X be a group, K a field and M a KX-module.

a) There is a module U and a short exact sequence $0 \to M \to U \to T \to 0$ with T a trivial module and $\phi\colon T \cong \mathrm{Ext}^1_{KX}(K,M)$, in the sense that if $a \in \mathrm{Ext}^1_{KX}(K,M)$, $t = \phi^{-1}(a)$ is the unique element of T such that $0 \to M \to M + Kt \to K \to 0$ realizes the module extension given by a cocycle in the class of t. Furthermore, U is unique: given $0 \to M \to U_1 \to T_1 \to 0$ with T_1 corresponding to a subspace of $\mathrm{Ext}^1_{KX}(K,M)$, the inclusion $T_1 \to T$ is induced by a module homomorphism $U_1 \to U$ which is 1 on M.

b) Dual statement for $\mathrm{Ext}^1_{KX}(M,K)$.

c) If M is self-dual and $H^1(X,K) = H^2(X,K) = 0$, there is a module N with the property that there are $N^o \le \{v \in N \mid v^x = v$, for all $x \in X\}$ and $N_o \ge [N,X]$, such that $N_o/N^o \cong M$ and $0 \to N^o \to N_o \to M \to 0$ is the sequence of (b) and $0 \to M \to N/N^o \to N/N_o \to 0$ is the sequence of (a). Furthermore N is unique: given any module L with $L^o \le \{v \in L \mid v^x = v$ for all $x \in X\}$ and $L_o \ge [L,X]$ such that $L_o/L^o \cong M$ and L^o, L/L_o correspond to quotient spaces, subspaces of $\mathrm{Ext}^1(M,K)$, $\mathrm{Ext}^1(K,M)$, respectively, there are $N^1 \le N^o \le N$, $N_o \le N_1 \le N$ with $N_1/N^1 \cong L$ and N_o/N^1, N^o/N^1 corresponding to L_o, L^o, respectively.

Proof. a) The "Baer sum" [Macl] underlies the idea of this proof. Let $\{e_i\}$ be a basis for $\mathrm{Ext}^1_{KG}(K,M)$. For each i, let $c_i \in e_i$ be a 1-cocycle and let the associated module extension be written $0 \to M_i \to U_i \to K \to 0$, with $\phi_i\colon M \xrightarrow{\cong} M_i$. Form $A = \coprod_i U_i$, $B = \{(\phi_i(m_i)) \in \coprod_i M_i \mid \Sigma m_i = 0\}$. Then $U = A/B$ with M identified with $(\coprod M_i)/B$ has the right properties. A Zorn's lemma argument gives the uniqueness statement.

b) Dualize.

c) First, build a short exact sequence as in (a): $0 \to M \to U \to T \to 0$. Let $F = \mathrm{Hom}_K(\cdot,K)$. Then $0 \to FT \to FU \to FM \to 0$ is exact and is isomorphic to $0 \to T \to FU \to M \to 0$. Let E^i be the functor

$\mathrm{Ext}^i_{KG}(K,\cdot)$. The long exact sequence for $\{E^i\}$ gives $0 = E^1T \to E^1(FU) \to E^1M \to E^2T = 0$, i.e. the epimorphism $FU \to M$ induces an isomorphism of E^1. Now let N be a module as in (a) for the module FU (in the role of M). Define N^o as the image of FT in $FU \leq N$ and define $N_o = FU$.

Now for uniqueness. There is an essentially unique embedding $\bar{L} = L/L^o \to N/N^o$, as in (a). Let P/N^o be the image. By the uniqueness of (b) applied to $\bar{L}$ (and the natural isomorphism induced on $\mathrm{Ext}^1_{KG}(\cdot,K)$ by $M \hookrightarrow \bar{L}$), there is an epimorphism $P \to L$. Define $N^1 = \ker(P \to L)$, $N_1 = P$.

Lemma 6. Let $H = \cdot 2$, λ a vector of type 2 in the Leech lattice, $\bar{\Lambda} = \Lambda/2\Lambda$, $V_1 = \mathbb{F}_2\bar{\lambda}$, $V_2 = \bar{\lambda}^{\perp}$, $M = V_2/V_1$. Then (a) $0 < V_1 < V_2 < \bar{\Lambda}$ is the unique H-composition series of $\bar{\Lambda}$; (b) $\dim_{\mathbb{F}_2} \mathrm{Ext}^1_{\mathbb{F}_2H}(\mathbb{F}_2,M) = 1$ and $\dim_{\mathbb{F}_2} \mathrm{Ext}^1_{\mathbb{F}_2H}(M,\mathbb{F}_2) = 1$.

Proof. a) Uniseriality of $\bar{\Lambda}$ is easy; see [Gr4], p. 31 for an argument.

b) Since M is a self-dual module, it suffices to prove the first statement.

Construct a module N for M as in Lemma 5(c): $0 \to N^o \to N \to N_o \to 0$. There are $N^1 \leq N^o$ and $N_1 > N_o$ with $N_1/N^1 \cong \Lambda/2\Lambda$. Define $U := N/N^1$, $\bar{\Lambda} := N_1/N^1 \leq U$, $V_1 = N^o/N^1$, $V_2 :=$ the subgroup of index 2 in $\bar{\Lambda}$ corresponding to the annihilator of the 1-space V_1.

Take any $U_1 \leq U$ which contains $\bar{\Lambda}$ with codimension 1. Since T is a trivial module, U_1 is a submodule. Let K be an M_{23} subgroup of H conjugate in $\cdot 0$ to a natural M_{23} in the subgroup N_{24} of $\cdot 0$. Then M has the same nontrivial constituents as the permutation module. Set $U_o := \{u \in U | K \text{ fixes } u\}$. By Lemma 4, U_o complements $[U_1,K] \cong M$.

Take a subgroup $\langle x,y\rangle$ of H conjugate to a subgroup of the natural M_{24} in $\cdot 0$ with $H \cap \langle x,y\rangle = \langle x\rangle \cong Z_{11}$ and $|y| = 2$ with $\mathrm{trace}_{\Lambda}(y) = 0$. Then y acts on $C_U(K) = C_U(x) = U_o \oplus C_{\bar{\Lambda}}(x)$. Since $C_{\bar{\Lambda}}(x)$ is a free $\mathbb{F}_2\langle y\rangle$-module, we may assume that U_o admits y. Then $[U_o,y] = 0$.

Suppose that $v \in C_U(x) - \bar{\Lambda}$. Then v is fixed by $K_1 := \langle K,y\rangle$. We study possibilities for K_1. We have $|K_1| \equiv 0 \pmod{2|K| = 2^8 3^2.5.7.11.23)}$ so that $i := |H:K_1|$ divides $2^{10}3^4 5^2$. Since K and

K_1 contain a Sylow 23-normalizer for $.2$ [CGP1], $i \equiv 1 \pmod{23}$. Since K_1 contains a subgroup of index 1 or 2 in a Sylow 11-normalizer for H, $i \equiv 1$ or $2 \pmod{11}$, whence $i \equiv 1$ or $24 \pmod{253}$. The only possible solutions are $i = 1, 24$ and $2025 = 3^4 5^2 \equiv 1 \pmod{253}$. If $i = 1$ or $3^4 5^2$, K_1 has odd index in H. Thus, splitting of the extension $0 \to M \to M + F_2 v \to F_2 \to 0$ with respect to K_1 implies splitting with respect to H, a contradiction to the definition of U. Thus, $i = 24$. But, H has no subgroup of index 24; if it did, any $U_4(2)$-subgroup would lie in every subgroup of index 24 [Di1], in contradiction to simplicity of H.

We conclude that no such v exists, i.e. $U_o = 0$ and $U = \bar{\Lambda}$, which proves the theorem.

Proposition 7. The Schur multiplier of F_2 has order 2.

Proof. Trivially, the p-part of the multiplier is trivial for $p \geq 11$. Lemmas 1, 2 and 3 and the facts that the subgroups $2_+^{1+22}(\cdot 2)$, $3^{1+8}2_-^{1+6}U_4(2)$, $5^{1+4}2_-^{1+4}A_5$, $7 \times 2.L_3(4).2$ have indices prime to 2, 3, 5 and 7, respectively, imply that the Schur multiplier of F_2 has order dividing 2. This suffices.

4 THE SCHUR MULTIPLIER OF F_1

We let G be a finite simple group having a 2-central involution with 2-constrained centralizer of shape $C = (2^{1+24})(\cdot 1)$. Then G has order $N = 2^{46}\cdot 3^{20}\cdot 5^9\cdot 7^6\cdot 11^2\cdot 13^3\cdot 17\cdot 19\cdot 23\cdot 29\cdot 31\cdot 41\cdot 47\cdot 59\cdot 71$. We summarize below other properties we need to prove that the multiplier of G is trivial.

Lemma 1. The normalizer in G of the center of a Sylow p-subgroup has shape

$(2^{1+24})(.1)$	for $p = 2$,
$(3^{1+12})(2.Suz.2)$	for $p = 3$,
$(5^{1+6})(2.HJ.4)$	for $p = 5$,
$(7^{1+4})(2A_7 6)$	for $p = 7$,
$11^2.(SL(2,5).10)$	for $p = 11$, and
$(13^{1+2})(SL(2,3).12)$	for $p = 13$, respectively.

For all these subgroups, H, if $x \notin O_p(H)$ generates a normal cyclic subgroup of H modulo $O_p(H)$, $C_{O_p(H)}(x) = O_p(H)'$. Also $O_p(H)/O_p(H)'$ is an absolutely irreducible module for $H/O_p(H)$.

<u>Proof</u>. [FLT]; also see [CN].

<u>Lemma 2</u>. Let G_1 be a group, M an FG_1-module for some field F. If there is a subnormal abelian subgroup A of G_1 with $\{x \in M | x^a = x$, for all $a \in A\} = 0$, then $H^n(G_1,M) = 0$.

<u>Proof</u>. [Cu] or [R].

<u>Proposition 3</u>. The p-part of the Schur multiplier of G is trivial, for $p \neq 2$.

<u>Proof</u>. If the Sylow p-group is cyclic, this is trivial. Now assume that $p \in \{3,5,7,11,13\}$. Let N_p be the normalizer of a Sylow p-center. In the notation of Lemma 2 of Section 3, we get $d = 0$ and the p-part of the multiplier of N_p is trivial, from Lemmas 1 and 2 and the facts that the multipliers of $N_p/O_p(N_p)$ are p'-groups, for all relevant p (see [Gr6] for references). The Proposition follows.

<u>Proposition 4</u>. The 2-part of the Schur multiplier of G is trivial.

<u>Proof</u>. Let $H = N_G(V)$, where $V \leq O_2(C)$ is a four-group with $V^{\#}$ consisting of conjugates of z. Then H is 2-constrained with shape $(2^{2+11+22})(M_{24}\times\Sigma_3)$ [Gr4]. Let θ be an element of order 3 in H which maps modulo $O_2(H)$ to a nontrivial element of the Σ_3 direct factor. Then $C_G(\theta) \cong 3.F_{24}'$ [Gr4] and $C_H(\theta)' = 2^{11}.M_{24}$ with $O_2(C_H(\theta))$ an irreducible and non-self-dual module for M_{24}.

Now let $\hat{G} \to G$ be a 2-fold cover of G. We derive a contradiction. For a subgroup $X \leq G$, let $\hat{X}$ denote the induced extension.

We claim that $\hat{V}$ splits. If false, the action of $\langle\theta\rangle$ on $V = Z(O_2(H))$ forces $\hat{V} \cong Q_8$. Then $\hat{H}''' \leq C(\hat{V})$ since $\mathrm{Aut}(\hat{V}) \cong \Sigma_4$. But $H''' = H'' \geq O_2(H)$, a contradiction to $\hat{V} \triangleleft \hat{H}$, $\hat{V} \leq O_2(\hat{H})$, $\hat{V}' \neq 1$. The claim follows.

Since the multiplier of F_{24}' is $\mathbb{Z}_3$, $\widehat{C_G(\theta)}$ splits. Since $O_2(H)' \cong \mathbb{Z}_2^{13}$, the previous sentence and paragraph imply that $(\widehat{O_2(H)'})$

splits.

We argue that it suffices to prove that $B := [\hat{V},\theta]\times((\widehat{C_G(\theta)})' \cap O_2(\hat{H}))$ is normal in $\hat{H}$. Suppose so. Assume that $\hat{H}/B \cong 2^{1+22}(M_{24}\times S_3)$. Since the Frattini factor of $O_2(\hat{H}/B)$ is the direct sum of two irreducible modules, each isomorphic to M, the Golay code over $\mathbb{F}_2$ modulo the universe [Gr4], this is a contradiction since these modules are not self-dual. Therefore, $O_2(\hat{H}/B)$ is elementary abelian. We must now consider extensions in the category of $\mathbb{F}_2M_{24}$-modules. We know that $\dim \mathrm{Ext}^1_{\mathbb{F}_2M_{24}}(M,\mathbb{F}_2) = 1$ [Gr3]. In case $O_2(\hat{H}/B)$ gives a nonsplit module extension of $M\oplus M$ over $\mathbb{F}_2$, it must be given by a cocycle invariant under the action of S_3. However, the action of our S_3 on this copy of $M\oplus M$ gives a faithful action on $\mathbb{F}_2\oplus\mathbb{F}_2 \cong \mathrm{Ext}^1_{\mathbb{F}_2M_{24}}(M\oplus M, \mathbb{F}_2)$, a contradiction. Therefore, $O_2(\hat{H}/B)$ is completely reducible as an $\mathbb{F}_2M_{24}$-module. The fact that M_{24} has trivial Schur multiplier [BF] implies that $M_{24}\times\Sigma_3$ has trivial Schur multiplier. The complete reducibility of $O_2(\hat{H}/B)$ as an $\hat{H}$-module implies that $\hat{H}$ splits over $Z(\hat{G})$. Since $|G:H|$ is odd, $\hat{G}$ splits, a contradiction.

Now to prove normality of B in $\hat{H} = O_2(\hat{H})\cdot N_{\hat{H}}(\langle\hat{\theta}\rangle)$. Evidently, B is normalized by $N_{\hat{H}}(\langle\hat{\theta}\rangle)$, so it suffices to show $B \triangleleft O_2(\hat{H})$. Since all trivial $\hat{H}''$ chief factors lie within $B\cdot Z(\hat{G})$, $B_o := [\hat{V},\theta] \leq Z(\hat{H}'')$. So, we consider $\hat{H}/B_o \cong (2^{1+11+22})(M_{24}\times S_3)$. Set $R = O_2(\hat{H}/B_o)$. Assuming B not normal in $\hat{H}$, we get $Z(R) \cong \mathbb{Z}_2$ and a nontrivial pairing of $R/R' \cong 2^{22}$ with $R'/Z(R)$ into $Z(\hat{H}/B_o)$ induced by commutation. Since the action of our S_3 stabilizes $R'/Z(R)$ but stabilizes no M_{24}-irreducible in R/R', we have a contradiction since the pairing must be stable under the action of $\hat{H}$. This contradiction proves the required normality.

The results of this section prove:

<u>Theorem 5.</u> F_1 has trivial Schur multiplier.

5 CONCLUDING REMARKS

Since [Gr6] was written, some articles mentioned in [Gr6] have appeared, namely Steinberg's work on Schur multipliers of the (untwisted) Chevalley groups [St2] and Mazet's article on Schur multipliers of the Mathieu groups [Maz1]. Also, Holt [Hol], using computer calculations, determined that F_3 has trivial multiplier. Independently, Lyons [Ly1] proved the same result with only pure group theory. In [Gr6],

I erroneously credited Thompson with settling the multiplier of F_3. Thus, the claim I made in [Gr6], that the Schur multiplier situation for the finite simple groups is settled, was not strictly correct. I now hope that this is so.

I have looked back on this subject and made the following observations. Work on Schur multipliers of finite simple groups began to get rather technical, starting with the work of Burgoyne and Fong. Proofs of triviality of the Schur multiplier of some simple groups are generally easy, but not always so (the Tits group, ${}^2F_4(2)'$, and the Thompson group F_3 are good examples), but there are many more hard proofs of nontriviality. At this point, many of the difficult proofs of nontriviality may be avoided by knowing where to find the right nonsplit extension. Steinberg recognized this in many cases [St1][St2][St3]. Also, we get the perfect group $2.2.{}^2E_6(2)$ simply by looking in F_1 [Gr4]. The original construction was a lengthy calculation with the "stable cocycle" method [Gr2], [C.E.1].

I will give an example of what one can do with the right observations. Let $G = L_3(4) = A_2(4)$. It is easy to bound the multiplier $M(G)$ from above by $Z_4 \times Z_4 \times Z_3$; see [G.S.1], for example. The "hard part" is to show that $|O_2(M(G))| \geq 16$. Let $G^* = SL(3,4)$, $G^{**} = G_2(4)$ and let H^{**} be the extension of G^{**} constructed by the "permutation trick" [Gr2]. Then H^*, the extension of the natural G^* in G^{**}, is perfect. Let $\langle z \rangle = Z(H^*)$. The generators and relations of [Gr2] show that the induced extension of P_1 is split, where P is any parabolic of H^* and $P_1 \leq P$ contains $O_2(P)$ with index 2. For any such P_1, write $P_1 = \langle z \rangle \times R$. Then $|H^*:R| \equiv 4 \pmod 8$. Thus the "permutation trick" applies again to give a perfect group $\hat{G}$ with $|Z(\hat{G})| = 4$ such that a preimage of z in G has order 4. This proves that $O_2(M(G))$ involves $\mathbb{Z}_4$. Since an element of order 3 in the outer automorphism group acts fixed point freely on $O_2(M(G))$ [G.S.1], $O_2(M(G))$ must involve $\mathbb{Z}_4 \times \mathbb{Z}_4$, at least, which suffices.

The above argument, that a multiplier of exponent $0 \pmod 2$ leads to a multiplier of exponent $0 \pmod 4$, works also for M_{22}.

The multiplier of $L_3(4)$ was settled originally by Thompson and Burgoyne (unpublished). A proof, using the "stable cocycle" method, was written up by Mazet [Maz1]. I remark that the existence of a double cover of M_{12} follows from the permutation trick [Gr1]; it is unnecessary to bring in facts about the ternary Golay code, as in [Maz1].

A number of interesting nonsplit extensions of finite groups G by $\mathbb{Z}_2$ may be made by embedding G in the orthogonal group $O(n, \mathbb{R})$, then taking the preimage in $pin(n, \mathbb{R})$. The interesting paper of Gagola and Garrison [Gag-Gar1] examines this question in some detail. If G is a finite simple group with multiplier of even order, a 2-fold covering of G may, in many cases, be exhibited by an embedding of G in some $O(n,\mathbb{R})$. One way to do this is by a permutation representation of G; see [Gr1] and [Gag-Gar1] for a sampling of applications (and see Section 2 of this article). However, if $G = Sz(8)$, Gagola and Garrison show that no orthogonal representation of G realizes a 2-fold cover of G, which has multiplier $\mathbb{Z}_2 \times \mathbb{Z}_2$. The only proof that F_2 has multiplier of order divisible by 2 requires existence of F_1 [Gr4]. It would be quite interesting to see an "easy" proof, such as with an orthogonal representation, but we haven't found one yet.

Are there any analogues of these tricks for odd primes?

REFERENCES

A.G.1 Alperin, J. & Gorenstein, D. (1966). The multiplicators of certain simple groups. Proc. Amer. Math. Soc. 17, 515-519.

BF Burgoyne, N. & Fong, P. (1966). Schur multipliers of the Mathieu groups. Nagoya Math. Jour. 27, 733-745. Correction, Ibid. 31, 297-304 (1968).

C.E.1 Cartan, H. & Eilenberg, S. (1956). Homological Algebra. Princeton: Princeton University Press.

CGP1 Conway, J., Guy, M., & Patterson, N. The character table and conjugacy classes of .0,.1,.2,.3 and Suz. Unpublished.

CN Conway, J. & Norton, S. (1979). Monstrous Moonshine. Bull. London Math. Soc. 11, 303-339.

Cu Curran, P. (1976). Fixed point free action on a class of abelian groups. Proc. Amer. Math. Soc. 57, 189-193.

Dem Dempwolff, U. (1974). A characterization of the Rudvalis simple group of order $2^{14}3^3 5^3 7.13.29$ by the centralizer of a noncentral involution. J. Algebra 32, 53-88.

Di1 Dickson, E. (1958). Linear Groups. New York: Dover. (reprint edition).

FLT Fischer, B., Livingston, D. & Thorne, M.P. The character table of F_1. Unpublished.

Gag-Gar1 Gagola, S. & Garrison, S. Real characters, double covers and the multiplier. J. Algebra 74 (1982), 20-51.

Gor1 Gorenstein, D. (1968). Finite Groups. New York: Harper & Row. 2nd ed. New York: Chelsea. (1980).

Gr1 Griess, R.L., Jr. (1970). A sufficient condition for a finite group of even order to have nontrivial Schur multiplier. Notices of the American Mathematical Society.

Gr2 Griess, R.L., Jr. (1973). Schur multipliers of finite simple groups of Lie type. Trans. Amer. Math. Soc. 183, 355-421.

Gr3 Griess, R.L., Jr. (1974). Schur multipliers of some sporadic simple groups. Jour. of Algebra 32, 445-466.

Gr4 Griess, R.L., Jr. (1982). The Friendly Giant. Inventiones Math. 69, 1-102.

Gr5 Griess, R.L., Jr. (1980). Schur multipliers of the known finite simple groups, II. In Santa Cruz Conference on Finite Groups, American Mathematical Society.

G.S.1 Griess, R.L., Jr. & Solomon, R. (1979). Finite groups and unbalancing 2-components of $\{L_3(4)^{\wedge},He\}$-type. Jour. of Algebra 60, 96-125.

Hol Holt, D. To appear in the Journal of Algebra.

Hul Huppert, B. (1967). Endliche Gruppen, I. Berlin & New York: Springer-Verlag.

Lyl Lyons, R. The Schur multiplier of F_3 is trivial. To appear in Comm.Alg.

Mac.1 Maclane (1975). Homology. Berlin, Heidelberg & New York: Springer-Verlag. Third Printing.

Mazl Mazet, P. (1982). Sur les multiplicateurs de Schur des groupes de Mathieu, 77, 552-576.

R Robinson, D. (1976). The vanishing of certain homology and cohomology groups. Jour. of Pure and Applied Algebra, 7, 145-167.

Rul Rudvalis, A. A rank 3 simple group of order $2^{14}3^{3}5^{7}7.13.29$. J. Algebra 86 (1984), 181-218.

Stl Steinberg, R. (1967). Lectures on Chevalley Groups. Yale Lecture notes.

St2 Steinberg, R. (1981). Generators, relations and coverings of algebraic groups II. Journal of Algebra 71, 525-543.

St3 Steinberg, R. (1959). Variations on a theme of Chevalley. Pacific Journal of Math. 9, 875-891.

WIELANDT'S FORMULA IMPLIES GOOD GENERATION OF GROUPS OF LIE TYPE

R. Lyons*
Department of Mathematics, Rutgers University, New Brunswick, N.J. 08903

Gary Seitz [4] has proved basic generation and balance properties for finite groups of Lie type. These results are needed for the signalizer functor method to be applied (see, e.g., [1, Part II, Chap. 2]) in finite simple groups whose local subgroups are K-groups (that is, are groups whose composition factors are of known type). In this note we show how a portion of his proofs can be eased, if one is willing to assume that subgroups of the group of Lie type in question are themselves K-groups. While this hypothesis is less than ideal, nevertheless our proof is at least of heuristic interest, and should be a useful revision for the purpose of the classification of finite simple groups.

The key observation is that an analogue of Wielandt's formula [7, (3.2)] for the action of a group $D \cong Z_p \times Z_p$ on a p-group can be proved when D acts on a Chevalley group G in characteristic $r \neq p$. One must restrict attention to the r-shares of G and centralizers of subgroups of D; in addition, the formula fails in certain cases when the Schur multiplier of $G/Z(G)$ has order divisible by p.

After notation is established in Section 1, we prove the relevant generalizations of Wielandt's formula in Section 2. Then in Section 3, we combine them with a fundamental lemma of Seitz to sketch a proof of a generation theorem for Chevalley groups. For reasons of space, we do not include details of this proof, which still unfortunately requires a certain number of ad hoc arguments in some special configurations. The details should appear elsewhere.

1 NOTATION

If E is a group of automorphisms of the group H, we write H_E for $C_H(E)$. We also use the following terminology throughout:

*Research partially supported by the National Science Foundation.

r, p = distinct primes

k = the algebraic closure of $GF(r)$

$\mathcal{G}$ = a quasisimple linear algebraic group over k

Π = the root system of $\mathcal{G}$

Σ = a fundamental system in Π

σ = a surjective endomorphism of $\mathcal{G}$ with finitely many fixed points

$G = O^{r'}(\mathcal{G}_\sigma)$

u = dimension of a maximal unipotent subgroup of G

u_E = dimension of a maximal unipotent subgroup of G_E (whenever E is a single automorphism or a group of automorphisms of the abstract group $\mathcal{G}$)

For any linear algebraic group $\mathcal{G}$, a <u>Borel-torus pair</u> $(\mathcal{B},\mathcal{T})$ for $\mathcal{G}$ consists of a Borel subgroup $\mathcal{B}$ of $\mathcal{G}$ and a maximal torus $\mathcal{T}$ of $\mathcal{B}$.

By [6, (10.10) and (11.1) and (11.6)], σ normalizes some Borel-torus pair $(\mathcal{B},\mathcal{T})$, and $(\mathcal{B}_\sigma, N(\mathcal{T}_\sigma))$ is a B-N pair of $\mathcal{G}_\sigma$. The $\mathcal{T}$-root subgroups $\mathcal{U}_\alpha$ of $\mathcal{G}$, $\alpha \in \Pi$, (i.e. the minimal $\mathcal{T}$-invariant unipotent subgroups of $\mathcal{G}$) lie in $\mathcal{B}$ or its opposite, and are clearly permuted by σ. Moreover, one of the following holds:

1. σ normalizes $\mathcal{U}_\alpha$ for at least one root α of each root length.

2. $\mathcal{G} = B_2$, F_4, or G_2; $r = 2$, 2, or 3, respectively, and σ interchanges long and short root groups.

In case (1), we define $q(\sigma,\mathcal{G}) = |(\mathcal{U}_\alpha)_\sigma|$ for any σ-invariant $\mathcal{T}$-root group $\mathcal{U}_\alpha$. When there is no danger of ambiguity, we write $q(\sigma)$ for $q(\sigma,\mathcal{G})$. By [6, 11.6], $q(\sigma)$ is independent of α. (It is also independent of the choice of the σ-invariant Borel-torus pair $(\mathcal{B},\mathcal{T})$, since that choice is unique up to $\mathcal{G}_\sigma$-conjugation, by Lang's theorem.) In case (2) we set $q(\sigma,\mathcal{G}) = q(\sigma^2,\mathcal{G})^{1/2}$, observing that the right side is defined since σ^2 satisfies (1) above. By [6,(11.9) and (11.12)], we have (as $u = |\Pi|/2$)

$$|G|_r = |\mathcal{G}_\sigma|_r = q(\sigma,\mathcal{G})^u \tag{1.1}$$

Thus by [6, (11.13)],

$$q(\sigma^m, G) = q(\sigma, G)^m \tag{1.2}$$

for any positive integer m.

2 THE WIELANDT FORMULA

Throughout this section, D is a group of type (p,p) acting on a group X (or G or G or U). We assume that X is finite. Let $D_0, D_1, \ldots, D_p$ be the subgroups of D of order p. We define

$$W_r(X,D) = \Big(\prod_{i=0}^{p} |C_X(D_i)|_r\Big)/|C_X(D)|_r{}^p$$

and

$$Q_r(X,D) = |X|_r/W_r(X,D).$$

For such a group D, Wielandt's formula [7, (3.2)] implies that if $p \nmid |X|$, then $|X|_r = W_r(X,D)$, i.e. $Q_r(X,D) = 1$. We generalize this in a few directions. First, a couple of trivial generalizations:

<u>Lemma 1</u>. Suppose $X = L_{p'}(X)$ and let $\bar{X} = X/O_{p'}(X)$. Let $\bar{X}_1, \bar{X}_2, \ldots, \bar{X}_n$ be the minimal D-invariant products of components of $\bar{X}$, so that $\bar{X}$ is the central product of $\bar{X}_1, \bar{X}_2, \ldots, \bar{X}_n$. Then

a) $Q_r(X,D) = \prod_{i=1}^{n} Q_r(\bar{X}_i, D)$.

b) If $Q_r(\bar{X}_i, D) \neq 1$ for any i, then $\bar{X}_i$ is quasisimple and $C_D(\bar{X}_i) = 1$.

<u>Proof</u>. For any subgroup E of D, $|C_X(E)| = |C_{\bar{X}}(E)|\cdot|C_{O_{p'}(X)}(E)|$, so $Q_r(X,D) = Q_r(\bar{X},D)Q_r(O_{p'}(X),D) = Q_r(\bar{X},D)$ by Wielandt's formula. So we may assume that $O_{p'}(X) = 1$. Then $O_p(X) = Z(X)$, and it follows that any r-element of $X/Z(X)$ centralized by E is the image of an r-element of $C_X(E)$. So $Q_r(X,D) = Q_r(X/Z(X))$, and we may assume that $Z(X) = 1$. Then $X = X_1 \times X_2 \times \ldots \times X_n$ and $C_X(E)$ decomposes similarly, so (a) holds. In (b), say $X_i = K_1 \times K_2 \times \ldots \times K_m$ with the K_j's simple. If $m = p$, then for $D_0 = N_D(K_1)$, we have $D_1C_{X_i}(D_0) \cong C_{K_1}(D_0) \wr Z_p$ and so $C_{K_j}(D_0) \cong C_{X_i}(D)$ for each $j = 1,\ldots,p$; as $C_{X_i}(D_j) \cong K_1$ for $j > 0$, we compute

$Q_r(X,D) = 1$. The cases $m = p^2$ and $m = 1$ $(C_D(X_i) \neq 1)$ are just as easy.

Lemma 2. Suppose D acts via semisimple automorphisms on the unipotent linear algebraic group $\mathcal{U}$ over k. Then if we set $v = \dim \mathcal{U}$ and $v_E = \dim \mathcal{U}_E$ for any $E \leq D$, we have

$$v = \sum_{i=0}^{p} v_{D_i} - pv_D.$$

Proof. Standard arguments (using [2, Corollary, p. 120]) reduce us to the case that $\mathcal{U}$ is connected and has no proper closed D-invariant normal subgroup. Then by [2, Proposition, p. 127], $\mathcal{U}$ is a k-vector space on which D acts by linear transformations. The result is then familiar.

To generalize to the case of a Chevalley group we must consider the following situation. Let E be a finite abelian r'-group acting on $\mathcal{G}$ via (semisimple) automorphisms, such that $[E,\sigma] = 1$. By [6, Section 8] and induction on $|E|$ we see that $\mathcal{G}_E^o$ is reductive and $\mathcal{G}_E/\mathcal{G}_E^o$ is an r'-group. Write $\mathcal{G}_E^{(\infty)} = \mathcal{G}_1 \cdots \mathcal{G}_n$ with the $\mathcal{G}_i$'s pairwise commuting and quasisimple. Then σ permutes the $\mathcal{G}_i$; say $\{\mathcal{G}_1,\ldots,\mathcal{G}_m\}$ is an orbit. Then σ^m is a surjective endomorphism of $\mathcal{G}_1$ with finitely many fixed points.

Lemma 3. In the situation just described,

a) $q(\sigma^m, \mathcal{G}) = q(\sigma^m, \mathcal{G}_1)$.

b) $|G_E|_r = q(\sigma, \mathcal{G})^{u_E}$.

Proof. Let $u_1,\ldots,u_n$ be the dimensions of maximal unipotent subgroups of $\mathcal{G}_1,\ldots,\mathcal{G}_n$, respectively. Set $X = \mathcal{G}_1 \cdots \mathcal{G}_m$ and $\tau = \sigma^m$, so that $X_\sigma/Z(X_\sigma) \cong (\mathcal{G}_1)_\tau/Z((\mathcal{G}_1)_\tau)$ and so $|X_\sigma|_r = |(\mathcal{G}_1)_\tau|_r = q(\sigma^m,\mathcal{G}_1)^{u_1}$ by (1.1). If (a) holds, then by (1.2), this equals $q(\sigma,\mathcal{G})^{mu_1} = q(\sigma,\mathcal{G})^{u_1+u_2+\cdots+u_m}$. Taking the product over all products X of σ-orbits we have $|G_E|_r = |O^{r'}(G_{\langle E,\sigma\rangle})|_r = |(\mathcal{G}_E^o)_\sigma|_r = q(\sigma,\mathcal{G})^{u_1+u_2+\cdots+u_n}$, which is (b). Thus it is enough to prove (a).

Observe that by (1.2), it suffices to prove (a) with m replaced by any multiple of itself. Hence it is also sufficient to prove (a) for any positive power of σ in place of σ itself. Let F be a subgroup of E of prime index p, assume $F \neq 1$, and set $\mathcal{H} = (\mathcal{G}_F)^o$, so that $(G_E)^o = (H_{E/F})^o$. Replacing σ by a power, we may assume that σ normalizes the components $H_1, \ldots$ of H as well as all G_i, and so $m = 1$. Choose notation so that $G_1 \leq \langle H_1^{E/F} \rangle$. Then either E/F normalizes H_1, in which case $q(\sigma, H_1) = q(\sigma, G_1)$ by induction; or E/F cycles p components, say $H_1, \ldots, H_p$, of which G_1 is a diagonal, so $G_1/Z(G_1) \cong H_1/Z(H_1)$ as σ-groups and again $q(\sigma, H_1) = q(\sigma, G_1)$. But by induction $q(\sigma, G) = q(\sigma, H_1)$, so (a) holds in this case. We are thus reduced to the case $F = 1$, i.e., $|E| = p$. Say $E = \langle x \rangle$. Again replacing σ by a power, we may assume that σ normalizes G_1 and must prove that $q(\sigma, G) = q(\sigma, G_1)$. Moreover, replacing σ by σ^2, we may assume that σ fixes root lengths. Without loss, G is simple. Suppose first that x is inner on G; we regard $x \in G$. Choose a σ-invariant maximal torus T of G_x. Then T is a maximal torus of G (since x lies in some maximal torus and all maximal tori are conjugate). Clearly $T_1 = T \cap G_1$ is a maximal torus of G_1, and $T = T_1 C_T(G_1)$, so T_1-root subgroups of G_1 are minimal T-invariant in G, and so are T-root subgroups of G. As σ fixes root lengths, there exist such root subgroups U_α which are σ-invariant; and then both sides of (a) equal $|(U_\alpha)_\sigma|$, so (a) holds in this case.

The only other possibility is that x is in the Inn(G)-coset of a graph automorphism of G. In this case G is A_ℓ, D_ℓ, or E_6, and we fix a σ-invariant Borel-torus pair (B, T) of G and a graph automorphism γ of G of order p (see [5, p. 157]) such that for the T-root groups U_α in B, parametrized as $U_\alpha = \{x_\alpha(t) | t \in k\}$, we have $\gamma: x_\alpha(t) \rightsquigarrow x_{\alpha'}(c_\alpha t)$, where $c_\alpha = 1$ or -1 (independent of t) and $\alpha \rightsquigarrow \alpha'$ is an isometry of Π. Let σ_q be the endomorphism $x_\alpha(t) \rightsquigarrow x_\alpha(t^q)$, where $q = q(\sigma)$. Then $[\sigma_q, \gamma] = 1$, and we may assume that $\sigma \in \sigma_q \langle \gamma \rangle$. Replacing σ by a power we may assume $\sigma = \sigma_q$, where $q = q(\sigma)$. Of course $x \in \langle \gamma \rangle \mathrm{Inn}(G)$. Now by inspection of the groups A_ℓ, D_ℓ, and E_6, since x has order p, it is G-conjugate to an element $x_1 = \gamma h$, where $h \in T$. (For $G = D_4$ and $p = 3$ this is proved in [1, (9-1), pp. 104-105]. For $G = A_\ell$ and $p = 2$, we may take $\gamma: g \rightsquigarrow f(g^{-1})^T f^{-1}$ where $g \in PGL_{\ell+1}(k)$,

$$f = \begin{pmatrix} & & & & 1 \\ & & & -1 & \\ & & 1 & & \\ & \cdot & & & \\ \pm 1 & & & & \end{pmatrix},$$

and the assertion follows from the fact that if any nonsingular bilinear symmetric or skew-symmetric form is placed on $k^{\ell+1}$, then it is the orthogonal sum of hyperbolic planes (together with a one-dimensional space if ℓ is even). For $G = D_\ell$ and $p = 2$, we have $G\langle\gamma\rangle = PSO_{2\ell}(k)$, and the assertion follows from the fact that every element g of $SO_{2\ell}(k)$ with $g^2 = 1$ or -1 and $\det g = -1$ is of order 2 and so conjugate to $\mathrm{diag}(-1,-1,\ldots,1,1,\ldots)$ with an odd number of -1's. For $G = E_6$ and $p = 2$, the embedding $D_5 * GL_1(k) \leq E_6$ permits verification of the assertion by reduction to the D_5 case.) Replacing σ by a power, we may assume that σ centralizes h, whence σ centralizes x_1. Then writing $x_1 = x^g$ $(g \in G)$, we may replace σ by a power and assume that σ centralizes g. So $q(\sigma,G_1) = q(\sigma,G_1^g)$ and it suffices to show that $q(\sigma,G) = q(\sigma,G_1^g)$. We may therefore assume that $x = x_1$.

But then by the uniqueness in the Bruhat decomposition (cf. [6, p. 53 (1)]), (B_x,T_x) is a Borel-torus pair for G_x. Since $x = \gamma h$, the action of γ on T-root subgroups shows that every T_x-root subgroup U_α of G_x in B_x satisfies one of the following:

1) U_α is a T-root subgroup of G;

2) U_α is the diagonal of p T-root subgroups of G cycled by x;

3) $p = 2$ and there exist non-commuting T-root groups U_1, U_2, interchanged by x, such that if we set $U^* = \langle U_1,U_2\rangle$, then U^* is of type A_2 and U_α maps isomorphically on a diagonal of the direct product $U^*/Z = U_1Z/Z \times U_2Z/Z$, where $Z = Z(U^*)$.

Now σ normalizes B_x and T_x and hence, replacing σ by a power if necessary, we can find a σ-invariant U_α. In any of the three cases, since $\sigma = \sigma_q$, it is clear that $|(U_\alpha)_\sigma| = q(\sigma,G)$. Hence $q(\sigma,G_1) = q(\sigma,G)$, and the proof is complete.

This has the following immediate consequence:

<u>Proposition 4</u>. Suppose that $D \cong Z_p \times Z_p$ acts on G by algebraic automorphisms commuting with σ. Suppose that D normalizes a Borel-torus pair (B,T) of G. Then $|G|_r = W_r(G,D)$, so that $Q_r(G,D) = 1$.

<u>Proof</u>. Since $p \neq r$, D acts via semisimple automorphisms. Since D normalizes B and T it again follows easily from the Bruhat decomposition that for any $E \leq D$, B_E is a Borel subgroup of G_E. Applying

Lemma 2 to the unipotent radical of B_E, raising $q(\sigma)$ to both sides of the resulting equation, and using (1.1) and Lemma 3(b), we get $|G|_r = W_r(G,D)$.

The assumption in Proposition 4 about (B,T) usually holds, and when it does not, we shall usually get generation by using [4, (2.3)] (see Lemma 9 below). Specifically, we have:

Proposition 5. Let D ($\cong Z_p \times Z_p$) act on G ($= O^{r'}(G_\sigma)$). Then one of the following holds:

1) $Q_r(G,D) = 1$.

2) $\mathcal{N}_G(D;r) \neq \{1\}$.

3) $G/Z(G) \cong L_p(q)$ $(p|q-1)$ or $U_p(q)$ $(p|q+1)$, and $W_r(G,D)=1$.

4) $G/Z(G) \cong B_2(q)$, $p = 2$, and $W_r(G,D) = q^3$.

In case (4), we have $|G|_r = q^4$, so the proposition immediately implies the following.

Corollary 6. Under the assumptions of Proposition 5, either

1') $Q_r(G,D) \geq 1$, or

2) $\mathcal{N}_G(D;r) \neq \{1\}$.

Proof of Proposition 5. Clearly we may assume that G is simple. As in Lemma 1, we may assume that D acts faithfully on G. We do not distinguish between an inner automorphism of G and the element of G which induces it. By the well-known structure of Aut(G) (see [5]), either $D = \langle x,\phi\rangle$ where x is an inner-diagonal-graph automorphism of G (i.e., a product of possibly trivial automorphisms of these three types) and ϕ is a field or graph-field automorphism of G; or else every element of D is an inner-diagonal-graph automorphism of G.

In the first case, there exists a semisimple automorphism y of G and a surjective endomorphism τ of G, with G_τ finite, inducing x and ϕ, respectively, on G; in particular, such that $[y,\tau] = 1$ and $\langle\tau^p\rangle = \langle\sigma\rangle$. Without loss, $(\tau y^i)^p = \sigma$ for each i. If we choose notation so that $D_0 = \langle x\rangle$, then for each $i > 0$, we have by (1.1) and (1.2) that

$$|G_{D_i}|_r = |G_{\tau y^i}|_r = (q(\tau y^i))^u = (q(\sigma)^{1/p})^u = |G|_r^{1/p}$$

and, using Lemma 3 for $E = \langle y \rangle$,

$$|G_D|_r = |G_{\langle y,\tau\rangle}|_r = q(\tau)^{u_y} = q(\sigma)^{u_y/p} = |G_{\langle y,\sigma\rangle}|_r^{1/p}$$

$$= |G_{D_0}|_r^{1/p}.$$

These relations immediately give $Q_r(G,D) = 1$. So we assume that the second case of the initial paragraph holds, whence there exist semisimple automorphisms x, y of G commuting with σ, such that $\langle x,y\rangle$ acts on G like D. The group of graph automorphisms has p-rank at most one, so we may assume without loss that $y \in G$.

By [6, (7.5)] x normalizes a Borel-torus pair (B,T) of G, and then as usual T_x is a maximal torus of G_x. If $y \in (G_x)^o$, then $y^g \in T_x$ for some $g \in G_x$ by [2, p. 139], and so $\langle x,y\rangle$ normalizes the Borel-torus pair $(B^{g^{-1}}, T^{g^{-1}})$, and (1) follows from Proposition 4.

We therefore may assume that $y \notin (G_x)^o$. Let $\hat{G}$ be the universal covering group of G, so σ lifts to an endomorphism of $\hat{G}$, which we also call σ. Let $\hat{y}$ be a p-element of $\hat{G}$ mapping on y via the covering projection $\hat{G} \to G$. We define $\hat{x}$ similarly if $x \in G$; while if $x \notin G$, there exists an automorphism $\hat{x}$ of $\hat{G}$ of order p lifting x, and we form the semidirect product $\hat{G}\langle\hat{x}\rangle$. In any case, $\langle\hat{x},\hat{y}\rangle$ is a p-group normalizing $\hat{G}$; $[\langle\hat{x},\hat{y}\rangle,\sigma] \leq Z(\hat{G})$; $[\hat{x},\hat{y}] \in C_{\hat{G}\langle\hat{x}\rangle}(\hat{G}) = Z(\hat{G})$; and $\hat{x}^p$, $\hat{y}^p \in Z(\hat{G})$.

Let $z = [\hat{x},\hat{y}]$, so that $z^p = 1$ and $[z,\sigma] = 1$. If $z = 1$ then $\hat{y} \in \hat{G}_{\hat{x}}$, which is connected by [6, (8.1)]; hence $y \in (G_x)^o$, contrary to assumption. Hence $z \neq 1$. In particular, $p \,|\, |Z(\hat{G})|$. (Indeed as $[z,\sigma] = 1$, $p \,|\, |Z(\hat{G}_\sigma)|$, so the Schur multiplier of G has order divisible by p.)

Next, observe that if the reductive group $(G_D)^o$ is not abelian, then the fixed points on it of σ form a Chevalley group of characteristic r, so in particular, $r \,|\, |G_D|$. Hence $\mathcal{N}_G(D;r) \neq \{1\}$, one of the desired conclusions. We may therefore assume that $(G_D)^o$ is abelian.

Case analysis now takes over. If p is odd, then we have $\hat{G} = SL_n(k)$ (with $p|n$) or $\hat{G} = E_6$ (with $p = 3$). In either case, G has no graph automorphism of order p, so $x \in G$.

Suppose that $\hat{G} = SL_n(k)$. Since $\hat{x}^{\hat{y}} = \hat{x}\hat{z}$, we see that $\hat{x}$ has p eigenvalues of equal multiplicity, and $\hat{y}$ cycles their eigenspaces.

Hence $\hat{G}_{\hat{x}}$ is the direct product of p copies of $SL_{n/p}(k)$ cycled by $\hat{y}$, so $\hat{G}_{<\hat{x},\hat{y}>}$ contains $SL_{n/p}(k)$. As $(G_D)^o$ is abelian, this forces $n = p$, so $\hat{G}_{\hat{x}}$ is a torus. The same argument applies to any element of $<\hat{x},\hat{y}> - Z(\hat{G})$, so conclusion (3) holds ($p|q \pm 1$ since $[z,\sigma] = 1$). For the remaining cases we use the following well-known fact.

Lemma 7. Suppose $a \in G$ has order p. Let Π_a be the root system of G_a. Let α^* be the highest root in Π, and write $\alpha^* = \Sigma c_\alpha \alpha$, the sum over Σ. Then:

a) There exists an additive homomorphism $\lambda: Z\Pi \to Z_p$ such that Π_a can be identified with $\Pi \cap \ker(\lambda)$; moreover, a fundamental system Σ_a in Π_a is one of the following:

1) $\Sigma_a = \Sigma \cap \ker(\lambda)$, or

2) $\Sigma_a = \Sigma \cup \{-\alpha^*\} - \{\alpha\}$ for some $\alpha \in \Sigma$ with $c_\alpha = p$.

b) If $p = 3$, then in case (a1), $\Sigma_a = \Sigma - \{\alpha\}$ where $c_\alpha \leq 2$, or $\Sigma_a = \Sigma - \{\alpha,\beta\}$ where $c_\alpha = c_\beta = 1$.

c) If $p = 2$, then in case (a1), $\Sigma_a = \Sigma - \{\alpha\}$ where $c_\alpha = 1$.

d) In case (a2), $(G_a)^o$ is semisimple and self-centralizing in G.

Now if $G = E_6$ and $p = 3$ in Proposition 5, then $<x,y> \leq G$. By Lemma 7(b), $(G_x)^{(\infty)}$ is A_5, $A_4 \times A_1$, D_5, or $A_2 \times A_2 \times A_2$. In the first three cases, $(G_{<x,y>})^{(\infty)} \neq 1$ by Lemma 7(b), contradiction. So the last case holds, and $(G_x)^o = (G_x)^{(\infty)}$ by Lemma 7(d). Since $(G_{<x,y>})^o$ is abelian, y must normalize the three components of $(G_x)^o$, and hence induces an inner automorphism on each component (A_2 has no graph automorphism of order 3). Hence by Lemma 7(d), $y \in (G_x)^o$, contradiction.

Finally, assume that $p = 2$. Since $(G_D)^o$ is abelian, we can use Lemma 7(c), as 7(b) was used above, to see that for each $d \in D^{\#}$, all components of $(G_d)^o$ are of type A_1 and normalized by D. Since A_1 has no graph automorphism of order 2, we see again that $(G_d)^o$ is not self-centralizing for any $d \in D^{\#} \cap G$, so by Lemma 7(d,c), the only possibilities for $(G,(G_d)^o)$ are (A_2,A_1), $(A_3,A_1 \times A_1)$, and (B_2,A_1). Now if $G = A_2$, the only graph automorphisms γ of G of order 2 satisfy $(G_\gamma)^o \cong A_1$, so $(G_d)^o \cong A_1$ for each $d \in D^{\#}$. Thus, since $(G_D)^o$ is abelian,

$$u = \sum_{d \in D^{\#}} u_d - 2u_D \tag{2.1}$$

(both sides equal 3). As in the proof of Proposition 4, this implies, by Lemma 3, that $Q_r(G,D) = 1$. Similarly, if $G = A_3$, then any graph automorphism γ of G of order 2 satisfies $(G_\gamma)^o \cong A_1 \times A_1$ or B_2; the latter is impossible if $\gamma \in D$ by the second sentence of this paragraph. So $u = 6$, $u_d = 2$ for each $d \in D^{\#}$, and again (2.1) holds, leading to $Q_r(G,D) = 1$. Finally, if $G = B_2$, then $u_d = 1$ for each $d \in D^{\#}$, so the right side of (2.1) equals 3 and $W_r(G,D) = q^3$. This gives Proposition 5.

For completeness we make some remarks on Lemma 7. A proof of (a) appears in [1, p. 176]. In (b), if $|\Sigma-\Sigma_a| > 2$, then we can find a connected subdiagram Δ of the Dynkin diagram of G containing exactly 2 or 3 roots in $\Sigma-\Sigma_a$ and such that if δ is the sum of the roots in Δ, then $\delta \in \ker(\lambda)$. This is a contradiction as Σ_a spans $\Pi \cap \ker(\lambda)$. So $|\Sigma-\Sigma_a| \leq 2$. Suppose that $\Sigma-\Sigma_a = \{\alpha,\beta\}$. If $\lambda(\alpha) \neq \lambda(\beta)$, we have $\delta \in \ker(\lambda)$ for δ = sum of all roots in Σ. If $\lambda(\alpha) = \lambda(\beta)$, there exists for each integer $n \leq c_\alpha+c_\beta$ a root $\delta_n = \Sigma d_\alpha \alpha$ with $d_\alpha+d_\beta = n$. So $c_\alpha+c_\beta < 3$, otherwise $\delta_3 \in \ker(\lambda)-Z\Sigma_a$. Thus (b) holds in this case. If $\Sigma-\Sigma_a = \{\alpha\}$, then similarly $c_\alpha < 3$, which proves (b). The proof of (c) is similar. Finally, (d) holds since a maximal torus of $(G_a)^{(\infty)}$ has dimension $|\Sigma_a|$, the same as the dimension $|\Sigma|$ of a maximal torus of G.

3 A GENERATION THEOREM

The following theorem covers a substantial part of Seitz's main generation result [4, Theorems 1 and 2], at the cost of adding a K-group hypothesis.

Theorem 8. (Seitz) Suppose r, p, G, σ, and $G = O^{r'}(G_\sigma)$ are as described in Section 1, p is odd, and $D \cong Z_p \times Z_p$ acts on G. Set $\Gamma = \langle O^{r'}(C_G(x)) \mid x \in D^{\#}\rangle$. Assume that every simple section of Γ is a K-group. Then either $G = \Gamma$, or else $C_D(G) = 1$ and one of the following holds:

a) $G/Z(G) \cong L_p(q)$ $(p|q-1)$ or $U_p(q)$ $(p|q+1)$ and D acts on G like the image of a nonabelian subgroup of $GL_p(q)$ (resp. $GU_p(q)$).

b) $p = 3$ and G is defined over $GF(2)$ $(q(\sigma,G) = 2$ or $\sqrt{2})$.

c) $G = {}^2F_4(2)$, ${}^2B_2(2^5)$, or $A_1(2^3)$, with $p = 5$, 5, or 3, respectively.

Our proof depends on the following lemma [4].

Lemma 9. Under the hypotheses of Theorem 8, either of the following conditions implies that $G = \Gamma$:

a) $\mathcal{N}_G(D;r) \neq \{1\}$.

b) $|G|_r = |\Gamma|_r$.

The sufficiency of (a) follows from [4, (2.3)] and the Borel-Tits theorem, and (b) implies (a) (or $G = \Gamma$) by [3, (1.6)].

Thus, in proving the theorem, we may assume that $\mathcal{N}_G(D;r) = \{1\}$ and $|G|_r > |\Gamma|_r$. Also, Proposition 5 allows us to assume that $Q_r(G,D) = 1$. Thus $W_r(G,D) = |G|_r > |\Gamma|_r$. But clearly for any $E \leq D$, $O^{r'}(C_G(E))$, lying in Γ, equals $O^{r'}(C_\Gamma(E))$. So by definition, $W_r(G,D) = W_r(\Gamma,D)$, whence $W_r(\Gamma,D) > |\Gamma|_r$. Therefore,

$$Q_r(\Gamma,D) < 1. \tag{3.1}$$

Now for each $x \in D^{\#}$, $O^{r'}(C_G(x))$ is the central product $G_1G_2\ldots G_n$ of Chevalley groups of characteristic r (see [4, (2.6)]). Since $\mathcal{N}_G(D;r) = \{1\}$, $O_{p'}(C_G(x))$ is an r'-group, so $p\,\big|\,|G_i|$ for each i. Unless some G_i is one of the eight exceptions to simplicity ($A_1(2)$, etc.), it follows that

$$O^{r'}(C_G(x)) \trianglelefteq L_{p'}(C_G(x)). \tag{3.2}$$

(In fact, equality obviously holds.) If some $G_i \cong A_1(2)$, ${}^2A_2(2)$, or ${}^2B_2(2)$, then (b) or (c) of the theorem must hold (cf. [1,(9-1) and (14-1)]). Since $p \neq r$ and p is odd, $G_i \not\cong A_1(3)$ for all i. The cases where some $G_i \cong B_2(2)$, $G_2(2)$, ${}^2F_4(2)$, or ${}^2G_2(3)$ must be dealt with directly, but they are straightforward, and the possibilities for G are at any rate extremely limited.

Now assuming (3.2), we get for each $x \in D^{\#}$ that $O^{r'}(C_G(x)) \leq L_{p'}(C_\Gamma(x)) \leq L_{p'}(\Gamma)$, the last by $L_{p'}$-balance applied to the K-group Γ [1, (18-1)]. Hence $\Gamma = L_{p'}(\Gamma)$. Writing $\bar{\Gamma} = \Gamma/O_{p'}(\Gamma) = \bar{\Gamma}_1 \times \ldots \times \bar{\Gamma}_n$ as in Lemma 1, we see by that lemma and (3.1) that $Q_r(\bar{\Gamma}_i,D) < 1$ for some i with $\bar{\Gamma}_i$ quasisimple and $C_D(\bar{\Gamma}_i) = 1$. Hence by Corollary 6,

$$\bar{\Gamma}_i \notin \mathrm{Chev}(r), \tag{3.3}$$

which is the main point of the whole argument.

Now (3.3) is essentially a contradiction, since for each $x \in D^{\#}$, $O^{r'}(C_{\bar{\Gamma}_i}(x))$ is isomorphic to a central factor group of a product of some components of $O^{r'}(C_G(x))$, which are in Chev(r). Surveying the possibilities for $\bar{\Gamma}_i$ -- known quasisimple groups outside Chev(r) -- and comparing $O^{r'}(C_{\bar{\Gamma}_i}(x))$ with $O^{r'}(C_G(x))$, we see that the possibilities for the components of $O^{r'}(C_{\bar{\Gamma}_i}(x))$ are extremely limited: Chevalley groups of characteristic r which are at the same time either of other characteristics, or alternating groups, or components of centralizers of p-elements in sporadic groups. It is rather simple then to reach a contradiction in all cases. It is convenient first to reduce to the case $m_p(C_{GD}(D)) = 2$ by applying the theorem inductively to $C_G(y)$ for appropriate y of order p in $C_G(D)$.

We give no further details, but close with some remarks about Theorem 8. First, it generalizes also to the case $p = 2$; the only new exception is $G \cong {}^2G_2(3)$. The same methods of proof may be used, but yield more exceptional cases which must be handled directly, namely, orthogonal groups of dimension at most 8, and groups of Chevalley rank at most 2. In fact, in all other cases, Lemma 7(c) gives $\mathcal{U}_G(D;r) \neq \{1\}$ anyway, so Proposition 4 is not of much use when $p = 2$. Second, to obtain Seitz's main result [4, Theorems 1 and 2] from Theorem 8, i.e., to show that in general

$$G = \langle C_G(E_o) \,\big|\, |E:E_o| = p\rangle \tag{3.4}$$

for an elementary abelian p-group E acting on G, one still needs to use his proof [4, pp. 373-379] for the troublesome case $G/Z(G) \cong L_p(q)$ or $U_p(q)$ (but not $L_3(4)$). The precise list of counterexamples G to (3.4) can then be obtained easily by induction, using Theorem 8, although a number of special cases still must be examined, particularly to construct examples of non-generation. Likewise, closer examination of particular groups narrows the groups that must be listed in Theorem 8(b) to only a small number.

REFERENCES

1. Gorenstein, D. & Lyons, R. (1983). The local structure of finite groups of characteristic 2 type. Memoirs A.M.S. #276.
2. Humphreys, J.E. (1975). Linear Algebraic Groups. New York: Springer-Verlag.

3. Seitz, G.M. (1973). Flag-transitive subgroups of Chevalley groups. Ann. Math. 97, 27-56.
4. ________. (1982). Generation of finite groups of Lie type. Trans. A.M.S. 271, 351-407.
5. Steinberg, R. (1967). Lectures on Chevalley groups, notes prepared by J. Faulkner & R. Wilson, Yale University.
6. __________. (1968). Endomorphisms of linear algebraic groups. Memoirs A.M.S. #80.
7. Wielandt, H. (1960). Beziehungen zwischen den Fixpunktzahlen von Automorphismengruppen einer endlichen Gruppe. Math. Zeit. 73, 146-158.

OVERGROUPS OF IRREDUCIBLE LINEAR GROUPS

G.M. Seitz
University of Oregon and Institute for Advanced Study

In this note we describe a current project concerning the subgroup structure and representation theory of finite groups of Lie type and corresponding algebraic groups. For such a group G, let V be a finite dimensional irreducible G-module in the natural characteristic. The ultimate goal of the project is to determine the overgroups of the image of G in $GL(V)$ (finite overgroups if G is finite and connected overgroups if G is a connected algebraic group). Such a determination would be of fundamental importance for the study of the subgroup structure of groups of Lie type and we begin our discussion with a recent result of Aschbacher indicating the relevance to the analysis of maximal subgroups of finite classical groups.

Let L be a finite classical group over a field of characteristic p and let V be the natural module for L. In [1] Aschbacher describes a family C of subgroups of L arising naturally from V. Most of these subgroups occur as stabilizers of subspaces or sets of subspaces of V, normalizers of classical groups over subfields or extension fields of the field of definition of V, normalizers of subgroups preserving a tensor product decomposition of V, or normalizers of certain extraspecial groups. He then proves

Theorem 1 ([1]). Let X be a subgroup of L. Then either X is contained in some element of C or $F^*(X) = A$ is quasisimple and absolutely irreducible on V.

So to determine the maximal subgroups of L we must consider quasisimple groups A embedded as absolutely irreducible subgroups of L and then determine whether or not $N_L(A)$ is maximal in L. Our approach to such a question is to find all overgroups of A in $GL(V)$ and then determine which of these lie in L. So when A is of Lie type in characteristic p, this is precisely the problem mentioned in the first paragraph, which we state formally.

Problem 1. For A of Lie type in characteristic p acting absolutely irreducibly on a finite vector space V over a field of characteristic p, determine all overgroups of A in GL(V).

A first step in a solution to Problem 1 is the following

Theorem 2 ([4]). There is a finite collection F of quasisimple groups such that whenever A and V are as in Problem 1 with A not isomorphic to a group in F and $A \leq Y \leq GL(V)$, then

i) $\langle A^Y \rangle = E(Y)$ and $F^*(Y) = E(Y)C$ with C cyclic.

ii) E(Y) is a commuting product of groups of Lie type in characteristic p and alternating groups.

The groups in F are of low Lie rank and defined over small fields. We refer the reader to [4] for the precise list. The appearance of alternating groups in (ii) is unavoidable as can be seen by considering groups A having 2-transitive permutation representations and reducing the permutation module mod p. However, there is a variation of Theorem 2 which avoids both the family F and the alternating group configuration at the cost of a dimension restriction on V. Write $A = A_d(p^a)$ to indicate that A is defined over a field of order p^a and that the associated algebraic group has rank d.

Theorem 2' ([4]). Let $A = A_d(p^a)$ and V be as in Problem 1, with $\dim(V) < \frac{1}{2}(p^{ad}-1)$. If $A \leq Y \leq GL(V)$, then

i) $\langle A^Y \rangle = E(Y)$ and $F^*(Y) = E(Y)C$ with C cyclic.

ii) E(Y) is a commuting product of groups of Lie type in characteristic p.

In view of the above results we now consider embeddings $A < Y < SL(V)$ where A is of Lie type in characteristic p, Y is a commuting product of such groups, and A is absolutely irreducible on V. Our goal is to determine such triples (A,Y,V).

Replace V by $\bar{V} = \bar{K} \otimes_K V$, where $\bar{K} = \bar{\mathbb{F}}_p$ and K is the underlying field for V. Two immediate reductions are possible. Writing $\bar{V}$ as a tensor product of irreducibles, one for each component of Y, we reduce to the case Y quasisimple. Moreover, from the Steinberg tensor product theorem we may assume that $\bar{V}$ is a restricted module.

Let $\bar{A}$ and $\bar{Y}$ be simply connected algebraic groups over $\bar{K}$ corresponding to A and Y, respectively. Then we may regard $\bar{V}$ as a rational module for $\bar{A}$ and for $\bar{Y}$, and it would be desirable to arrange things so that the image of $\bar{A}$ in $SL(\bar{V})$ is contained in the image of $\bar{Y}$. We have a result that establishes this containment for most cases when Y is a classical group, but a completely general result has yet to be established. In any case we are led to the following important problem.

<u>Problem 2</u>. Determine all triples $(\bar{A},\bar{Y},\bar{V})$ where $\bar{A}$ and $\bar{Y}$ are quasisimple algebraic groups, $\bar{A} \leq \bar{Y} \leq SL(\bar{V})$, $\bar{V}$ is restricted for $\bar{Y}$ and irreducible under the action of $\bar{A}$.

A solution to the above problem requires a list of all pairs $(\bar{A},\bar{Y})$ together with all modules $\bar{V}$ identified by high weights for maximal tori in both $\bar{A}$ and $\bar{Y}$. This is the main problem we wish to discuss and it is currently under investigation by the author and Donna Testerman, a doctoral student at the University of Oregon.

In two remarkable papers ([2], [3]) Dynkin completely solved Problem 2 in characteristic 0. The papers are long and complicated, but the end result is a surprisingly short list of possible configurations. In particular, he determines all maximal (connected) subgroups of classical groups (in characteristic 0).

The corresponding problem in characteristic p is substantially more difficult. Some of Dynkin's key reductions fail and there are embeddings here that Dynkin did not have to contend with. Nevertheless, there is now strong evidence to suggest that a complete solution is possible, and in the following we outline our approach to such a solution.

Fix $\bar{A}$, $\bar{Y}$, $\bar{V}$ as in Problem 2. Let $\bar{P}_1$ be a proper parabolic subgroup of $\bar{A}$, $\bar{P}_1 = \bar{Q}_1\bar{L}_1$, where $\bar{Q}_1 = R_u(\bar{P}_1)$ and $\bar{L}_1$ is a Levi factor of $\bar{P}_1$. We can embed $\bar{P}_1$ in a parabolic subgroup $\bar{P}_2 = \bar{Q}_2\bar{L}_2$ of $\bar{Y}$ in such a way that $\bar{Q}_1 \leq \bar{Q}_2$. For $i = 1,2$ let $\bar{V}_i$ be the fixed points of $\bar{Q}_i$ in $\bar{V}$.

An easy consequence of the result in [5] is the following

<u>Lemma A</u>. $\bar{V}_1 = \bar{V}_2$ and each of $\bar{L}_1$ and $\bar{L}_2$ acts irreducibly on this fixed point space.

Hence, we obtain a smaller configuration of our problem. So our method is induction, and to get things started we first look at the minimal case, where $\bar{A}$ has rank 1.

Theorem 3. Assume that $\bar{A}$ has rank 1 ($\bar{A} = SL_2$ or PSL_2) and that $\bar{Y}$, $\bar{V}$ are as above. Then one of the following holds:

i) $\bar{Y} = SL(\bar{V})$, $Sp(\bar{V})$, or $SO(\bar{V})$.

ii) $\bar{Y} = G_2(\bar{K})$, $p \geq 7$, and $\dim V = 7$.

iii) $p = 2$, $\dim \bar{V} = 2^n$, and either

a) $\bar{Y} = B_n(\bar{K})$, $\bar{A}$ is diagonally embedded in the commuting product of n copies of $B_1(\bar{K})$, and as a module for $\bar{Y}$, $\bar{V}$ has high weight with labeling 1 0 0 ... 0 0.

b) $\bar{Y} = D_{n+1}(\bar{K})$, $\bar{A} \leq B_n(\bar{K})$, $\bar{V}$ has high weight with label (1, 0; 0 0 ... 0), and the pair $(\bar{A}, B_n(\bar{K}))$ is as in (a).

Our proof of Theorem 3 is fairly easy when $\bar{V}$ is also restricted as a module for $\bar{A}$. But the general case is complicated and involves some additional results of independent interest and of significance for the general case of Problem 2. One begins by observing that the weight spaces for a fixed maximal torus of $\bar{A}$ are necessarily 1-dimensional, and so this must also be true for $\bar{Y}$. We then, more or less, determine those pairs $(\bar{Y},\bar{V})$ satisfying this condition. There are very few such pairs, especially if $\bar{Y}$ is an exceptional group. A little additional work settles the case where $\bar{Y}$ is exceptional. For $\bar{Y}$ a classical group with natural module $\bar{W}$, the universal cover of $\bar{A}$ acts on $\bar{W}$ and it is necessary to describe this action. We establish the following theorem which does not depend on $\bar{A}$ being a rank 1 group.

Theorem 4. Let $\bar{A}$, $\bar{Y}$, $\bar{V}$ be as in Problem 2 and assume that $\bar{Y}$ is a classical group with natural module $\bar{W}$. Regard $\bar{W}$ as a module for the universal covers $\hat{A}$, $\hat{Y}$ of $\bar{A}$, $\bar{Y}$ respectively and assume that $\bar{V}$ is not isomorphic to $\bar{W}$ or $\bar{W}^*$ (as a $\hat{Y}$-module). Then one of the following holds:

i) $\hat{A}$ acts irreducibly on $\bar{W}$ and $\bar{W}$ is a conjugate of a restricted $\hat{A}$-module.

ii) $\bar{Y} = C_n(\mathbb{F}_2)$ and $\bar{V} \longleftrightarrow$ 1 0 0 ... 0 0.

iii) $\bar{Y} = D_n$, $\hat{A}$ stabilizes a nondegenerate subspace of odd dimension and $\bar{V} \longleftrightarrow$ (1, 0; 0 0 ... 0 0)

iv) $\bar{Y} = D_n$, $\hat{A}$ fixes a unique 1-space of $\bar{W}$, and $\bar{V}_{\bar{Y}}$ is determined.

v) $\bar{Y} = D_8$, $\hat{A}$ acts irreducibly on $\bar{W}$ and $\bar{V}$ is a spin module for $\bar{Y}$.

While it is not surprising that in the context of Theorem 4, $\hat{A}$ is usually irreducible on $\bar{W}$, it is quite significant that the action of $\hat{A}$ on $\bar{W}$ is given by a conjugate of a restricted module. For one thing this makes it fairly easy to compare weights of $\bar{W}$ and $\bar{V}$ for a maximal torus of $\hat{Y}$ with their restriction to a maximal torus of $\hat{A}$ (especially if $\hat{A}$ is rank 1). In addition, it is possible to establish useful results regarding the embedding of parabolic subgroups of $\bar{A}$ in those of $\bar{Y}$. We'll come back to this later.

Our only comment about the proof of Theorem 4 is that it makes use of the following result which represents the opposite of the situation in Theorem 3.

<u>Theorem 5</u>. Let $\bar{A}, \bar{Y}, \bar{V}$ be as in Problem 2 and assume that $\text{rank}(\bar{A}) = \text{rank}(\bar{Y})$. Then

i) $\bar{Y} = B_n, C_n, F_4$, or G_2 with $p = 2,2,2,3$, respectively.

ii) $\bar{A}$ is generated by a subsystem of root subgroups in a fixed system of $\bar{Y}$ and $\bar{A}$ contains either all long root subgroups or all short root subgroups in this system.

iii) As a module for $\bar{Y}$, $\bar{V} = \bar{V}(\lambda)$, where λ is a dominant weight with support on those fundamental dominant weights corresponding to long (resp. short) fundamental roots.

The proof of Theorem 3 is completed using Theorem 4, Theorem 5, and a few additional arguments. It is curious that one settles the maximal case (Theorem 5) of Problem 2 in order to deal with the minimal case (Theorem 3). We note that the configurations appearing in Theorem 5 do not occur in characteristic 0, so are not on Dynkin's list.

With Theorem 3 established we are in a position to proceed toward a solution to the general case of Problem 2 using Lemma A and induction. Unfortunately, the inductive information is fairly weak and additional methods are required. Before describing these we point out two major differences between the characteristic 0 and characteristic p versions of Problem 2.

1) Let $\bar{A}, \bar{Y}, \bar{V}$ be as in Problem 2 and write $\bar{V} = V(\lambda)$ for λ a dominant weight of a maximal torus of $\bar{Y}$, and suppose $\lambda = \mu+\nu$ for μ, ν dominant. If $\hat{A}, \hat{Y}$ denote the universal covers of $\bar{A}, \bar{Y}$,

respectively, then $\hat{A}$ acts on the irreducible $\hat{Y}$-modules corresponding to μ and ν. When the underlying field has characteristic 0, it is not difficult to show that $\hat{A}$ is also irreducible on each of these modules. Consequently, Dynkin first considers the case where λ is a fundamental dominant weight and then considers cases where λ is compound. But in characteristic p the modules corresponding to μ and ν may fail to be irreducible for $\hat{A}$, so this reduction is not available.

2) Let $\bar{P}_1, \bar{P}_2$ be parabolics of $\bar{A}, \bar{Y}$, as before, and choose $\bar{P}_1$ corresponding to an end node of the Dynkin diagram of $\bar{A}$. Then $\bar{L}_1'$ is quasisimple. If the underlying field is of characteristic 0, then nontrivial $\bar{L}_1'$ modules cannot have irreducible tensor product and this implies that at most one of the quasisimple constituents of $\bar{L}_2$ acts nontrivially on the fixed point space $\bar{V}_2$ of Lemma A. However, in characteristic p arbitrary irreducible modules for $\bar{L}_1'$ are tensor products of conjugates of restricted modules, and so, a priori, $\bar{L}_2$ could have any number of constituents acting irreducibly on $\bar{V}_2$.

The methods used to overcome the above difficulties (not to mention the absence of Freudenthal's formula, the Weyl character formula, and complete reducibility of modules) involve a detailed study of the embedding of $\bar{A}$ in $\bar{Y}$, as well as the action of these groups on $\bar{V}$. It would appear that these ideas have application to other problems in representation theory as they yield general information on the embedding of $\bar{A}$ in $SL(\bar{V})$, which can then be used to obtain further information about $\bar{V}$.

Recall the parabolic subgroups $\bar{P}_1 \leq \bar{P}_2$ and unipotent radicals $\bar{Q}_1 \leq \bar{Q}_2$. For $i = 1,2$ set $\bar{Z}_i = Z(\bar{L}_i)^0$. Of particular interest are the following:

a) The embedding of $\bar{Q}_1$ in $\bar{Q}_2$.

b) The projections of the quasisimple components of $\bar{L}_1\bar{Q}_2/\bar{Q}_2$ to the components of $\bar{P}_2/\bar{Q}_2 \cong \bar{L}_2$.

c) For $i = 1,2$, the commutator series $\bar{V} > [\bar{V},\bar{Q}_i] > [\bar{V},\bar{Q}_i,\bar{Q}_i] > \dots > 0$.

d) The group $C_{\bar{Y}}(\bar{Z}_1)$ and the action of $\bar{Z}_1$ on the series in (c).

The composition factors of $\bar{L}_i$ on $\bar{Q}_i$ have a natural module structure. The fact that $\bar{Q}_1 \leq \bar{Q}_2$ and that both $\bar{A}$ and $\bar{Y}$ are irreducible on $\bar{V}$ imposes very strong restrictions on the various composition factors. In turn, this information feeds back to yield more information regarding the action of $\bar{L}_i$ on the series in (c). This is an extremely tight situation and is a key feature of our approach to Problem 2. Details will appear elsewhere. Here we have space for only a few general remarks.

For $r \geq 1$ set $[\bar{V},\bar{Q}_i^r] = [\bar{V},\overbrace{\bar{Q}_i,\ldots,\bar{Q}_i}^{r}]$ and set $[\bar{V},\bar{Q}_i^0] = \bar{V}$. In addition, for $r \geq 1$ we define $\bar{V}_{\bar{Q}_i}^r = [\bar{V},\bar{Q}_i^{r-1}]/[\bar{V},\bar{Q}_i^r]$. If there is no ambiguity regarding $\bar{Q}_i$ we simply write $\bar{V}^r$.

The following lemma is easy but quite useful.

<u>Lemma B</u>. i) $[\bar{V},\bar{Q}_1] = [\bar{V},\bar{Q}_2]$.

ii) $\bar{V}_{\bar{Q}_1}^1 = \bar{V}_{\bar{Q}_2}^1$ and is an irreducible module for both $\bar{L}_1$ and $\bar{L}_2$.

iii) As an $\bar{L}_i$-module, $\bar{V}_{\bar{Q}_i}^1 \cong (\bar{V}_i^*)^*$, where $\bar{V}_i^*$ is the fixed space of $\bar{Q}_i$ on V^*.

In view of (i) we have $\bar{V}_{\bar{Q}_2}^2$ an $\bar{L}_1$-invariant section of $\bar{V}_{\bar{Q}_1}^2$ and this section is of great importance in what follows.

In order to illustrate the methods without introducing unnecessary technicalities we will make the following hypotheses for the remainder of the paper.

<u>Hypotheses</u>. 1) $\bar{P}_1$ is a maximal parabolic corresponding to an end node α.

2) $\bar{Q}_1$ is the product of those root subgroups of $\bar{A}$ corresponding to negative roots with non-zero coefficients of α.

3) Neither $\bar{A}$ nor $\bar{Y}$ is of type B_n, C_n, F_4, or G_2, with $p = 2,2,2,3$ respectively.

We take $\bar{Q}_1$ to be a product of root subgroups for negative roots in order that the high weight of $\bar{L}_1$ on $\bar{V}_{\bar{Q}_1}^1$ is just the high weight of $\bar{V}$ restricted to $\bar{L}_1$ (really, to a maximal torus of $\bar{L}_1$).

<u>Lemma C</u>. We may choose $\bar{P}_2$ and Levi factor $\bar{L}_2$ of $\bar{P}_2$ such that

$\bar{Z}_1 \leq \bar{Z}_2$. Hence, $\bar{L}_2$ acts on the weight spaces of $\bar{Z}_1$ on $\bar{V}^2_{\bar{Q}_2}$.

To describe the weight spaces of $\bar{Z}_1$ on $\bar{V}^2_{\bar{Q}_1}$ we introduce the following notation. Write $\bar{V}|_{\bar{A}} = \bar{V}_1{}^{q_1} \otimes \ldots \otimes \bar{V}_k{}^{q_k}$, where each $\bar{V}_i$ is restricted and $q_i \neq q_j$ for $i \neq j$. Let $\bar{W}_j = \bar{V}_j{}^{q_j}$ for $1 \leq j \leq k$. Finally, we set $\bar{V}^2_{\bar{Q}_1,q_i} = \bar{W}^1_1 \otimes \ldots \otimes \bar{W}^1_{i-1} \otimes \bar{W}^2_i \otimes \bar{W}^1_{i+1} \otimes \ldots \otimes \bar{W}^1_k$ (recall that we write $\bar{W}^r_i$ for $(\bar{W}_i)^r_{\bar{Q}_1}$).

Each of the spaces $\bar{V}^2_{\bar{Q}_1,q_i}$ is an $\bar{L}_1$-module and we have

Proposition D. With the above notation

i) $\bar{V}^2_{\bar{Q}_1} \cong \bar{V}^2_{\bar{Q}_1,q_1} \oplus \ldots \oplus \bar{V}^2_{\bar{Q}_1,q_k}$.

ii) Each summand $\bar{V}^2_{\bar{Q}_1,q_i}$ of $\bar{V}^2_{\bar{Q}_1}$ is a weight space of $\bar{Z}_1$, with corresponding weight $(\lambda-q_i\alpha)|_{\bar{Z}_1}$ (λ is the high weight of $\bar{V}$).

iii) For $1 \leq i \leq k$, $\dim(\bar{V}^2_{\bar{Q}_1,q_i}) \leq \dim(\bar{V}^1_{\bar{Q}_1})\dim(\bar{Q}_1)$.

In order to get any use out of the above decomposition we need a decomposition of $\bar{V}^2_{\bar{Q}_2}$ into weight spaces for $\bar{Z}_1$.

Choose $\bar{P}_2$ as in Lemma C. So $\bar{Z}_1 \leq \bar{Z}_2$ and we label root subgroups of $\bar{P}_2$ (for suitable maximal torus) so that $\bar{Q}_2$ corresponds to a product of root subgroups for negative roots. For each fundamental root δ not in the root system of $\bar{L}_2$ we set $\bar{V}^2_{\bar{Q}_2,\delta}$ to be the sum of all weight spaces $\bar{V}_{\lambda-\mu-\delta}$ such that $\lambda-\mu-\delta$ is a sum of positive roots in the root system of $\bar{L}_2$.

Proposition E. With notation as above,

i) $\bar{V}^2_{\bar{Q}_2} \cong \bigoplus_{\delta} \bar{V}^2_{\bar{Q}_2,\delta}$.

ii) For each δ there is a unique i such that $\bar{V}^2_{\bar{Q}_2,\delta}$ is isomorphic to an $\bar{L}_1$-invariant image of $\bar{V}^2_{\bar{Q}_1,q_i}$.

iii) With δ and q_i as in (ii), $\delta|_{\bar{Z}_1} = q_i\alpha|_{\bar{Z}_1}$ and $\dim(\bar{V}^2_{\bar{Q}_2,\delta}) \leq \dim(\bar{V}^1_{\bar{Q}_2})\dim(\bar{Q}_1)$.

The bound in Proposition E(iii) is already quite restrictive and leads to a contradiction in many situations, as follows. We produce a weight μ such that $\bar{V}^2_{\bar{Q}_2,\delta}$ has an $\bar{L}_2$-composition factor with high weight μ (really μ restricted to a suitable maximal torus of $\bar{L}_2$) and then establish certain lower bounds for the dimension of this factor. Comparison with Proposition E(iii) gives the contradiction.

At this stage we point out ways in which group theoretical information involving the embedding $\bar{A} \leq \bar{Y}$ can be used in conjunction with the above methods.

For each δ with $\bar{V}^2_{\bar{Q}_2,\delta} \neq 0$ it is possible to produce a certain normal subgroup $\bar{Q}_{2,\delta}$ of $\bar{Q}_2$ such that the quotient $\bar{Q}_2/\bar{Q}_{2,\delta}$ is a chief factor of $\bar{P}_2$ and such that $\bar{Q}_1 \not\leq \bar{Q}_{2,\delta}$. Moreover, this chief factor has a natural module structure and the image of $\bar{Q}_1$ becomes an $\bar{L}_1$-invariant submodule. The action of $\bar{L}_1$ on $\bar{Q}_1$ is well understood and the action of $\bar{L}_1$ on the chief factor is partly determined by the inductive information contained in Lemma A. This analysis leads to some very precise information. For example one can argue that δ must adjoin a simple root existing in the root system of $\bar{L}_2$ and that with few exceptions $\bar{L}_2$ must have commuting factors each with such a simple root. Moreover, one of these factors is usually trivial on $\bar{V}^1_{\bar{Q}_2}$ but nontrivial on $\bar{V}^2_{\bar{Q}_2,\delta}$. So this is information not available from induction alone and the information feeds in to give better lower bounds for $\dim(\bar{V}^2_{\bar{Q}_2,\delta})$.

Everything so far has been based on the pair of parabolic subgroups $\bar{P}_1$, $\bar{P}_2$. At some point it is necessary to consider other parabolic subgroups of $\bar{A}$ (containing a fixed Borel subgroup of $\bar{P}_1$) and corresponding parabolic subgroups of $\bar{Y}$. To do this most effectively one would like to have the parabolics of $\bar{Y}$ so obtained to be built on the same system of root subgroups that determine $\bar{P}_2$. In other words, the parabolics should at least contain a common maximal torus. The following result settles this when $\bar{V}$ is a conjugate of a restricted $\bar{A}$-module.

<u>Proposition F.</u> Assume that $\bar{V}$ is an algebraic conjugate of a restricted $\bar{A}$-module. Then it is possible to choose $\bar{P}_2$ such that

i) $\bar{Q}_1 \leq \bar{Q}_2$ and $\bar{Z}_1 \leq \bar{Z}_2$.

ii) If $\bar{T}_{\bar{A}}$ is any maximal torus of $\bar{L}_1$ and if $\bar{T}_{\bar{Y}}$ is any maximal torus of $\bar{Y}$ containing $\bar{T}_{\bar{A}}$, then $\bar{T}_{\bar{Y}} \leq \bar{P}_2$.

iii) Fix $\bar{T}_{\bar{A}} \leq \bar{T}_{\bar{Y}}$ as in (ii). There is a group $\bar{Q}$ such that $\bar{Q}_1 \leq \bar{Q} \leq \bar{Q}_2$, $\bar{Q}$ is a product of $\bar{T}_{\bar{Y}}$-root subgroups of $\bar{Y}$, and $[\bar{V},\bar{Q}_1^d] = [\bar{V},\bar{Q}^d]$ for each $d > 0$.

More can be obtained than indicated above. For example, one can show that the parabolic $\bar{P}_2$ is contained in a parabolic $\bar{P}_3$ such that $\bar{L}_1 \leq \bar{L}_3 \geq \bar{T}_{\bar{Y}}$ for all tori $\bar{T}_{\bar{Y}}$ as in (ii). While it is clearly desirable to have such a parabolic $\bar{P}_3$, we do not necessarily have $\bar{Z}_1 \leq Z(\bar{L}_3)$. Consequently, for such a parabolic the conclusions of Proposition E may fail to hold.

Parts (i) and (ii) of Proposition F are the most essential parts and these are usually available even if $\bar{V}$ is not a conjugate of a restricted $\bar{A}$-module. For example, if $\bar{Y}$ is a classical group with natural module $\bar{W}$, then Theorem 4 shows that in most cases $\bar{W}$ is an algebraic conjugate of a restricted module for $\bar{A}$ (or suitable covering group). So we can use $\bar{W}$ to obtain the desired parabolic $\bar{P}_2$.

We conclude the discussion of Problem 2 with a rough indication of our basic strategy for the general case. Fix parabolics $\bar{P}_1 \leq \bar{P}_2$ (hopefully $\bar{P}_2$ satisfies the conditions of Proposition F) and maximal tori, $\bar{T}_{\bar{A}}$, $\bar{T}_{\bar{Y}}$ of $\bar{P}_1$, $\bar{P}_2$, respectively, such that $\bar{T}_{\bar{A}} \leq \bar{T}_{\bar{Y}}$. Write $\bar{V}_{\bar{A}} = \bar{V}_1^{q_1} \otimes \ldots \otimes \bar{V}_k^{q_k}$, as before.

Assume that $\bar{Y}$ is a classical group (things are somewhat different when $\bar{Y}$ is an exceptional group). Then Theorem 4 tells us how $\bar{A}$ (or a suitable cover of $\bar{A}$) acts on the usual module associated with $\bar{Y}$. If this action is reducible then $\bar{A}$ is contained in a product of smaller classical groups and we can use induction. So assume the action is irreducible. Then we also know that the action is given by a twist, say by q, of a restricted module (except for a case when $\bar{Y}$ is of type D_4).

One can then argue that $\bar{P}_2$ can be chosen with the following additional property: if δ is a fundamental root for $\bar{T}_{\bar{Y}}$ adjoining a factor of $\bar{L}_2$ which acts non-trivially on $\bar{V}/[\bar{V},\bar{Q}_2]$, then $\delta|_{\bar{Z}_1} = q\alpha|_{\bar{Z}_1}$. On the other hand, there is a technical result based on induction and a detailed analysis of the embedding $\bar{Q}_1 \leq \bar{Q}_2$, which can usually be applied to show that for each $1 \leq i \leq k$ there is such a fundamental root δ satisfying $\delta|_{\bar{Z}_1} = q_i\alpha|_{\bar{Z}_1}$. Consequently, $k = 1$

and $q_i = q$.

Of course, this shows that there is just one term in the decomposition of Proposition D(i). Also we see that there is at most one factor of $\bar{L}_2$ acting nontrivially on $\bar{V}/[\bar{V},\bar{Q}_2]$. In turn, this can be used to show that the decomposition of Proposition E(i) has only a few (almost always 1 or 2) non-zero terms. By this time we are getting close to a precise description of the high weight λ associated with $\bar{V}$. For example, if $\bar{L}_2$ is nontrivial on $\bar{V}/[\bar{V},\bar{Q}_2]$ (if $\bar{L}_2$ is trivial the analysis is easier) and $\bar{L}_0$ the above mentioned factor of $\bar{L}_2$, then the only fundamental roots δ of $\bar{Y}$ satisfying $\langle\lambda,\delta\rangle \neq 0$ are those in the root system of $\bar{L}_0$ or adjacent to $\bar{L}_0$ in the Dynkin diagram. Moreover, we already know the labeling of $\bar{L}_0$ by induction. Using various additional techniques we eventually arrive at an explicit description of $\lambda|_{\bar{T}_{\bar{A}}}$ and $\lambda|_{\bar{T}_{\bar{Y}}}$. Of course we must then decide which configurations actually occur, but in most cases this is not too difficult since the high weight λ is not very complicated.

At this writing the above procedure has been used to settle most configurations with $\bar{Y}$ a classical group and it seems that methodology exists for handling the remaining cases. We are therefore hopeful of a complete solution to Problem 2.

The following are open problems relating to the topics of this paper. Each of the problems appears to be difficult, with the last problem especially challenging.

Problem 3. In the context of Theorem 2, determine all cases where Y has a component which is a cover of an alternating group.

Problem 4. Determine all triples (A,Y,V) with both A and Y covers of alternating groups, $A < Y \leq SL(V)$, and A irreducible on V.

Problem 5. Study Problem 1 when A is quasisimple but not of Lie type in characteristic p.

REFERENCES

1. Aschbacher, M. On the maximal subgroups of the finite classical groups. (To appear, Invent. Math.)
2. Dynkin, E.B. (1957). Maximal subgroups of classical groups. A.M.S. Translation, 111-244.

3. Dynkin, E.B. (1957). Semisimple subalgebras of semisimple Lie algebras, A.M.S. Translation, 245-378.
4. Seitz, G. Overgroups of irreducible subgroups of GL(V). (manuscript)
5. Smith, S. (1982). Irreducible modules and parabolic subgroups. J. Alg. 75, 286-289.

TOWARD THE CONSTRUCTION OF AN ALGEBRA FOR O'NAN'S GROUP

D. Frohardt
Wayne State University, Detroit, Michigan 48202

In [3], O'Nan provided evidence for the existence of a simple group of order 460, 815, 505, 920. Sims subsequently proved the existence of such a group using computer calculations. The uniqueness of this group was established by Sims and Andrilli, also using computer calculation. I am not aware of any other existence or uniqueness proofs for this group. The goal of the work I discuss here is to provide computer-free proofs for the existence and uniqueness of a group satisfying the original conditions given by O'Nan. This work will also provide a concrete setting for O'N analogous to the setting for the monster provided by Griess [2] and the setting for J_3 given in [1].

These remarks are intended to give a sketch of the general methods employed in my work in progress particularly as they differ from analogous parts of [1] and [2]. At the end, I shall mention possible approaches to other groups.

We let G denote a group satisfying the conditions given by O'Nan [3]. In that paper he proved that G has a triple cover $\hat{G}$, and he also derived the character tables for G and $\hat{G}$. In producing the character tables, O'Nan obtained a complete list of p-local subgroups as well as information about G-fusion. He also showed that G contains two non-conjugate subgroups isomorphic to $L_3(7)$*. O'Nan's results constitute the basic hypotheses under which we proceed.

The smallest faithful representation of G has degree 10944, but $\hat{G}$ has faithful representations of degrees 342 and 495. The representations of degree 342 admit no non-trivial trilinear forms, while each representation of degree 495 admits a non-trivial form which is essentially unique and also symmetric. We let M be a module for a 495-dimensional representation of $\hat{G}$. The basic plan is to construct a trilinear form ϕ on M and identify $\hat{G}$ as a subgroup of $GL(M)$ preserving ϕ.

To carry out this plan, the first step is to obtain an explicit description of the action of $\hat{G}$ on M. This will have as a byproduct a proof of the uniqueness of G. The second step is to use this description to define the form ϕ. This implicitly defines an algebra structure on $M \oplus M^*$, where M^* is the dual of M. The description of ϕ is then used in the final step to recover $\hat{G}$ and obtain an existence proof.

More specifically, the procedure is the following. Let H be an $L_3(7)^*$ subgroup of G and let $\hat{H}$ denote the preimage of H in $\hat{G}$. Since $Z(\hat{H})$ has order 3, it is clear that $\hat{H} \not\cong SL_3(7)^*$. Therefore $\hat{H}$ has a subgroup of index 3 which we shall call H, by abuse of notation. The two classes of $L_3(7)^*$ subgroups of G contain different classes of elements of order 8 and these elements act with different traces on M. We can choose H and M so that M is the sum of irreducible H-submodules M_1 and M_2 of degrees 343 and 152 respectively.

M_1 is in fact the Steinberg module for H, and the action of H on M_1 can be described by choosing a basis for M_1 parametrized by $\mathbb{F}(7) \times \mathbb{F}(7) \times \mathbb{F}(7)$. For later use in defining trilinear forms, it is convenient to have as well a description of M_1 as a submodule of the permutation module V of H on its Borel subgroups. This can be obtained easily from the Bruhat decomposition.

To give a description of M_2, we use a 456-dimensional monomial representation of $H_0 := [H,H]$. The module U for this representation is induced from a line on which the elements of a Borel subgroup of H_0 act as cube roots of unity. As an H_0-module U is the sum of three irreducible submodules of degree 152. Under the action of H on U, two of these submodules are interchanged and one, which is isomorphic to M_2, is left invariant. To find a basis for M_2, we restrict to a subgroup K of H of type $E_{49} \rtimes SL_2^{\pm}(7)$. As a K-module, M_2 is the sum of three irreducible submodules R, S, and T. Two of these submodules cannot be described as monomial modules without introducing primitive 7th roots of unity. To avoid such complications, we induce to K from 6-dimensional representations in which elements of order 7 act fixed-point-free. To identify M_2 as a submodule of U and thereby find the action of H on M_2, we first identify T which is the only K-submodule of U of its isomorphism type. Then we determine the action of a particular element in the Borel subgroup of H lying outside K.

At this point, we have found a basis for $M = M_1 \oplus M_2$ and we have also identified M_1 and M_2 as submodules of U and V. To establish the uniqueness result, it suffices to show that the action of $\hat{H}$ on M can be extended to the action of $\hat{G}$ on M in an essentially unique way. The approach of [1] and [2] is not practical here because no element of $\hat{G}\setminus\hat{H}$ normalizes a large subgroup of H. We use instead a non-conjugate $L_3(7)$*-subgroup H' which intersects H in $O^3(B)$ where B is a Borel subgroup of H. This requires a description of M as an H'-module. We have $M = M_0' \oplus M_2' \oplus M_3'$ where M_0', M_2', and M_3' are irreducible H'-modules of degrees 1, 152, and 342. The action of H' on M_0' is trivial, and a description of the action of H' on M_2' is easily obtained from the description of the action of H on M_2. In order to describe the action of H' on M_3', or equivalently of H on M_3, we induce a 6-dimensional irreducible representation of K up to H using the Bruhat decomposition again. (We note in passing that M_3 is the only irreducible representation for H_0 supporting a rational-valued character that is not involved in a monomial module of degree 456, the index of a Borel subgroup. This would complicate the task of describing trilinear forms on M_3.)

Since $M = M_1 \oplus M_2 = M_0' \oplus M_2' \oplus M_3'$, and $\hat{G} = \langle H,H' \rangle$, we can establish the uniqueness of $\hat{G}$ (and G) by showing that the basis for $M_1 \oplus M_2$ can be identified in an essentially unique way inside $M_0' \oplus M_2' \oplus M_3'$. This work is still in progress and uses relations obtained first from the identification B = B', a Borel subgroup of H', then from trace values of elements lying in 2-local and 3-local subgroups generated by their intersections with H and H'.

To obtain a description of ϕ, we shall use an embedding of $S^3M \cong S^3M_1 \oplus S^2M_1 \otimes M_2 \oplus S^2M_2 \otimes M_1 \oplus S^3M_2$ as an H-submodule of $(S^2M_1 \oplus S^2M_2) \otimes (V \oplus U)$. There are 20 parameters required to describe the general H-invariant trilinear form on M. The action of H' will be used to reduce the number of parameters to one. The most difficult part of the existence proof will undoubtably be showing that the automorphisms of H' preserve ϕ. It seems likely that various identities involving the arithmetic of $\mathbb{F}_7$ will play an important role.

The other sporadic groups for which only computer proofs are available are the fourth Janko group and the Lyons group. It would be desirable to have algebras, or equivalently trilinear forms, for these groups as well. My suggestions for the most promising candidates for modules to use are the following. J_4 has an irreducible representation R of degree 1333 and there is a unique non-trivial trilinear form on Λ^2R, which is irreducible of degree 887778, but not self-dual. This form is not symmetric. If Q is an irreducible representation of degree 2480 for Ly, then S^2Q has an irreducible constituent of degree 48174 which admits a unique trilinear form, and this form is symmetric. It should be relatively straightforward to describe the actions of these groups on R and Q. In fact, given the relatively small size of these modules compared to the groups, this may well require much less work than in the present case. Nonetheless, each of these groups offers new challenges, with S being a "small" submodule of a more easily accessible module for Ly, and with J_4 providing a non-symmetric form.

REFERENCES

1. Frohardt, D. (1983). A trilinear form for the third Janko group. J. Algebra, 83, 349-379.
2. Griess, R. (1982). The friendly giant. Invent. Math., 69, 1-102.
3. O'Nan, M. (1976). Some evidence for the existence of a new simple group. Proc. London Math. Soc., 32, 421-479.

COMMUTATIVE ALGEBRAS ASSOCIATED WITH PERMUTATION GROUPS

K. Harada
Ohio State University, Columbus, Ohio

Let G be a permutation group on a set $\Omega = \{0,1,\ldots,n\}$ and V be a permutation module of (G,Ω) over a field k with permutation basis $\{v_0,v_1,\ldots,v_n\}$. Moreover, let U be a one-dimensional submodule of V generated by $v_0+v_1+\ldots+v_n$ and put $A = V/U$. Thus A is an n-dimensional G-module. If G is doubly transitive on Ω and the characteristic (denoted char(k) hereafter) of k is greater than $n+1$, then A is known to be irreducible.

The tensor product $A \otimes A$ is a G-module by the canonical diagonal action of G. Let $f \in \mathrm{Hom}_G(A\otimes A, A)$ and define $ab = f(a \otimes b)$ for $a,b \in A$. Then the product ab on A is bilinear (distributive) and G-invariant. Thus A possesses a G-invariant algebra structure, which is not necessarily associative.

Let $\mathrm{Alt}(A)$ be the submodule of $A\otimes A$ generated by all elements of shape $u\otimes v - v\otimes u$, $u,v \in A$. If $\mathrm{char}(k) \neq 2$, then $A\otimes A = \mathrm{Sym}(A) \oplus \mathrm{Alt}(A)$, where $\mathrm{Sym}(A)$ is the (2nd) symmetric product of A generated by all elements of shape $u\otimes v + v\otimes u$, $u,v \in A$.

Suppose f vanishes on $\mathrm{Alt}(A)$, then $ab = ba$ for all $a,b \in A$ and so the algebra A is commutative. In this note <u>we restrict ourselves to considering only commutative algebras</u> A. The algebra structure depends on choice of f and so A_f will be more descriptive but f will be suppressed for simplicity. We define $\mathrm{Aut}(A)$ as follows:

$$\mathrm{Aut}(A) = \{g \in GL(n,k)) \mid (ab)^g = a^g b^g \text{ for all } a,b \in A\}.$$

If $f = 0$, then $A^2 = 0$ and A is the trivial algebra and $\mathrm{Aut}(A) = GL(n,k)$. More interesting cases will arise if $f \neq 0$. For this purpose, we need to know the following fact:

<u>Theorem 1</u>. Let A be an n dimensional G module associated with a permutation group (G,Ω) with $|\Omega| = n+1$. Then

$$\mathrm{Hom}_G(A\otimes A/\mathrm{Alt}(A),\ A) \neq 0.$$

In fact, the following map f is an element of $\mathrm{Hom}_G(A\otimes A,\ A)$ which vanishes on $\mathrm{Alt}(A)$.

$$f: \begin{cases} x_i\otimes x_i \to (n-1)x_i, & 1 \leq i \leq n \\ x_i\otimes x_j \to -x_i - x_j, & 1 \leq i < j \leq n. \end{cases}$$

Here x_i is an element of $A = V/U$ defined by $x_i = v_i + U$. f is bilinear and vanishes on $\mathrm{Alt}(A)$. One can show easily that for $x_0 = -x_1-x_2-\ldots-x_n$, we have

$$f(x_0\otimes x_0) = (n-1)x_0$$
$$f(x_0\otimes x_i) = -x_0-x_i, \quad 1 \leq i \leq n$$

and so f is a G-homomorphism. We call f the <u>canonical</u> homomorphism and the algebra $A = A_f$ the <u>canonical</u> algebra of (G,Ω). We also call the set $\{x_i | 0 \leq i \leq n\}$ the <u>canonical</u> basis elements. We note that any n— element subset of $\{x_i | 0 \leq i \leq n\}$ form a basis of A.

The automorphism group of the canonical algebra A has been determined in [3] and in [1] (see [6] also).

<u>Theorem 2</u>. Let A be the canonical algebra of a permutation group (G,Ω) of dimension n with $n \geq 2$. Then

$$\mathrm{Aut}(A) \cong \begin{cases} \Sigma_{n+1} & \text{if } \mathrm{char}(k) \nmid n+1 \\ k^{n-1}\cdot GL(n-1,k) & \text{if } \mathrm{char}(k) \mid n+1. \end{cases}$$

Here Σ_{n+1} is the symmetric group of degree $n+1$ and k^{n-1} is a vector space over k of dimension $n-1$. If $n = 1$, then $A^2 = 0$ and, so $\mathrm{Aut}(A) \cong GL(1,k) \cong k\backslash\{0\}$.

If (G,Ω) is triply transitive, we have the following theorem [3]

Theorem 3. Suppose (G,Ω) is triply transitive, then $\dim_k(\mathrm{Hom}_G(A\otimes A/\mathrm{Alt}(A), A)) = 1$.

Thus up to scalar multiplication of basis elements, the canonical algebra is the only commutative algebra structure for A.

Next assume that G is doubly transitive on Ω. If S is a subset of Ω, then G_S denotes the pointwise stabilizer of S and $G_{\{S\}}$ denotes the setwise (global) stabilizer of S. Let Ω_{ij}^{ℓ} $(0 \leq \ell \leq r)$ be the orbits of $G_{\{ij\}}$ on Ω. Without loss, we may assume that

$$\Omega_{ij}^{0} = \{i,j\} \text{ and}$$
$$(\Omega_{ij}^{\ell})^{g} = \Omega_{i^g j^g}^{\ell}, \quad g \in G.$$

Then the following holds

Theorem 4. Let G, Ω, Ω_{ij}^{ℓ} $(0 \leq \ell \leq r)$ be as above. Furthermore suppose $\mathrm{char}(k) \nmid 2$. Then the commutative algebra structure of A is described as follows:

$$\begin{cases} x_i x_i = a x_i, & 1 \leq i \leq n \\ x_i x_j = b(x_i + x_j) + \sum_{\ell=1}^{r-1} \left(c_\ell \sum_{m \in \Omega_{ij}^{\ell}} x_m\right), & 1 \leq i < j \leq n. \end{cases}$$

where a, b, c_ℓ $(1 \leq \ell \leq r-1)$ are constants satisfying

$$a + (n-1)b = \sum_{\ell=1}^{r-1} c_\ell |\Omega_{01}^{\ell}|.$$

In particular, the theorem shows

$$\dim_k(\mathrm{Hom}_G(A\otimes A/\mathrm{Alt}(A), A)) = r$$

and the structure identity of A involves r independent parameters. The same theorem applies even when $\mathrm{char}(k) = 2$ if $G_{\{ij\}}$ does not possess any orbit other than $\{i,j\}$ which is the union of two orbits of G_{ij}.

Example 1.

$G = PSL(m,q)$, $m \geq 3$,

Ω = the set of $\frac{q^m-1}{q-1}$ projective points,

A = the commutative algebra associated with (G,Ω),

$$\dim A = \frac{q^m-1}{q-1} - 1.$$

Then

$$\begin{cases} x_i x_i = a x_i \\ x_i x_j = b(x_i + x_j) + c \sum_{m \in L'_{ij}} x_m \end{cases}$$

where $L'_{ij} = L_{ij} \setminus \{i,j\}$ and L_{ij} is the projective line spanned by i and j. Moreover, a, b, and c satisfy $a + (n-1)b = (q-1)c$ where $n+1 = q^m-1/q-1$.

Example 2.

$G = Co.3$, dotto 3,

$\Omega = \{0,1,\ldots,275\}$,

A = the algebra associated with (G,Ω) of dimension 275.

Then

$$\begin{cases} x_i x_i = a x_i \\ x_i x_j = b(x_i + x_j) + c \sum_{m \in \Delta_{ij}} x_m \end{cases}$$

where Δ_{ij} is the orbit of G_{ij} of length 112 and $a + 274b = 112c$.

K. Narang, a graduate student at Ohio State University has shown:

Theorem 5. Let A be a commutative algebra for $PSL(m,q)$, $m \geq 3$. Suppose $b = 0$ and $\mathrm{char}(k) > n+1$. Then $\mathrm{Aut}(A) \cong P\Gamma L(m,q)$.

For the proof, Narang made use of minimal polynomial $\phi(X)$ satisfied by the canonical basis elements as elements in $\mathrm{End}(A)$. If $a \in A$ then $x \to ax$ defines an endomorphism ϕ_a of A. The minimal polynomial of ϕ_a is by definition the minimal polynomial of a. If A

is a commutative algebra associated with $PSL(m,q)$, then

$$\phi(X) = (X+c)(X-(q-1)c)$$

is the minimal polynomial of x_i's, $0 \le i \le n$, if $b = 0$.

If σ is an element of $Aut(A)$, then $\sigma(x_i) = \sum_{i=1}^{n} a_i x_i$ must have the same minimal polynomial. This fact yields many quadratic equations in a_i's. We also have $(\sigma(x_i))^2 = a\sigma(x_i)$. This also yields another set of quadratic equations in a_i's. Using those equations, Narang was able to show that σ permutes the canonical basis elements among themselves. He then shows that the projective lines are preserved. Thus the fundamental theorem of projective geometry shows that $Aut(A) \cong P\Gamma L(m,q)$.

If the assumption $b = 0$ is dropped, then the minimal polynomial $\phi(X)$ of x_i's is $(X-a)(X-b+c)(X-b-(q-1)c)$ (see [5]). Narang is now working on it.

D. Parrott is now trying to determine the automorphism group of the commutative algebra associated with Co.3. The dimension of A is 275 and the structure of A was given in Example 2. Suppose $b = 0$. Then $a = 112c$. $c \neq 0$ as otherwise A is the trivial algebra. The minimal polynomial $\phi(X)$ of x_i, $0 \le i \le 275$, is $(X-a)(X-2c)(X+28c)$. Parrott is now trying to classify all elements $x \in A$ such that $xx = 112x$ and the minimal polynomial of x is $(X-a)(X-2c)(X+28c)$. The dimension of eigenspace for a, $2c$, or $-28c$ is 1, 252, or 22 respectively. Classification of all such elements appears very involved.

After the work of Narang, I employed the idea of minimal polynomial to obtain an alternative proof of Theorem 1. I found that the use of minimal polynomial is much more effective than the use of a relation $xx = (n-1)x$. For example one of the theorems of [6] states:

<u>Theorem 6</u>. Let A be the canonical algebra of dimension n for (G,Ω). Let x be a nonzero element of A with a minimal polynomial dividing $X^2-(n-2)X-(n-1)$. Then the following holds.

1) If $n \ge 3$ and $p \nmid (n+1)(n-2)$, then $x \in \{x_i \mid 0 \le i \le n\}$.

2) If $n \ge 3$ and $p \nmid n+1$, $p \mid n-2$, then $x \in \{\pm x_i \mid 0 \le i \le n\}$.

In the canonical algebra A, x_i's have the minimal polynomial $X^2-(n-2)X-(n-1)$. I believe that the proof is more transparent than the earlier papers of Allen [1] or of myself [3].

I can even conjecture that the canonical basis elements are the only elements that satisfy the same minimal polynomial as $\{x_i \mid 0 \leq i \leq n\}$ in the commutative algebra A of dimension n associated with a permutation group of degree n+1. Of course, we must exclude some "degenerate" cases. Narang is now working on this question for the commutative PSL(m,q) algebra.

Finally we turn our attention to a little more general permutation groups. Let G = HS be the Higman-Sims simple group. View G as a rank 3 permutation group on a set Ω of 100 letters. Let A be a 99 dimensional commutative algebra associated with (G,Ω). Then the algebra structure of A is described as follows.

$$\begin{cases} x_i x_i = a x_i + b \sum_{m \in \Omega_i} x_m & \\ x_i x_j = c(x_i+x_j) + d \sum_{m \in \Delta_{ij}} x_m & \text{if } p(ij) = 1 \\ x_i x_j = e(x_i+x_j) + f \sum_{m \in \Gamma_{ij}} x_m + g \sum_{m \in \Sigma_{ij}} x_m & \text{if } p(ij) = 2 \end{cases}$$

Notation used above is as follows.

a, b, c, d, e, f, g are constant in k,

Ω_i is the orbit of G_i of length 22,

p(ij) is the distance between i and j in the standard graph of rank 3 permutation group G,

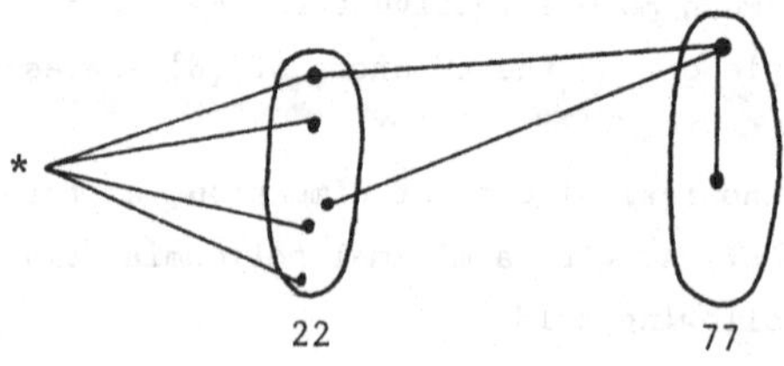

Δ_{ij} is the orbit of length 42 of G_{ij} with $p(ij) = 1$,

Γ_{ij}, Σ_{ij} is the orbit of length 6, or 32 respectively of G_{ij} with $p(ij) = 2$.

The constants satisfy the following equations

$$a + 22c + 77e = b + c + 21d + 21f + 56g = 6d + e + 16g. \qquad (*)$$

Thus there are at most five independent parameters. On the other hand, one can show that G acts on the algebra A if (*) is satisfied. Thus (*) is the only relation satisfied by the constants. If $k = C$, the number of independent parameters is equal to the value of the inner product $\langle S^2(\chi),\chi\rangle_G$ where χ is the character of degree 99 associated with (G,Ω) and $S^2(\chi)$ is the symmetric product of χ. Using the character table of $G = HS$, one computes that $\langle S^2(\chi),\chi\rangle = 5$ as expected.

(Added in Proof)

David Parrott has observed that the minimal polynomial $\phi(X)$ for the .3 algebra A is quadratic under some conditions.

Let A be the .3 algebra described in Example 2. The minimal polynomial $\phi(X)$ of x_i for the generic case is

$$(X - a)(X - (b + 2c))(X - (b - 28c)).$$

Suppose

1) $a = b + 2c$, or

2) $a = b - 28c$.

Then the minimal polynomial $\phi(X)$ is reduced to

1)' $(X - a)(X - (b - 28c))$, or

2)' $(X - a)(X - (b + 2c))$.

Using the relation

$$a + 274b = 112c,$$

we obtain

1)" $a = 6b$, $c = 5b/2$, or

2)" $a = -54b$, $c = 55b/28$.

Hence $\phi(X)$ becomes

1)"' $(X - 6)(X + 69)$, or

2)"' $(X + 1512)(X - 138)$

respectively.

Parrott says this quadratic case is more hopeful than the cubic case.

REFERENCES

1. Allen, Harry. Non-associative algebra associated with doubly transitive permutation groups: S_{n+1}. To appear in J. Alg.
2. Allen, Harry. A note on automorphism group of non-associative algebra associated with doubly transitive permutation group, to appear.
3. Harada, Koichiro. (1982). On a commutative nonassociative algebra associated with a multiply transitive group. J. Fac. Sci., Univ. of Tokyo, Vol. 28, 843-849.
4. Harada, Koichiro. On a commutative nonassociative algebra associated with a doubly transitive group. To appear in J. Alg.
5. Harada, Koichiro. On commutative nonassociative algebras associated with the doubly transitive permutation groups $PSL_m(q)$, $m \geq 3$, to appear.
6. Harada, Koichiro. Symmetric groups as automorphisms of some commutative algebras. To appear.

NILPOTENT π-SUBGROUPS BEHAVING LIKE p-SUBGROUPS

H. Bender
Christian-Albrechts-Universität Zu Kiel
2300 Kiel 1, Federal Republic of Germany

1 Let us begin with the simple question how some basic properties of p-subgroups of solvable or, more generally, of p-constrained groups, generalize from p to a given set π of primes.

The most important one of these properties is

$$O_{p'}(N_G(X)) \subseteq O_{p'}(G), \; X \text{ a p-subgroup.} \tag{1}$$

A glance at the proof of this result reveals that (1) remains true for π in place of p if, with $\bar{G} := G/O_{\pi'}(G)$,

$$C_{\bar{G}}(F(\bar{G})) \subseteq F(\bar{G}), \quad \text{and} \tag{2a}$$

$$F(\bar{G})\bar{X} \text{ is nilpotent.} \tag{2b}$$

Call G π-constrained if (2a) holds, and then let $F_\pi(G)$ denote the set of (nilpotent) π-subgroups X satisfying (2b). What turns out is that the family $F_\pi(G)$ and the set $F^*_\pi(G)$ of its maximal elements behave formally like the p-subgroups and the Sylow p-subgroups of any group:

$$X \subseteq Y \in F_\pi(G) \Longrightarrow X \in F_\pi(G) \tag{3a}$$

$$X, Y \in F_\pi(G),\; X^Y = X \Longrightarrow XY \in F_\pi(G) \tag{3b}$$

$$N \triangleleft G \Longrightarrow N \text{ is } \pi\text{-constrained,} \tag{3c}$$

$$F_\pi(N) = F_\pi(G) \cap N, \quad \text{and}$$

$$F^*_\pi(N) = F^*_\pi(G) \cap N.$$

$$F^*_\pi(G) \text{ is a class of conjugate subgroups} \tag{3d}$$

$$X \in F_\pi(G),\; H = N_G(X) \;\; (\text{or } C_G(X)) \Longrightarrow \tag{4}$$

$$H \text{ is } \pi\text{-constrained,} \tag{4a}$$

$$F_\pi(H) = F_\pi(G) \cap H, \tag{4b}$$

$$O_{\pi'}(H) = O_{\pi'}(G) \cap H, \tag{4c}$$

$$\pi(F(\bar{H})) = \pi(F(\bar{G})). \tag{4d}$$

Whereas a p-subgroup of G lying in some other subgroup H of G is still a p-subgroup of H, the analogous statement is not true for "F_π-subgroups". Hence we need results identifying nilpotent π-subgroups as F_π-subgroups, such as (4b) and the following. More can be found in [8], Theorem 2.

A nilpotent π-subgroup of G lies in $F_\pi(G)$ (5)
if and only if

for every $p \in \pi$ there exists $P \subseteq F_p$ such that (5a)

$$F_{p'} \in F_\pi(C_G(P)),$$

or, equivalently, $F_{p'} \in F_\pi(N_G(P))$.

$F \in F^*_\pi(G) \Longleftrightarrow$ (6)

F is a nilpotent π-subgroup of G containing (6a)
every F-invariant nilpotent π-subgroup of G.

The subgroups in $F^*_\pi(G)$ we call Fischer π-subgroups, after B. Fischer who first investigated such objects, and proved their conjugacy, within the general framework of his theory of injectors of solvable groups [5].

They can also be characterized as maximal nilpotent π-subgroups F of G such that $\bar{F} \supseteq F(\bar{G})$.

Injectors for non-solvable groups have first been introduced by A. Mann [8]. The above results are essentially contained in [8] or are easy exercises, mainly on Thompson's P×Q-Lemma. Mann has proved other generalizations of theorems on Sylow p-subgroups (Alperin's fusion theorem, Glauberman's ZJ-theorem), and it seems justified to regard Fischer π-subgroup as a further generalization of the Sylow subgroups, like the Hall π-subgroups, but along a different line.

The problem to generalize the ZJ-theorem from p to π actually goes back to Z. Arad and G. Glauberman [2]. They prove $ZJ(\bar{H}) \lhd \bar{G}$ for a Hall π-subgroup H of a solvable group G of odd order. In this context they study abelian subgroups A of maximal order of H

(generating $J(H)$) and prove a most remarkable and deep theorem, namely that AB is nilpotent for every A-invariant nilpotent subgroup B of H (Proposition 1). Equivalently, with G π-constrained as before and $\mathcal{A}_\pi(G)$ = set of abelian π-subgroups of G of maximal order,

$$\mathcal{A}_\pi(G) \subseteq \mathcal{F}_\pi(G), \text{ provided } 2 \notin \pi. \tag{7}$$

A theorem of J. Thompson [9] on p-constrained groups G states that if $p \neq 2$ and A is a p-subgroup containing every p-subgroup of its centralizer, then every A-invariant p'-subgroup of G lies in $O_{p'}(G)$. The $\mathcal{F}_\pi$-analogue (and easy corollary) of this is

$$A \in \mathcal{F}_\pi(G),\ \mathcal{F}_\pi(C_G(A)) \subseteq A,\ 2 \notin \pi \Rightarrow \mathcal{N}_G(A,\pi') \subseteq O_{\pi'}(G), \tag{8}$$

which together with (7) yields that

$$A \in \mathcal{A}_\pi(G),\ 2 \notin \pi \Longrightarrow \mathcal{N}_G(A,\pi') \subseteq O_{\pi'}(G). \tag{9}$$

A case of particular interest is when $\pi = 2'$. Here the conclusion is that $\mathcal{N}_G(A,\pi') \subseteq O_2(G)$. I have the feeling that abelian 2'-subgroups of maximal order could be interesting objects in arbitrary finite groups.

2 From (1) it is immediate that $O_{p'}$ is an A-signalizer functor for every abelian p-subgroup A of any finite group G with all p-local subgroups p-constrained. Can this be generalized from p to π? Suppose all π-local subgroups of a finite group G are π-constrained, and F is a nilpotent π-subgroup of G satisfying

$$F \in \mathcal{F}_\pi(N_G(F_p)) \text{ for every } p \in \pi(F). \tag{10}$$

Assume in addition that $\pi = \pi(F)$.

We shall see that

$$N_F(X) \in \mathcal{F}_\pi(N_G(X)) \text{ for every subgroup } X \neq 1 \text{ of } F. \tag{11}$$

To prove this, assume first that $X \triangleleft F$. Then $F_p \subseteq N_G(X)$ for every $p \in \pi$. By (10) and (4b), the latter applied to $N_G(F_p)$ in place of G, $F \in \mathcal{F}_\pi(N_{N_G(F_p)}(X))$. Thus, $F \in \mathcal{F}_\pi(N_{N_G(X)}(F_p))$. Now apply (5) to $N_G(X)$ in place of G to get $F \in \mathcal{F}_\pi(N_G(X))$, as required. This proves that $F \in \mathcal{F}_\pi(N_G(Z(F_p))$ for every $p \in \pi$. Now, for arbitrary $X \subseteq F$ repeat the above argument with $Z(F_p)$ in place of F_p and $N_F(X)$ in place of F.

As an immediate corollary of (11) and (4c), any two commuting non-identity subgroups X and Y of F satisfy

$$O_{\pi'}(C_G(X)) \cap C_G(Y) \subseteq O_{\pi'}(C_G(Y)). \tag{12}$$

Thus $O_{\pi'}$ is an F-signalizer functor, in the following sense.

3 The concept of an A-signalizer functor θ is usually only defined for abelian subgroups A (of rank at least 3), but it is straightforward to generalize everything to the case when A is nilpotent. Just require $\theta(C_G(X))$ to be defined for every subgroup $X \neq 1$ of A with $r(N_A(X)) \geq 3$, assume $\theta(C_G(X))$ to be a $N_A(X)$-invariant π'-subgroup (where $\pi = \pi(A)$), and assume the balance condition

$$\theta(C_G(X)) \cap C_G(Y) \subseteq \theta(C_G(Y))$$

only for commuting subgroups X,Y.

Then, for each $q \in \pi'$ and with $\mathcal{N}_\theta(A), \mathcal{N}_\theta(A,q), \mathcal{N}_\theta^*(A,q)$ defined as usual, we have the

Transitivity Theorem: $\theta(C_G(A))$ is transitive on $\mathcal{N}_\theta^*(A,q)$, and in addition every subgroup $A_o \subseteq A$ with $r(A_o) \geq 3$ satisfies

$$\mathcal{N}_\theta^*(A,q) \subseteq \mathcal{N}_\theta^*(A_o,q).$$

Moreover, $\theta(C_G(X)) \in \mathcal{N}_\theta(N_A(X))$ for every X.

Thus, if there exists just one A_o as above such that $\mathcal{N}_\theta^*(A_o,q) = 1$, then this is true for every A_o and $\theta(C_G(X))$ is a q'-group for every X.

4 Among the characterizations of Fischer π-subgroups of π-constrained groups there is one, namely (6), by a property which makes sense for an arbitrary finite group G.

So assume F to be a subgroup of G satisfying (6a).

Since this condition is inherited by every subgroup $G_o \supseteq F$ it follows from (6) that $F \in \mathcal{F}_\pi(G_o)$ for every π-constrained such G_o.

In particular, condition (10) holds if $N_G(F_p)$ is π-constrained for every $p \in \pi(F)$. Thus, if $N_G(X)$ is $\pi(F)$-constrained for every subgroup $X \neq 1$ of F, it follows from the final remarks of sections 2 and 3 that $O_{\pi(F)'}(N_G(X))$ is a q'-group for every prime $q \in \pi - \pi(F)$ and for every X with $r(N_F(X)) \geq 3$. Note here that by (6a) F does not normalize a non-identity q-subgroup of G. Thus any such X satisfies

$$O_{\pi(F)'}(N_G(X)) = O_{\pi'}(N_G(X)) \tag{12a}$$

whence $N_G(X)$ is π-constrained, $\mathcal{F}_{\pi(F)}(N_G(X)) = \mathcal{F}_\pi(N_G(X))$ and therefore, by (11), applied to $\pi(F)$ in place of π,

$$N_F(X) \in \mathcal{F}_\pi(N_G(X)). \tag{12b}$$

5 The above arguments can be applied to a wider class of subgroups F. Suppose all π-local subgroups of G are π-constrained. Let M be a π-constrained subgroup of G such that

$$H \subseteq G,\ H \ \pi\text{-constrained},\ M = (H \cap M)O_{\pi'}(M) \Longrightarrow H = (H \cap M)O_{\pi'}(H). \tag{13}$$

It is easy to see that every nilpotent π-subgroup of G lies in such a subgroup H.

Let F_o be a (nilpotent) Hall π-subgroup of $F(M \bmod O_{\pi'}(M))$ and let F be a maximal nilpotent π-subgroup of G containing F_o. Assume $M = N_G(F_o)$.

Applying (13) to $H = N_G(F_{op})$ for $p \in \pi(F_o)$ and using (4b) again, we conclude that $F_{p'} \in \mathcal{F}_\pi(C_G(F_p))$ which is equivalent to (10). Assume $r(F_o) \geq 3$ and $N_G(X)$ is $\pi(F)$-constrained for $1 \neq X \leq F$, $r(N(X)) \geq 3$.

Then by the final remarks of 3 and 4 we again get (12b) for such subgroups X provided $\mathcal{N}^*_\theta(F_o,q) = 1$ for every $q \in \pi - \pi(F)$. For $Q \in \mathcal{N}^*_\theta(F_o,q)$ the Transitivity Theorem yields that

$$M = N_G(F_o) = N_M(Q)O_{\pi(F_o)'}(C_G(F_o)) = N_M(Q)O_{\pi'}(M)$$

and then (13), applied to $H = N_G(Q)$, shows that $Q = 1$, as required.

An easy special case is that of a maximal subgroup M of a minimal simple group G, with $\pi = \Pi$. Here the conclusion is that if $r(F(M)) \geq 3$, then for every subgroup $X \neq 1$ of a maximal nilpotent subgroup $F \supseteq F(M)$ of G such that $r(N_F(X)) \geq 3$, we have

$$O_{\pi(F(M))'}(C_G(X)) = 1.$$

6 Let us recall the essentials of the above discussion. First we noticed that in a π-constrained group a certain class of nilpotent π-subgroups behaves quite nicely and shares a number of properties with the class of p-subgroups. Then we have seen that certain well known applications of these p-properties can be generalized from p to π. Here it was important to select certain nilpotent subgroups which were to play the role of the p-subgroups.

Now, given an arbitrary finite group G, in what direction shall we proceed? Motivated by (1), it is natural to concentrate first on generalizations of (1) of the following kind

$$f(N_G(X)) \subseteq f(G), \tag{14}$$

where f is a suitable functor and X is a p-subgroup of G.

An important example of such a functor is $f = L_{p'}$, studied by D. Gorenstein and J. Walter [7] and defined by

$$L_{p'}(G) = O^{p'}(O_{p',E}(G)).$$

In this definition we may replace p by π and then we can search for nilpotent π-subgroups X satisfying (14) with $f = L_{p'}$, or, to be more precise, search for maximal nilpotent π-subgroups F of G such that (14) holds for every subgroup X of F. In addition, $N_F(X)$ should be embedded in $N_G(X)$ like the subgroups of F are in G.

There is some evidence that the Fischer π-subgroups F of certain in some sense maximal π-constrained subgroups M (defined in [4] by some generalization of (13)) show such a behavior. These include the subgroups satisfying (6a).

Along a different line, A. Bialostocki [4] has proved some remarkable results on certain generalizations of the Fischer π-subgroups.

Among nilpotent π-subgroups A of nilpotence class at most 2 choose A of maximal order. Let F be a maximal nilpotent π-subgroup containing A. By [1], Proposition 1.4, $F \in F_{\pi}^{*}(G)$ if G is π-constrained (in general, F satisfies (6a)). This requires a theorem of Glauberman [6] (Theorem B) stating that AB is nilpotent for every A-invariant nilpotent π-subgroup B (compare this with the Arad-Glauberman theorem discussed in 1 and note that now π can be arbitrary). Bialostocki conjectures that all such subgroups F are conjugate in G and he proves this conjecture for the symmetric groups $G = S_n$.

REFERENCES

1. Arad, Z. & Chillag, D. (1979). Injectors of finite solvable groups. Comm. in Algebra 7(2), 115-138.
2. Arad, Z. & Glauberman, G. (1975). A characteristic subgroup of a group of odd order. Pac. J. Math. 56, 305-319.
3. Bender, H. (1974). Large sections of finite groups. Int. Symp. Theory of finite groups. (Sapporo), 17-26.
4. Bialostocki, A. (1982). Nilpotent injectors in symmetric groups. Israel J. Math. 41, 261-273.
5. Fischer, B. (1966). Klassen konjugierter Untergruppen in endlichen auflösbaren Gruppen. Habilitationsschrift, Univ. Frankfurt (M).
6. Glauberman, G. (1975). On Burnside's other $p^a q^b$ Theorem. Pac. J. Math. 56, 469-476.
7. Gorenstein, D. & Walter, J. (1975). Balance and generation in finite groups. J. Algebra 33, 224-287.
8. Mann, A. (1971). Injectors and normal subgroups of finite groups. Israel J. Math. 9, 554-558.
9. Thompson, J. (1964). Fixed points of p-groups acting on p-groups. Math. Z. 86, 12-13.

THE CLASSIFICATION OF N-GROUPS: TEST CASE FOR A REVISIONIST STRATEGY

D. Gorenstein[1]
Department of Mathematics, Rutgers University
New Brunswick, New Jersey 08903

1 Introduction

The classification of simple N-groups[2] can be viewed as a test case for any global strategy for determining the finite simple groups, for most (but not all) of the major features of the general case arise (in simplified form) in the course of the analysis.

For some time now, Richard Lyons and I (and more recently Ronald Solomon) have been developing a three-part revisionist strategy for classifying the finite simple groups, which can be compared schematically with the three parts of a chess game. Thus the following diagram indicates our overall "game plan":

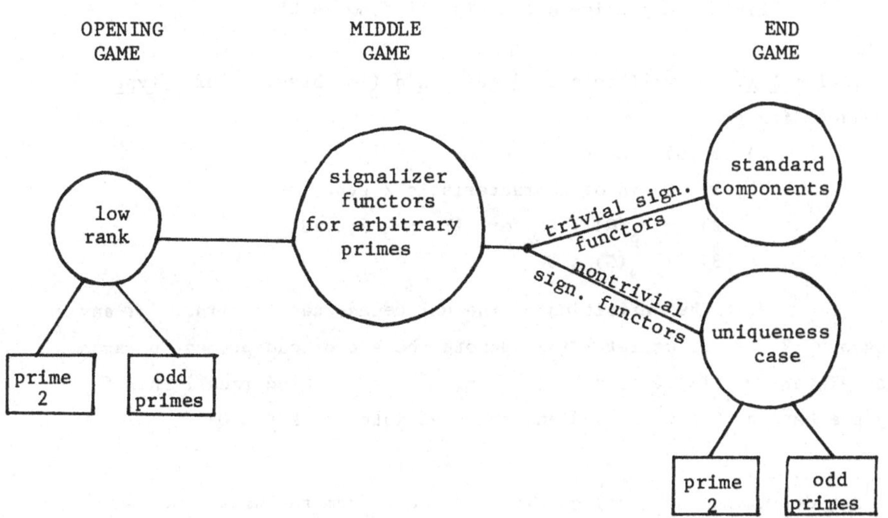

[1] Supported in part by National Science Foundation Grant MCS81-02394.

[2] By definition, an N-*group* is a group in which all local subgroups are solvable.

In this note, we shall show how to "play" the middle N-group game, thus reducing the complete classification of N-groups to a solution of their opening and end games.

It is important to observe that as the centralizer of every nonidentity element of an N-group is solvable, the standard component portion of the end game does not arise, so that segment of the diagram is excluded here.

2 Statement of the main result

We need a number of definitions before we can state the main theorem. Our notation will be standard. In addition, for the simple group G under investigation, we set $C_X = C_G(X)$ and $N_X = N_G(X)$ for any subgroup (or subset) X of G, and we denote by $m_{2,p}(X)$ the maximal p-rank $m_p(X)$ of a 2-local subgroup H of X. We also use the bar convention for homomorphic images.

We restrict the key definitions here to simple N-groups G. In this case G is said to be of characteristic 2 type if $O(H) = 1$ for every 2-local subgroup H of G. Also, in distinguishing the prime 2 case from the odd prime case, we shall use the suggestive symbols E and O, respectively.

First the precise definition of "low rank".

Definition 1.1. G will be called quasithin (for brevity, of Q-type) provided either

E) $m_2(G) \leq 2$; or

O) 1) G is of characteristic 2 type;

2) $m_{2,p}(G) \leq 2$ for all $p \geq 5$; and

3) $m_{2,3}(G) \leq 3$.[3]

Next, the definition of the uniqueness case. First, for any subgroup X of G, we let $\sigma(X)$ denote the set of odd primes p such that either $m_{2,p}(X) \geq 3$, $p \geq 5$, or $m_{2,3}(X) \geq 4$. Also recall that for any p-subgroup Q of X and any positive integer $k \leq m(Q)$,

[3] Our definition of (odd) quasithin differs from the usual one, which assumes $m_{2,p}(G) \leq 2$ for all odd p, including $p = 3$. There is reason to believe that a revision of the existing classification of quasithin groups, based on the Goldschmidt amalgam method, is possible, which covers the added case $m_{2,3}(G) = 3$. However in the case of N-groups, it is not difficult to work with the customary definition of quasithin (the argument is outlined in section 12).

$\Gamma_{Q,k}(X) = \langle N_X(R) \mid R \leq Q, m(R) \geq k \rangle$. Furthermore, an 0-local subgroup is the normalizer in G of a nonidentity subgroup of odd order.

Definition 1.2. G will be called of uniqueness type (for brevity, of U-type) provided either

E) $m_2(G) \geq 3$ and $\Gamma_{P,2}(G)$ is contained in an 0-local subgroup, where $P \in Syl_2(G)$; or

0) 1) G is of characteristic 2 type and $\sigma(G) \neq \emptyset$;

2) If M is a maximal 2-local subgroup, then for any $p \in \sigma(M)$ and $P \in Syl_p(M)$, $\Gamma_{P,2}(G) \leq M$;

3) If H is a 2-local subgroup such that $m_p(H \cap M) \geq 2$ for some $p \in \sigma(M)$, then M is the unique maximal 2-local subgroup containing H; and

4) If $\sigma(M) \neq \emptyset$ and $T \in Syl_2(M)$, then $N_T \leq M$. In particular, $T \in Syl_2(G)$.

If conditions 0(1)-(4) hold for M and a given prime $p \in \sigma(M)$, it will be convenient to call M a uniqueness subgroup for p. (Note that $\sigma(M) \subseteq \sigma(G)$.)

We shall prove the following result.

Theorem A. Every simple N-group is either of Q-type or of U-type.

The proof depends heavily on the signalizer functor method as developed in [7], including the Goldschmidt-Glauberman solvable signalizer functor theorem [3,4,5]. Theorem D of Klinger-Mason [9] is also critical for the proof as well as a few lemmas from Thompson's N-group paper [12]. In addition, we use the latter's odd prime signalizer result [13], as extended by Bender [1], as well as Glauberman's analysis of failure of Thompson factorization in solvable groups [2]. We also need some information concerning simple groups with Sylow 2-subgroups of type J_2, derived in [6], as well as results of Harada [9] on nonconnected 2-groups and of Konvisser [11] on 3-groups of rank ≥ 4.

Throughout G will denote a simple N-group. We assume G is not of Q-type or U-type. In particular, $m_2(G) \geq 3$. We set $\tau(G) = \{2\} \cup \sigma(G)$, so that $\tau(G) \neq \emptyset$.

3 Assumed results

In our present terminology, [10, Theorem D] assumes the following form (as stated above, G denotes a simple N-group throughout).

<u>Proposition 3.1</u>. If G is of characteristic 2 type and p is any prime in $\sigma(G)$, then there exists a maximal 2-local subgroup M which contains an elementary abelian p-subgroup A of rank $m_{2,p}(G)$ such that $|O_{p'}(C_a)|$ is even for some $a \in A^{\#}$.

<u>Remark</u>. It is this result which eliminates the "O_2 of symplectic type" case of Thompson's N-group analysis (except in so far as it is needed when G is of Q-type or U-type). Indeed, Klinger and Mason show first [10, Theorem A] that if the desired conclusion fails for a given M and A, then necessarily $p = 3$ (as $m(A) \geq 3$) and $O_2(M)$ is of symplectic type.

Then, assuming the desired conclusion fails for <u>each</u> such maximal 2-local subgroup M and corresponding p-subgroup A of M (whence each such $O_2(M)$ is of symplectic type) and choosing M so that the width n of $T = O_2(M)$ is <u>maximal</u>, they argue that for some involution x of T, there exists a maximal 2-local subgroup M^* containing C_x as well as an elementary abelian p-subgroup A^* of rank $m_{2,p}(G)$ (whence $T^* = O_2(M^*)$ is of symplectic type of width $n^* \leq n$) with the additional property that A^* is normalized, but not centralized by $\langle z \rangle = \Omega_1(Z(T))$. In particular, $z \notin T^*$. But $C_T(x)$ contains an extra-special group R with width $n-1$ with center $\langle z \rangle$. Thus $R \leq M^*$ and as $R \cap T^* \triangleleft R$ with $z \notin T^*$, it follows that $R \cap T^* = 1$. Thus M^*, T^*, and R satisfy the assumptions of [12, Lemma 5.13], which implies that $n \leq 2$. However, this is impossible as $m(A) \geq 3$ and A acts faithfully on T.

Recall next that if $P \in Syl_p(G)$, $\mathcal{U}(P)$ is the set of normal $Z_p \times Z_p$ subgroups U of P with the restriction that $U \leq Z(P)$ if $Z(P)$ is non cyclic. In [12, Lemma 6.1] Thompson has proved

<u>Lemma 3.2</u>. If G is of characteristic 2 type, $P \in Syl_2(G)$, and $U \in \mathcal{U}(P)$, then U centralizes every subgroup of G of odd order which it normalizes.

If X is a simple group with Sylow 2-subgroup P of type J_2 and $\langle z \rangle = Z(P)$, it is shown in [6] by an analysis of the fusion of involutions that $C_X(z)/O_{2',2}(C_X(z)) \cong A_5$. In particular, $C_X(z)$ is nonsolvable. Since centralizers of involutions in G are solvable, we thus obtain

<u>Proposition 3.3</u>. G does not have Sylow 2-subgroups of type J_2.

Proposition 3.3 will be used in conjunction with the following result of Harada [9]:

Lemma 3.4. Let P be a nonconnected 2-group of 2-rank ≥ 3 and U a normal four subgroup of P. If $R = C_P(U)$ admits an automorphism of order 3^n cyclically permuting the involutions of $U^{\#}$, then R is of type $L_3(4)$ and P is of type J_2.

For any group X and any prime p, $SC_p(X)$ denotes the set of elementary abelian p-subgroups B of X such that B contains every element of order p in $C_X(B)$. In [1], Bender has proved

Lemma 3.5. If X is a solvable group and $B \in SC_p(X)$, p an odd prime, then $O_{p'}(X)$ contains every B-invariant p'-subgroup of X.

Next, by Glauberman [2], we have

Lemma 3.6. Let X be a solvable group with $F(X) = O_2(X)$, C the largest normal subgroup of X which centralizes $Z(T)$, and $Y = \langle J(T)^X \rangle$. Then $X = CYN_X(J(T))$ and either $Y \leq C$ or else $p = 3$ and CY/C is a nontrivial direct product of Σ_3's.

Furthermore, Konvisser [11] has proved

Lemma 3.7. If P is a 3-group of rank ≥ 4, then P contains an elementary abelian normal subgroup of rank 4. In particular, P is connected.

Thompson's well-known dihedral lemma [12, Lemma 5.34] asserts

Lemma 3.8. If X is a solvable group with $O_2(X) = 1$ and A is a nontrivial elementary abelian 2-subgroup of X, then $F(O(X))$ contains an A-invariant subgroup Y such that $AY = A_1Y_1 \times A_2Y_2 \times \ldots \times A_rY_r$, where $Z_2 \cong A_i \leq A$, $Z_{p_i} \cong Y_i \leq Y$, p_i an odd prime, with A_i inverting Y_i, $1 \leq i \leq r$.

Finally, we have the following easily established property of solvable groups (a variation of this result appears in [12, Lemmas 5.19, 5.20]).

Lemma 3.9. Let X be a solvable group with $O_2(X) = 1$ which is faithfully represented on a vector space V over $GF(2)$. If t is an involution of X which centralizes a hyperplane of V and we set $Y = \langle t^X \rangle$, then $Y \cong \Sigma_3$ and $[V,Y] \cong Z_2 \times Z_2$.

4 Balanced groups, signalizer functors, and continuous conjugacy functors

We shall summarize the signalizer functor method as it applies to our N-group G.

First, if x,y are any two commuting elements of G of order p for $p \in \tau(G)$, then as C_x and C_y are solvable, it is immediate from the $A\times B$ lemma that

$$O_{p'}(C_x) \cap C_y = O_{p'}(C_y) \cap C_x.$$

Hence, according to the definition of balance, we have

Proposition 4.1. G is balanced with respect to p for any prime $p \in \tau(G)$.

In particular, if $E_{p^3} \cong A \leq G$ for $p \in \tau(G)$ and for $a \in A^{\#}$ we set $\theta(C_a) = O_{p'}(C_a)$, then θ is an A-signalizer functor on G. Hence by the solvable signalizer functor theorem [2,3,4] we have

Proposition 4.2. $\langle O_{p'}(C_a) | a \in A^{\#}\rangle$ is a p'-group.

Now for continuous conjugacy functors (cf. [7], where the notions were introduced for the prime 2 only).

For $p \in \tau(G)$, let $\Delta = \Delta_p(G)$ be the graph whose vertices are the set of $Z_p\times Z_p$ subgroups of G, with two vertices connected by an edge if and only if they commute elementwise. Let $\Delta^o = \Delta_p^o(G)$ be the subset of nonisolated points of Δ.

Clearly G acts by conjugation on both Δ and Δ^o.

For any connected component ψ of Δ^o, we denote by $N_G(\psi)$ the stabilizer of ψ in G.

Next, for $V \in \Delta^o$, we set

$$\alpha(V) = \alpha_p(V) = \langle O_{p'}(C_v) | v \in V^{\#}\rangle.$$

Clearly for $g \in G$, and $V \in \Delta^o$, we have

$$\alpha(V^g) = (\alpha(V))^g,$$

so α is an example of a conjugacy functor on Δ^o, as this term is defined in [7].

Furthermore, using the fact that G is balanced for p, one argues easily that if V and W are adjacent vertices of Δ^o, then

$$\alpha(V) = \alpha(W).$$

It follows that α is constant on the connected component of Δ^o containing V. Since this holds for every $V \in \Delta^o$, α is therefore <u>continuous</u>, as this term is defined in [7].

We see then that α is a continuous conjugacy functor on Δ^o, and consequently the general formalism developed in [7] applies to α.

To describe it, let $P \in Syl_p(G)$ and let U be a normal $Z_p \times Z_p$ subgroup of P. Since $U \triangleleft P$ and $m(P) \geq 3$, $U \leq A$ for some $E_{p^3} \cong A \leq P$. In particular, $U \in \Delta^o$. Furthermore, by Proposition 4.2 and the definition of $\alpha(U)$, we have

<u>Proposition 4.3</u>. $\alpha(U)$ is a p'-group.

We let ψ_U be the connected component of Δ^o containing U. The following elementary general results are proved in [7, section 1], for the prime 2. The identical proofs apply for arbitrary p.

<u>Proposition 4.4</u>. The following conditions hold:

i) ψ_U is determined independently of the normal subgroup U of P (thus ψ_U depends only on P);

ii) Δ^o is the disjoint union of G-conjugates of ψ_U;

iii) $N_G(\psi_U) \geq \langle \Gamma_{P,3}(G), \Gamma_{E,2}(G) \mid E_{p^k} \cong E \leq P$, with either $k \geq 3$ or $k = 2$ and $m(EC_P(E)) \geq 3 \rangle$;

iv) If P is connected, then $N_G(\psi_U) = \Gamma_{P,2}(G)$; and

v) $N_G(\psi_U) \leq N_{\alpha(U)}$.

In view of (iv), we shall write $\Gamma^o_{P,2}(G)$ for the stabilizer $N_G(\psi_U)$ of ψ_U in G. Hence in this terminology, (v) reads:

<u>Proposition 4.5</u>. We have $\Gamma^o_{P,2}(G) \leq N_{\alpha(U)}$.

We preserve the notation $p \in \tau(G)$, $\Delta^o = \Delta^o_p(G)$, P, U, ψ_U, and $\alpha = \alpha_p$ introduced above.

5 The case $p = 2$

Using the results of [7], we can prove

Proposition 5.1. If $p = 2$, then $\alpha(U) = 1$.

Proof. If not, then $N_{\alpha(U)}$ is an O-local subgroup. Since G is not of Q-type, we have $\Gamma_{P,2}(G) \not\leq N$. In particular, P is not connected by Proposition 4.4(iv), whence $U \not\leq Z(P)$. Set $R = C_P(U)$. The results of [7, section 3] imply that N_R contains a 3-element cycling the involutions of $U^{\#}$. But now as $m(P) \geq 3$, Lemma 3.4 shows that P is of type J_2, contrary to Proposition 3.3.

As an immediate corollary, we obtain

Proposition 5.2. G is of characteristic 2 type.

Proof. Since α is constant on ψ_U, Proposition 5.1 implies that $\alpha(V) = 1$ for every $V \in \psi_U$. Since Δ^o is a union of G-conjugates of ψ_U by Proposition 4.4(ii) and α is a conjugacy functor, it follows that $\alpha(V) = 1$ for all $V \in \Delta^o$.

But now if x is an involution of G, then as G is simple, Thompson's transfer lemma [12, Lemma 5.38] implies that a conjugate of x centralizes U, whence $x \in B \cong E_8$ for some $B \leq G$. But then if $Z_2 \times Z_2 \cong X \leq B$ with $x \in X$, we have $X \in \Delta^o$. Thus $O(C_x) \leq \alpha(X) = 1$.

This holds for every involution x of G. Using the A×B lemma, one easily concludes now that $O(H) = 1$ for every 2-local subgroup H, so G is of characteristic 2 type.

This in turn implies

Proposition 5.3. We have $\sigma(G) \neq \emptyset$.

If not, then by Proposition 5.2, G would be of Q-type, contrary to assumption.

6 The order of $\alpha(U)$

Henceforth we assume p is odd, whence $p \in \sigma(G)$. Proposition 3.1 quickly yields:

Proposition 6.1. $\alpha(U)$ is of even order.

Proof. By Proposition 3.1, there exists a maximal 2-local subgroup M of G with $E_{p^r} \cong A \leq M$ such that $r = m_{2,p}(G)$ and $|O_{p'}(C_a)|$ is even for some $a \in A^{\#}$. Without loss, $A \leq P$. Let $a \in V \leq A$ with $V \cong Z_2 \times Z_2$. Since $O_{p'}(C_a) \leq \alpha(V)$, $|\alpha(V)|$ is even. However, as $m(A) \geq 3$ and $U \triangleleft P$, clearly $V \in \psi_U$, whence $\alpha(V) = \alpha(U)$. Thus $|\alpha(U)|$ is even, as asserted.

7 Partial uniqueness

We now set $N = N_{\alpha(U)}$ and fix this notation as well. Thus $\Gamma^o_{P,2}(G) \leq N$. In this section we prove

Proposition 7.1. The following conditions hold:

i) $\Gamma_{P,2}(G) \leq N$;

ii) If H is a local subgroup such that $m_p(H \cap N) \geq 3$, then $H \leq N$; and

iii) If H is a local subgroup containing $V \in \Delta^o$ with $V \leq N$, then $H \leq N$.

The proof depends on a preliminary result:

Lemma 7.2. The following conditions hold:

i) If $E_{p^3} \cong E \leq N$, then N contains every E-invariant p'-subgroup of G; and

ii) If H is a p-local subgroup and $O_p(H)$ contains a non-trivial p-subgroup Q such that $m_p(N_Q \cap N) \geq 3$, then $O_{p'}(H) \leq N$.

Proof. Since $\Gamma_{E,2}(G) \leq N$ by Proposition 4.4(iv), (i) is immediate. As for (ii), it follows easily by the A×B lemma that $O_{p'}(H) \leq O_{p'}(N_Q)$. By assumption, there is $E_{p^3} \cong E \leq N_Q \cap N$. Since E normalizes $O_{p'}(N_Q)$, (ii) thus follows from (i).

We next prove

Lemma 7.3. We have $\Gamma_{P,2}(G) \leq N$.

Proof. Assume false. If $p = 3$, then P is connected by Lemma 3.7, in which case Proposition 4.4(iv) contradicts our assumption, so $p \geq 5$.

Let Q be a p-subgroup of N with $m(Q) \geq 2$ and $N_Q \not\leq N$, chosen so that $R \in Syl_p(N_Q \cap N)$ has maximal order; and subject to this condition so that $|Q|$ is maximal. Without loss, $R \leq P$. Since $N_Q \not\leq N$,

$m(Q) = 2$ by Proposition 4.4(iv), whence $Q < R$, and consequently $N_R \leq N$ by our choice of Q. Since $R \in Syl_p(N_Q \cap N)$, it follows that $R \in Syl_p(N)$.

We claim that $Y = O_{p'}(N_Q) \leq N$. If $m(R) \geq 3$, this holds by Lemma 7.2(i). On the other hand, if $m(R) = 2$, and we set $Z = \Omega_1(Z(P))$, then $Z \leq R$ and as $m(Q) = m(R) = 2$, we must have $Z \leq Q$. But then as $P \leq N_Z$, $Y \leq N$ by Lemma 7.2(ii).

Setting $T = R \cap O_{p'p}(N_Q)$, we have $Q \leq T$ and $N_Q = YN_{N_Q}(T)$, whence $N_{N_Q}(T) \not\leq N$. Since $R \leq N_T$, $T = Q$ by our choice of Q. Hence if we set $\bar{H} = H/YQ$, then $O_p(\bar{H}) = 1$ and $\bar{H}$ acts faithfully on Q. Since $m(Q) = 2$, it follows that $\bar{H} \leq GL_2(p)$. However, as $\bar{H}$ is solvable with $O_p(\bar{H}) = 1$ and $p \geq 5$, this is possible only if $\bar{R} = 1$, whence $Q = R$, contradiction.

Thus Proposition 7.1(i) holds. Let H be as in Proposition 7.1(ii) or (iii). In view of (i), N contains a Sylow p-subgroup R of H. Set $Y = O_{p'}(H)$ and $Q = R \cap O_{p'p}(H)$, so that $H = YN_H(Q)$. Since $m(R) \geq 2$ and H is solvable with p odd, $m(Q) \geq 2$, whence $N_H(Q) \leq N$ by (i).

Thus it suffices to show that $Y \leq N$. If $m(R) \geq 3$, this again follows from Lemma 7.2(i), so we can assume $m(R) = 2$. In particular, (ii) holds. Let then V be as in (iii). Since $V \in \Delta^o$, $m_p(C_V) \geq 3$. Also $C_V \leq N$ by (i). Thus for $v \in V^{\#}$, $m_p(C_v \cap N) \geq 3$ and consequently $C_v \leq N$ by (ii). Since $Y = \langle C_Y(v) \mid v \in V^{\#} \rangle$, we conclude that $Y \leq N$, so (iii) holds as well.

8 Nontriviality of $O_2(N)$.

We next prove

Proposition 8.1. We have $O_2(N) \neq 1$.

Proof. This argument appears in the N-group paper. Let $T \in Syl_2(N)$ and expand T to $S \in Syl_2(G)$. It will suffice to prove that T contains an element $W \in \mathcal{U}(S)$. Indeed, then W acts on $O(F(N))$, in which case W centralizes $O(F(N))$ by Lemma 3.2. Since N is solvable, it follows that $O(F(N)) < F(N)$, whence $O_2(N) \neq 1$, as asserted.

Without loss, T is permutable with P. Set $Q = O_2(TP)$. Since $\alpha(U) \triangleleft N$ with $\alpha(U)$ a p'-group, $Q \cap \alpha(U) \in Syl_2(\alpha(U))$. But $|\alpha(U)|$ is even by Proposition 6.1, so $Q \neq 1$. Since $P \leq N_Q$, Proposition

7.1(ii) implies now that $N_Q \leq N$. Since $T \leq N_Q$ and $T \in Syl_2(N)$, $T \in Syl_2(N_Q)$ and consequently $Q = T \cap O_{p'}(N_Q) \in Syl_2(O_{p'}(N_Q))$. Hence $Q = O_2(N_Q)$. But now as $Z(S) \leq T$ and $O(N_Q) = 1$, it follows that $Z(S) \leq Q$, whence every element of $\mathcal{U}(S)$ is contained in T. Moreover, $\mathcal{U}(S) \neq \emptyset$ as $m(S) \geq 3$, and the proposition is proved.

Thus there exists a maximal 2-local subgroup M containing $N_{O_2(N)}$. Since $N \leq M$, the conclusions of Proposition 7.1 therefore hold with M is place of N. We shall show that M is a uniqueness subgroup for the prime p.

9 2-local subgroups

In this section we prove

Proposition 9.1. If H is a 2-local subgroup such that $m_p(H \cap M) \geq 2$, then M is the unique maximal 2-local subgroup containing H.

Again by Proposition 7.1(i), M contains a Sylow p-subgroup R of H. If $m(R) \geq 3$ or if R contains a vertex of Δ^o, Propositions 7.1(ii) and (iii) imply that any 2-local subgroup containing R necessarily lies in M, so M is the unique maximal local subgroup containing H in these cases.

Thus it remains to consider the case in which $m(R) = 2$ and every vertex B of Δ contained in R is isolated in Δ. It follows that $m(C_B) = 2$ and thus B contains every element of order p in its centralizer, whence $B \in SC_p(G)$. Without loss, $B \leq P$. Then $Z = \Omega_1(Z(P)) \leq B$.

The proof in this case appears (under more general conditions) in [8, Part II, Chapter 3, section 6].

We therefore limit ourselves to an outline of the argument.

Lemma 9.2. If W is a maximal B-invariant 2-subgroup of M, then we have

i) $W \in Syl_2(O_{p'}(M))$;

ii) $N_{Z(W)}$ and $N_{J(W)}$ are contained in M; and

iii) W is a maximal B-invariant subgroup of G.

Proof. First, (i) follows from Lemma 3.5. Since $N_M(W)$ contains a Sylow p-subgroup of M, (ii) follows from Proposition 7.1(ii), and then (iii) follows from (ii).

We now fix a B-invariant Sylow 2-subgroup T of $O_{p'}(H)$ and next prove:

Lemma 9.3. If $\langle N_{Z(T)}, N_{J(T)} \rangle \leq M$, then $H \leq M$.

Proof. Assume false and set $Y = O_{p'}(H)$. Then $H = YN_H(T)$, so $Y \not\leq M$. Set $X = \langle J(T)^Y \rangle$, let C be the largest normal subgroup of Y that centralizes $Z(T)$, and set $K = N_Y(J(T))$, so that $CK \leq M$. Hence $CK < Y$ and so by Lemma 3.6, $p = 3$, $Y = CXK$ and if we set $\bar{Y} = Y/C$, then $\bar{X}$ is a direct product of Σ_3's with M not covering $\bar{X}$. On the other hand, Z acts on $\bar{X}$ and $N_Z \leq M$ by Proposition 7.1(ii). However, considering the action of Z on $\bar{X}$, we easily check that $\bar{X} = \langle \bar{T} \cap \bar{X}, C_{\bar{X}}(Z) \rangle$, so M covers $\bar{X}$, contradiction.

To apply the lemma, we assume the proposition fails, and choose H with $H \not\leq M$, and subject to this condition so that $S = T \cap M$ has maximal order. We prove

Lemma 9.4. We have $S \neq 1$.

Proof. Suppose false. Since B char $C_P(B)$, $N_P(B) > C_P(B)$. Hence if $Z = B_o, B_1, \ldots, B_p$ denote the $p+1$ subgroups of B of order p, there is $x \in N_P(B)$ transitively permuting $\{B_1, B_2, \ldots, B_p\}$. Since $B\langle x \rangle$ acts faithfully on $W = O_2(M)$, it follows that $W_i = C_W(B_i) \neq 1$ for all i, $1 \leq i \leq p$. On the other hand, as B acts faithfully on T, $T_j = C_T(B_j) \neq 1$ for some j, $1 \leq j \leq p$. By Lemma 3.5, $\langle T_j, W_j \rangle \leq Y = O_{p'}(C_{B_j})$.

Expand T_j, W_j to B-invariant Sylow 2-subgroups T^*, W^* of Y, respectively. Maximality of H, T (together with $S = T \cap M = 1$) implies that $W^* \leq M$. But $T^* = W^{*y}$ for some $y \in C_Y(B) \leq C_Z \leq M$, so $T_j \leq T^* \leq M$ and the lemma follows.

Now we complete the proof of Proposition 9.1. We choose H so that in addition to the above conditions, $|O_2(H)|$ is maximal; and subject to this condition so that $|H|$ is maximal. By Lemma 9.4, $S \neq 1$. Also $H = N_{O_2(H)}$ by maximality of $|H|$. Furthermore, $S \leq W$ for some B-invariant $W \in Syl_2(O_{p'}(M))$.

If $S < T$, then $S < W$ by Lemma 9.2(iii), whence $N_W(S) > S$. Hence $N_S \leq M$ by our maximal choice of H. But then $N_T(S) \leq M$, forcing $T = N_T(S)$, and we conclude that $S = T$.

If $T = W$, then $\langle N_{Z(T)}, N_{J(T)} \rangle \leq M$ by Lemma 9.2(ii), whence $H \leq M$ by Lemma 9.3; while if $T < W$, then $N_W(T) > T$ and it follows from our maximal choice of H that $\langle N_{Z(T)}, N_{J(T)} \rangle \leq M$, whence again $H \leq M$. The proposition is proved.

10 Sylow 2-normalizers

Let M be as above. Now we can obtain our objective.

Proposition 10.1. M is a uniqueness subgroup for p.

In view of Propositions 7.1(i) and 9.1, it remains only to show that $N_T \leq M$ for $T \in Syl_2(G)$. The proof appears in the N-group paper [12, Lemmas 13.7, 13.8], so again we limit ourselves to an outline of the argument.

We assume false. Let V be the normal closure of $Z = \Omega_1(Z(T))$ in M and let V^* be the weak closure of V in T with respect to G. Also set $C = C_M(V)$, so that $C \triangleleft M$.

Lemma 10.2. V^* does not centralize V. In particular, $m(V) \geq 2$.

Proof. If false, then $M = CN_M(V^*)$. Since $m_p(M) \geq 3$ and $C \triangleleft M$, one easily checks that either $m_p(C) \geq 2$ or $m_p(N_M(V^*)) \geq 2$. Since $C \leq N_Z$, Proposition 9.1 implies that correspondingly $N_Z \leq M$ or $N_{V^*} \leq M$. But $N_T \leq N_Z \cap N_{V^*}$, so $N_T \leq M$, contrary to assumption.

Thus there is $g \in G-M$ such that $X = V^g \leq T$ and $X \not\leq C$. Set $\bar{M} = M/C$. It is immediate that $O_2(\bar{M}) = 1$, so $1 \neq \bar{X}$ acts faithfully on $O(\bar{M})$. Hence by Lemma 3.8, there is $\bar{A} \leq F(O(\bar{M}))$ invariant under $\bar{X}$ with $\bar{X}\bar{A} = \bar{X}_1\bar{A}_1 \times \bar{X}_2\bar{A}_2 \times \ldots \times \bar{X}_r\bar{A}_r$ and such that $Z_2 \cong \bar{X}_i \leq \bar{X}$, $Z_{p_i} \cong \bar{A}_i \leq \bar{A}$ for some odd prime p_i, and $\bar{X}_i$ inverting $\bar{A}_i$, $1 \leq i \leq r$. Let X^i be the subgroup of index 2 in X such that $\bar{X}^i$ centralizes $\bar{A}_i$. By the A×B lemma, $V_i = [C_V(X^i), \bar{A}_i] \neq 1$, $1 \leq i \leq r$. Then X acts on each V_i and, in particular, on V_1. Moreover, we can choose $v \in V_1 - C_{V_1}(X)$. Finally set $L = M^g$, $D = C^g = C_L(X)$, and $\tilde{L} = L/D$, so that $v \notin D$ and $O_2(\tilde{L}) = 1$. We fix this notation.

Lemma 10.3. The following conditions hold:

i) $v \in L$ and $\langle v^L \rangle \cong \Sigma_3$;

ii) $\langle x\rangle = [X,v] \cong Z_2$ and $x \in V_1$;

iii) L is the unique maximal 2-local subgroup containing C_x;

and

iv) $V \leq L$.

Proof. Since P^g acts on X and $m(X) = m(V) \geq 2$, $m_p(C_L(W)) \geq 2$ for some four subgroup W of X and consequently $W_1 = W \cap X^1 \neq 1$. But then $v \in C_{X^1} \leq C_{W_1} \leq L$ by Proposition 9.1, applied to M^g. Hence v acts on X centralizing the hyperplane X^1 of X and $\tilde{v} \neq 1$. In particular, the first assertion of (ii) follows, and the second also holds as X normalizes V_1. Furthermore, as $\tilde{L}$ is solvable with $O_2(\tilde{L}) = 1$, Lemma 3.9 implies that $\tilde{K} = \langle \tilde{v}^{\tilde{L}}\rangle \cong \Sigma_3$ and that $Y = [X,\tilde{K}] \cong \Sigma_3$. In particular, (i) holds and $x \in Y$. But also $Y \triangleleft L$ and so $m_p(C_L(Y)) \geq 2$. Now (iii) follows by another application of Proposition 9.1. Finally as V is abelian, (iv) follows from (ii) and (iii).

By the lemma, V acts on X, so $[V,X] \leq X$ and hence $[V,X,X] = 1$. We use this quadratic action to prove

Lemma 10.4. We have $r = 1$.

Proof. Suppose false and set $W = [V,\bar{A}_1]$ and $W_1 = [W,\bar{X}_1]$, so that $V_1 \leq W$. Furthermore, it is immediate that $W = \langle W_1^{\bar{A}_1}\rangle$. Now $[W_1,\bar{X}_2] \leq [V,X,X] = 1$, so $\bar{X}_2$ centralizes W_1. But W_1 admits $\bar{A}_2$, which is inverted by $\bar{X}_2$, so $\bar{A}_2$ centralizes W_1. Since $C_W(\bar{A}_2)$ is $\bar{A}_1$-invariant and $W = \langle W_1^{\bar{A}_1}\rangle$, it follows that $\bar{A}_2$ centralizes W. In particular, $\bar{A}_2$ centralizes $x \in V_1 \leq W$. But then if A_2 denotes the preimage of $\bar{A}_2$ in M, we have $A_2 \leq C_x \leq L$, by Lemma 10.3(iii), whence $[A_2,X] \leq X$, contrary to the fact that $\bar{A}_2 = [\bar{A}_2,\bar{X}_2]$ is nontrivial of odd order.

By the lemma, $X^1 \leq C$, whence $|X/C_X(V)| = 2$. Since $V \leq L$, it follows by symmetry that also $|V/C_V(X)| = 2$. These equalities in turn imply that $|[V,X]| = 2$, whence $\langle x\rangle = [V,X]$. But again by symmetry, Lemma 10.3(iii) must therefore hold for x relative to M as well as to L. Hence $M = L$ and so $g \in N_M \leq N_{O_2(M)} = M$, which is not the case. This completes the proof of Proposition 10.1.

11 Theorem A

We have thus shown that G possesses a uniqueness subgroup

M_p for every $p \in \sigma(G)$. But now if M is an arbitrary maximal 2-local subgroup with $\sigma(M) \neq \emptyset$ and if for $q \in \sigma(M)$ we choose M_q so that $M \cap M_q$ contains a Sylow q-subgroup of M, the unicity of M_q implies that $M \leq M_q$, whence $M = M_q$. Thus M is a uniqueness subgroup for every $q \in \sigma(M)$. Since G is of characteristic 2 type with $\sigma(G) \neq \emptyset$, G is therefore of U-type, contrary to assumption. This completes the proof of Theorem A.

12 The case $m_{2,p}(G) = 3$

Finally we describe the additional argument required to establish the stronger form of Theorem A in which our present definition of (odd) quasithin is replaced by the customary definition $m_{2,p}(G) \leq 2$ for all odd primes p. One begins by redefining $\sigma(G) = \{p | m_{2,p}(G) \geq 3\}$.

Now, assuming the stronger theorem is false, one argues as before that G is of characteristic 2 type. Since G is not (odd) quasithin, the redefined $\sigma(G) \neq \emptyset$. Hence to prove the theorem, one must again argue that G possesses a uniqueness subgroup for each $p \in \sigma(G)$. Clearly the arguments of sections 6-10 apply without change if either $p \geq 5$ or if $p = 3$ and $m_{2,3}(G) \geq 4$. Thus it remains to prove the existence of such a uniqueness subgroup when $p = 3$ and $m_{2,3}(G) = 3$.

One easily checks that the assumption $m_{2,3}(G) \geq 4$ was used at only one place in those five sections: namely, in showing that $\Gamma_{P,2}(G) \leq N = N_{\alpha(U)}$ (Lemma 7.3), where the connectivity of P was invoked to obtain the desired conclusion. Note that at that point in the argument, we already had $\Gamma^{o}_{P,2}(G) \leq N$ and $|\alpha(U)|$ even (whence also $|O_{3'}(N)|$ is even).

Thus the following single result will suffice to establish the desired stronger form of Theorem A:

Proposition 12.1. Assume $m_{2,3}(G) = 3$ and let $P \in Syl_3(G)$. If $\Gamma^{o}_{P,2} \leq N$ for some local subgroup N with $|O_{3'}(N)|$ even, then $\Gamma_{P,2}(G) \leq N$.

Likewise this argument appears (under more general conditions) in [8, Part III, Chapter 3, section 2], so again we limit ourselves to an outline.

Note first that if H is any local subgroup containing N, then as $P \leq N$ and N is solvable, it is immediate that $O_{3'}(N) \leq O_{3'}(H)$, so also $|O_{3'}(H)|$ is even. Hence without loss we can assume that N is a maximal local subgroup, whence $N = N_{O_{3'}(N)}$.

Assume the proposition fails. Then exactly as in Lemma 7.3, G contains a 3-subgroup Q of rank 2 with $N_Q \not\leq N$ satisfying the following conditions: $R = P \cap N_Q \in Syl_3(N_Q)$, $O_{3'}(N_Q) \leq N$, and $O_{3',3}(N_Q) = O_{3'}(N_Q) \times Q$. We set $Z = \Omega_1(Z(P))$ and $\bar{N}_Q = N_Q/O_{3',3}(N_Q)$ and fix this notation. Note that, in particular, $Z \leq Z(Q)$.

We separate the cases $R = P$ and $R < P$. Thus in Lemmas 12.2-12.6 we <u>assume</u> $R = P$.

<u>Lemma 12.2</u>. The following conditions hold:

i) Q is a central product of an extra-special group of order 27 and a cyclic group;

ii) $\bar{N}_Q \cong SL_2(3)$ or $GL_2(3)$;

iii) $Z = \Omega_1(Z(Q)) \cong Z_3$;

iv) If U is a normal $Z_3 \times Z_3$ subgroup of P with $U \leq Q$, then $Z \leq U$ and $E = \Omega_1(C_P(U)) \cong E_{27}$;

v) If $u \in U - Z$, then $N_{N_Q}(E)$ contains a 2-element t with $\langle \bar{t} \rangle = Z(\bar{N}_Q)$ such that t inverts u and centralizes $E/\langle u \rangle$; and

vi) $Q \in Syl_3(O_{3',3}(N_Z))$ and $N_Z/O_{3',3}(N_Z) \cong \bar{N}_Q$.

<u>Proof</u>. If Q contains a noncyclic elementary abelian characteristic subgroup V, then $m(V) = m(Q) = 2$ and $V \triangleleft P$, so $Z_2 \times Z_2 \cong V \in \Delta^o$. Hence $N_Q \leq N_V \leq \Gamma^o_{P,2}(G) \leq N$, contradiction, so no such V exists. Since $m(Q) = 2$, (i) follows at once now from P. Hall's well-known result. Since $m(P) = 3$ and $\bar{N}_Q$ acts faithfully on Q with $O_3(\bar{N}_Q) = 1$, (ii) is immediate from (i). Clearly (i) also implies (iii). Furthermore, again as $m(P) = 3$, (iv) and (v) follow easily from the structure of P and N_Q, respectively. Since $N_Q \leq N_Z$ and N_Z is solvable, likewise (vi) is easily verified.

With U as in Lemma 12.2, we next prove

<u>Lemma 12.3</u>. For any $u \in U^{\#}$, $U \leq O_{3',3}(C_u)$.

<u>Proof</u>. If $\langle u \rangle = Z$, the lemma holds by the choice of U, so we can assume $u \in U - Z$. Set $X = C_P(u) = C_P(U)$ and expand X to $Y \in Syl_2(C_u)$. If

$X = Y$, then $U \leq Z(Y)$ and the desired conclusion follows as C_u is 3-constrained. Hence we can assume that $X < Y$. Since $|P:X| = 3$, it follows that $Y \in Syl_3(G)$. Thus $\langle u \rangle = \Omega_1(Z(Y))$ and so $\langle u \rangle = Z^g$ for some $g \in G$. Setting $H = N_{\langle u \rangle}$, we have $H = (N_Z)^g$, so the structure of H is determined from Lemma 12.2(vi). Hence $Y \in Syl_3(H)$, $D = Y \cap O_{3',3}(H) \cong Q$ and $\tilde{H} = H/O_{3',3}(H) \cong SL_2(3)$ or $GL_2(3)$.

On the other hand, if $E = \Omega_1(C_P(u))$ and t is determined as in Lemma 12.2(v), then $E_{27} \cong E \leq X \leq Y$, $t \in H$, and t centralizes $E/\langle u \rangle$. In particular, t centralizes $\tilde{E} \cong Z_3$, which implies that $\langle \tilde{t} \rangle = Z(\tilde{H})$, whence t inverts $D/Z(D)$. But $F = E \cap D \cong Z_2 \times Z_2$ and as $u \in Z(D)$, it follows that t inverts $F/\langle u \rangle$, contrary to the fact that t centralizes $E/\langle u \rangle$.

As an immediate corollary, we obtain

Lemma 12.4. N contains every U-invariant 3'-subgroup of G.

Proof. Let X be such a subgroup of G. Since $U \in \Delta^o$, $N_U \leq N$, so $C_X(U) \leq N$. Hence we need only show that $X_u = [C_X(u), U] \leq N$ for each $u \in U^{\#}$. However, by the previous lemma, $X_u \leq O_{3'}(C_u)$, so it will suffice to prove that each $O_{3'}(C_u) \leq N$. But $U \leq E \leq P$ for some $E \cong E_{27}$ and E acts on each $O_{3'}(C_u)$, so the desired conclusion follows from Lemma 7.2(i).

Now we can prove

Lemma 12.5. $O_{3'}(N)$ is the unique maximal Q-invariant 3'-subgroup of G.

Proof. If X is any Q-invariant 3'-subgroup of G, then as $U \leq Q$, the preceding lemma implies that $X \leq N$. Therefore it will suffice to prove that X is, in fact, contained in $O_{3'}(N)$. However, as $C_P(Q) \leq Q$, this is a consequence of Thompson's result in [13].

Now we can obtain our objective.

Lemma 12.6. We have $R < P$.

Proof. If false, then the preceding lemma is applicable. However, clearly N_Q permutes the set of all maximal Q-invariant 3'-subgroups of G, so N_Q normalizes $O_{3'}(N)$, whence $N_Q \leq N_{O_{3'}(N)} = N$, contrary to the choice of Q.

This yields the following corollary.

Lemma 12.7. If H is a local subgroup containing P, then $H \leq N$. In particular, $N_Z \leq N$.

Proof. Setting $D = P \cap O_{3',3}(H)$, we have $H = O_{3'}(H)N_H(D)$ with $m(D) \geq 2$. As usual, $O_{3'}(H) \leq N$ by Lemma 7.2(i). Furthermore, as $P \leq N_D$, Lemma 12.6 and our choice of Q implies that $N_D \leq N$. Thus $H \leq N$, as asserted.

Now we can prove

Lemma 12.8. We have $Q \cong Z_3 \times Z_3$ with $Q \in SC_3(G)$.

Proof. Set $E = \Omega_1(Z(Q))$, so that $Z \leq E$. Since $N_Q \leq N_E$, $N_E \not\leq N$. Since $N_Z \leq N$, this forces $Z < E$. But $m(Q) = 2$, so $Z_3 \times Z_3 \cong E = \Omega_1(Q)$. Furthermore, $m(C_E) = 2$, otherwise $E \in \Delta^o$, in which case $N_E \leq N$, contradiction. Thus $E \in SC_3(G)$ and so it remains only to show that $Q = E$.

Since $m(C_P(E)) = 2$ and $m(P) = 3$, it is immediate that $E \not\leq \Phi(P)$. Thus $E \not\leq \Phi(Q)$. But again as $m(Q) = 2$, $\bar{N}_Q = N_Q/O_{3',3}(N_Q) \cong SL_2(3)$ or $GL_2(3)$. Furthermore, $N_{N_Q}(R)$ covers $\bar{N}_Q/O_2(\bar{N}_Q)$ and $N_R \leq N$ by the choice of Q. Hence N does not cover $\bar{Y} = O_2(\bar{N}_Q) \cong Q_8$ and consequently $\bar{Y}$ does not centralize Z, thus forcing $\bar{Y}$ to act irreducibly on E. Since $\bar{Y}$ leaves $E \cap \Phi(Q)$ invariant and $E \not\leq \Phi(Q)$, this in turn implies that $E \cap \Phi(Q) = 1$. Since $E = \Omega_1(Z(Q))$, we conclude that $\Phi(Q) = 1$, whence $Q = E$ and the lemma is proved.

We are now in a position to repeat the argument of section 9 with N, Q in the roles of M, B, respectively, for we have established the three conditions on which that argument depends: namely, (a) $N_Z \leq N$; and if W is a Q-invariant Sylow 2-subgroup of $O_{3'}(N)$, then (b) $W \neq 1$ (as $|O_{3'}(N)|$ is even), and (c) $N_W \leq N$ (by Lemma 12.7, as $N_N(W)$ contains a Sylow 3-subgroup of N).

That argument yields that every 2-local subgroup containing Q lies in N. In particular, if T is a Q-invariant 2-subgroup of G, then as N_T is a 2-local containing Q, $T \leq N$, whence by Lemma 3.5, $T \leq O_{3'}(N)$. We thus conclude:

Lemma 12.9. If T is a maximal Q-invariant 2-subgroup of G, then $T \in Syl_2(O_{3'}(N))$.

Now we can quickly derive a final contradiction. Choose $x \in N_Q$ with $x \notin N$ and let T be a maximal Q-invariant 2-subgroup of G. Then so is T^x and hence by Lemma 12.9, T and T^x are each Q-invariant Sylow 2-subgroups of $Y = O_{3'}(N)$. It follows that $T^{xy} = T$ for some $y \in C_Y(Q)$. Then $xy \in N_T \leq N$ (again by Lemma 12.7). But also $y \in N$ (as $Z \leq Q$), so $x \in N$, contrary to our choice of x.

Thus Proposition 12.1 holds, and hence so does the stronger form of Theorem A.

REFERENCES

1. Bender, H. (1967). Über die grössten p'-normal Teiler in p-auslösbaren Gruppen. Arch. Math. 18, 474-478.
2. Glauberman, G. (1973). Failure of factorization in p-solvable groups. Quart. J. Math. 24, 71-77.
3. ____________. (1976). On solvable signalizer functors in finite groups. Proc. Lon. Math. Soc. 33, 1-27.
4. Goldschmidt, D. (1972). Solvable signalizer functors on finite groups. J. Algebra 21, 137-148.
5. ____________. (1972). 2-signalizer functors on finite groups. J. Algebra 21, 321-340.
6. Gorenstein, D. & Harada, K. (1970). A characterization of Janko's two new simple groups. J. Fac. Sci. Univ. Tokyo 16, 331-406.
7. Gorenstein, D. & Lyons, R. (1982). Signalizer functors, proper 2-generated cores, and nonconnected groups. J. Algebra 75, 10-22.
8. ____________________. The local structure of finite groups of characteristic 2 type. Memoirs of Amer. Math. Soc., 42, #276, 1-731.
9. Harada, K. (1981). Groups with nonconnected Sylow 2-subgroups revisited. J. Algebra 70, 339-349.
10. Klinger, K. & Mason, G. (1975). Centralizers of p-groups in groups of characteristic 2,p-type. J. Algebra 37, 362-375.
11. Konvisser, M. (1971). Embedding of abelian subgroups in p-groups. Trans. Amer. Math. Soc. 153, 469-481.
12. Thompson, J. (1968). Nonsolvable finite groups all of whose local subgroups are solvable. I. Bull. Amer. Math. Soc. 74, 383-437; (1973). IV Pac. J. Math. 48, 511-592.
13. __________. (1964). Fixed points of p-groups acting on p-groups. Math. Z. 86, 12-13.

SIMPLE GROUPS OF SMALL NON-EVEN TYPE

R. Solomon
Ohio State University and Rutgers University

In 1970 Helmut Bender introduced a viewpoint for studying finite simple groups which extended to groups of even order the techniques he had developed for his proof of the "Uniqueness Theorem" for simple groups of odd order. He applied his method at the time to simple groups with abelian Sylow 2-subgroups and subsequently to simple groups with dihedral Sylow 2-subgroups. As a method for proceeding from certain 2-local data for an unknown simple group G to the dichotomy: either G has a "strongly embedded subgroup" or the centralizer of some involution in G is of "standard type", Bender's method bears comparison with the "covering p-local" methods of Alperin, Gorenstein, and Harada and the "signalizer functor" methods of Gorenstein and Walter. With the current techniques, Bender's method seems to work best when G (or at least the initial 2-local H) has 2-rank at most 4 and the signalizer functor method increasingly prevails as the 2-rank of G increases.

From this perspective our efforts have focussed on determining the optimal range of applicability for Bender's methods and writing up a unified reduction theorem for these cases. This seems to be nearing completion and we shall outline the anticipated results below.

<u>Definitions</u>. Let H be a finite group. A subgroup L of H is a <u>2-component</u> of H if L is a perfect subnormal subgroup of H with $L/O(L)$ quasisimple. A <u>2-component</u> L of H is a <u>component</u> of H if $O(L) \subseteq Z(L)$.

For G a finite group, <u>$\mathcal{L}(G)$</u> is the set of all 2-components of H as H ranges over all centralizers of involutions of G.

An element $L \in \mathcal{L}(G)$ is <u>2-terminal</u> if for every involution t in $C_G(L/O(L))$,

$$\langle L(C_L(t))^{L(C_G(t))}\rangle = O(\langle L(C_L(t))^{L(C_G(t))}\rangle)L(C_L(t)).$$

An element $L \in \mathcal{L}(G)$ is <u>terminal</u> if L is quasisimple and for every involution t in $C_G(L)$, L is normal in $L(C_G(t))$. By results of Aschbacher [2] and Gilman [3], if L is terminal and $m_2(L) > 1$, then $C_G(L)$ is a tightly embedded subgroup of G, hence of quite restricted structure. Indeed if $L/Z(L)$ is a Chevalley group of odd characteristic, an alternating group or a sporadic simple group, one can show fairly easily that a Sylow 2-subgroup of $C_G(L)$ is cyclic or of maximal class.

<u>Definition</u>. Let G be a finite simple group all of whose p-local subgroups are K-groups. We shall say that G is of <u>small non-even type</u> if one of the following holds:

1) $m_2(G) \leq 2$, or

2) a) If $L \in \mathcal{L}(G)$ and $L/Z^*(L) \cong A_n$, then $n \leq 11$.

b) If $L \in \mathcal{L}(G)$ and $L \in \text{Chev}(q)$, q odd, then $L/Z^*(L) \cong L_2(q), U_3(3), L_3(3), G_2(3), \Omega_5(3)$ or $\Omega_6^{\pm}(3)$.

c) There exists a 2-terminal $L \in \mathcal{L}(G)$ with $L/O(L)$ isomorphic to one of: $L_2(q)$ or $SL(2,q)$ (q odd), A_7, $\hat{A}_n$ $(7 \leq n \leq 11)$ or $\hat{L}_3(4)$ (center of exponent 4).

<u>Theorem A</u>. Let G be a finite simple group of small non-even type. Then either

a) G is of 2-uniqueness type; or

b) G is of 2-standard type.

By "2-uniqueness type", we shall mean that G has a primitive permutation representation in which some 2-central involution fixes exactly one point. Then we may invoke Holt's theorem [4] to identify G. In fact this result will be used only when the structure of $M = G_\alpha$ yields an immediate contradiction. Bender's "strongly embedded subgroup" theorem will serve to identify $L_2(4)$ and $PSU(3,4)$.

By "2-standard type" in this context, we shall mean roughly the following. Either

1) A Sylow 2-subgroup S of G is dihedral, semidihedral, wreathed or of type $PSU(3,4)$. If t is an involution in $Z(S)$ and $H = C_G(t)$, then $O(H) = BA$ where B is inverted by every involution of $H-\{t\}$ and $N_G(X) \subseteq H$ for all $1 \neq X \subseteq B$. A is cyclic and $A = C_{O(H)}(s)$ for involutions $s \in H-\{t\}$. If $A \neq 1 \neq B$, then BA is a Frobenius group with kernel B; or

2) G has a standard subgroup L isomorphic to one of $L_2(q)$ or $SL(2,q)$ (q odd), A_7, $\hat{A}_n$ ($7 \le n \le 11$) or $\hat{L}_3(4)$ (center of exponent 4); or

3) There is 2-central involution $t \in L$ with $O_2(C_G(t))E(C_G(t)) = LL^x$ with $L/Z(L) \cong L_2(q)$, $L \neq L^x$, $x \in C_G(t)$.

More will in fact be established in reaching standard types (2) and (3). Indeed $O^{2'}(C_G(t)/O(C_G(t)))$ will be determined precisely and $O(C_G(t))$ will be shown to be cyclic of bounded order. From this point one will proceed to the identification stage and prove:

Theorem B. Let G be a finite simple group of small non-even type. Then G is isomorphic to one of:

1) A_n, $5 \le n \le 11$

2) A group of Lie type of odd characteristic and BN rank 1 or 2, excluding $PSU(4,q)$

3) $P\Omega^{\pm}(6,q)$, $q \equiv \pm 3 \pmod 8$

4) $PSU(3,4)$ or M_{11}

5) J_1, M_{12} or HJ

6) M^c or LyS

7) O'NS

or 8) G has a standard subgroup $L \cong L_2(9)$, $|C_G(L)|_2 = 2$, $m_2(N_G(L)) = 4$ and $|G:N_G(L)|$ is even.

We remark that the standard subgroup problem listed in (8) is particularly unpleasant to solve and leads only to groups of even type which are more naturally identified from their "parabolic" structure. It appears that it will be possible to complete the revised proof of the classification without ever treating case (8).

The proof of Theorem A is accomplished in two phases. The first phase entails bounding the size of a Sylow 2-subgroup of G. In the cases when $m_2(G) = 2$ or $|Z^*(L)|$ is even, this is accomplished by 2-fusion analysis and leads to the precise Sylow 2-structure of the target groups. On the other hand, when $L/O(L) \cong L_2(q)$ or A_7, signalizer functor methods will be employed in conjunction with fusion analysis to study $C_G(L/O(L))$ and force certain 2-locals to lie in a common maximal subgroup of G.

Once the 2-structure is under control, phase 2 begins with the study of maximal subgroups M containing $C_G(t)$ where t is a 2-central involution of G if $L/O(L) \not\cong L_2(q)$ or A_7. If $L/O(L) \cong L_2(q)$ or A_7, then t is chosen with $|C_G(t)|_2$ maximal subject to $C_G(t)$ having a non-quasisimple 2-terminal 2-component L with $L/O(L) \cong L_2(q)$ or A_7. A very general argument shows that $F^*(M)$ has odd order if G is not of 2-standard type. A somewhat more delicate argument establishes that $F^*(M)=P$ is a p-group and if $1 \neq R \subseteq P'C_P(P')$ and $N_G(R) \subseteq N$, a maximal subgroup of G, then $F^*(N)$ is a p-group as well. In the "classical" Bender cases (abelian or dihedral Sylow 2-subgroups) G is p-stable and Glauberman's ZJ-Theorem may be employed to yield an easy contradiction.

In case, $L/O(L) \not\cong L_2(q)$ or A_7, we let $S \in Syl_2(M)$ and choose a 4-subgroup $U = \langle t,u \rangle \subseteq \{1\} \cup (t^G \cap S)$ with $C_S(U) \in Syl_2(C_G(U))$. We let $T = N_S(U)$ and study $\mathcal{N}_G^*(T,p)$, establishing first that $M = N_G(ZJ(R))$ for some $R \in \mathcal{N}_G^*(T,p)$. We next suppose that there exists a non-trivial $N_G(U)$-invariant p-subgroup X of M. This is immediate if $m(S) \geq 3$, as then $1 \neq C_P(U) \subseteq O_p(C_G(U))$. Studying maximal subgroups N of G containing $N_G(X)$ reduces to the case $m(S) = 2$ and $O^{2'}(N/O_p(N)) \cong (S)L(3,p^n)$. This seems to be an unavoidable obstruction first encountered by Alperin, Brauer and Gorenstein [1] and handled by modular character theory. Aschbacher has provided an argument via weak closure techniques to eliminate this configuration. It may also be amenable to the amalgam method. In any event, we arrive at the case $\mathcal{N}_M(N_G(U),p) = \{1\}$ and $m(S) = 2$. Aschbacher has provided an elegant argument in this case that G acts 2-transitively on a finite projective plane, whence $G \cong PSL(3,q)$ by the Ostrom-Wagner Theorem.

If $L/O(L) \cong L_2(q)$ or A_7, we reduce easily to the case $p = 3$ and there exists a non-3-stable 3-local subgroup N of G containing $N_G(X)$ for some non-trivial S-invariant p-subgroup X of $O(M)$. $N/O_3(N)$ must have a normal subgroup K isomorphic to $SL(2,3)$ or $SL(3,3)$. If $m_2(G) = 3$, analysis of 2-fusion forces a contradiction. If $m_2(G) > 3$, the earlier signalizer analysis will force K to lie in M, yielding a contradiction.

The value of this approach is that it consolidates and simplifies the treatment of 2-signalizers for a large family of groups, hitherto treated by ad hoc methods. In terms of the nature of the target groups and the place in the analysis of simple groups of non-even type, this chapter of the classification is analogous to the chapter on thin and quasi-thin groups in the analysis of simple groups of even type.

Unfortunately the method of attack has but few and apparently superficial similarities to the parabolic approach being developed by Goldschmidt, Stellmacher, Stroth and others, although in both cases the groups of Lie rank 2 will be ultimately identified via their parabolic geometries. It is a challenge for the future to find a line of attack that leads more directly from the hypotheses that G is a finite simple group of "small type" in a suitable sense to the conclusion that G is a flag-transitive automorphism group of one of a list of specified geometries.

REFERENCES

1. Alperin, J.L., Brauer, R. & Gorenstein, D. (1970). Finite groups with quasi-dihedral and wreathed Sylow 2-subgroups. Trans. Amer. Math. Soc. 151, 1-261.
2. Aschbacher, M. (1975). On finite groups of component type. Illinois J. Math. 19, 78-115.
3. Gilman, R. (1976). Components of finite groups. Comm. Alg. 4, 1133-1198.
4. Holt, D. (1978). Transitive permutation groups in which a 2-central involution fixes a unique point. Proc. London Math. Soc. 37, 165-192.

AUTOMORPHISMS OF GROVES

A.L. Chermak
Kansas State University, Manhattan, Kansas

INTRODUCTION

My aim in this article is to present, as briefly as possible, some ideas and results that arose from an attempt to extend Goldschmidt's work on automorphisms of graphs* to a more general setting. Proofs have necessarily had to be deleted, and will appear elsewhere.

1 GROVES AND INDEX-p SYSTEMS

Let Γ be a tree. Denote the vertex set of Γ by Vert Γ, and for any $\alpha \in \text{Vert}\,\Gamma$ denote by $\Delta(\alpha)$ the set of vertices of Γ which are incident to α. A subset X of Vert Γ is an orientation of Γ if $\Delta(\alpha) \cap X = \Delta(\beta) \cap X' = \emptyset$ for all $\alpha \in X$ and all $\beta \in X' = (\text{Vert}\,\Gamma) - X$. We say that X generates Γ if no proper subtree of Γ contains X.

A grove consists of a tree Γ (called the base tree), a family of trees $(\Gamma_\alpha)_{\alpha \in \text{Vert}\,\Gamma}$, and a family $(h_\alpha)_{\alpha \in \text{Vert}\,\Gamma}$ of injective mappings

$$h_\alpha : \Delta(\alpha) \hookrightarrow \text{Vert}\,\Gamma_\alpha$$

such that the image of any h_α is an orientation of Γ_α which also generates Γ_α. In particular, if β and β' are distinct vertices in $\Delta(\alpha)$ then $\text{dist}(h_\alpha(\beta), h_\alpha(\beta')) \geq 2$. When equality holds, we say that β and β' are linked (at α).

Let $G = (\Gamma; (\Gamma_\alpha, h_\alpha)_{\alpha \in \text{Vert}\,\Gamma})$ be a grove, and let g be an automorphism of Γ which sends linked vertices to linked vertices. Then the maps

---

*D.M. Goldschmidt, Automorphisms of trivalent graphs, Annals of Math., 111 (1980), 377-406.

$$\bar{g}_\alpha : \Delta(\alpha) \to \Delta(\alpha^g)$$
$$\beta \mapsto \beta^g$$

"extend" in a unique way to isomorphisms

$$g_\alpha : \Gamma_\alpha \to \Gamma_{(\alpha^g)}$$

via the maps h_α and $h_{(\alpha^g)}$. Thus Aut G may be identified with the subgroup of Aut Γ consisting of link-preserving automorphisms.

Notice that for any subgroup G of Aut G and $\alpha \in$ Vert Γ, G_α induces a group of automorphisms of the "fiber" Γ_α.

<u>Example</u>. Let Γ be the tree of valance two, and identify Vert Γ with $\mathbb{Z}$, so that $\Delta(n) = \{n-1, n+1\}$ for all n. Let Γ_n and h_n be given by

x————o————x

$h_n(n-1)$ $h_n(n+1)$

for any $n \in \mathbb{Z}$. We denote this grove by $\|\mathbb{Z}\|$. Its automorphism group is $\mathbb{D}_\infty$.

Whenever G is a grove, T is a subgrove of G isomorphic to $\|\mathbb{Z}\|$, and τ is an element of Aut G such that τ induces a 2-translation on $\|\mathbb{Z}\|$, we say that (T,τ) is a <u>track</u> in G. If G is a subgroup of Aut G, we say that (T,τ) is G-<u>generating</u> if, for any edge $\{\alpha,\beta\}$ of Γ which is also an edge of the underlying tree of T, there exist elements $r_\alpha \in G_\alpha$ and $r_\beta \in G_\beta$, such that $\tau = r_\alpha r_\beta$, and such that $G = \langle G_{\alpha,\beta}, r_\alpha, r_\beta \rangle$.

Let P_1 and P_2 be subgroups of some group, and set $B = P_1 \cap P_2$. One may then form the free amalgamated product

$$G = P_1 \underset{B}{*} P_2$$

and there is an associated tree Γ, on which G induces a group of automorphisms. Namely, take Vert Γ to be the (disjoint) union $P_1\backslash G \,\dot\cup\, P_2\backslash G$ of the right cosets of P_1 and P_2 in G, and define incidence by $P_1g_1 \sim P_2g_2$ if and only if $P_1g_1 \cap P_2g_2 \neq \emptyset$ (equivalently, $P_1g_1 \cap P_2g_2 = Bg$ for some $g \in G$). Then G acts on Γ by right translation.

Now suppose that also P_1 and P_2 are themselves presented as free amalgamated products:

$$P_i = B \underset{B_i}{*} C_i, \quad (i = 1,2)$$

and let Γ_1 and Γ_2 be the associated trees. There are natural maps

$$h_i : B\backslash P_i \hookrightarrow \text{Vert } \Gamma_i$$

which, by transport of structure, yield a grove $\mathcal{G} = (\Gamma;(\Gamma_\alpha,h_\alpha))$, and then G induces a subgroup of Aut $\mathcal{G}$ in the natural way.

An <u>index</u>-p <u>system</u> consists of a finite p-group B, a pair of maximal subgroups B_1 and B_2 of B, and a pair of automorphisms r_1 and r_2, of B_1 and B_2, respectively. We may abuse notation and denote such a system briefly by $S = (B;r_1,r_2)$. Set

$$P_i = B \underset{B_i}{*} B_i\langle r_i\rangle, \quad (i = 1,2), \quad \text{and}$$

$$G = P_1 \underset{B}{*} P_2.$$

As just described, there is an associated grove $\mathcal{G} = \mathcal{G}(S) = (\Gamma;(\Gamma_\alpha,h_\alpha))$, and a natural action by G on $\mathcal{G}$. Moreover it is possible to fix a G-generating track (T,τ) in $\mathcal{G}$, such that $\tau = r_1r_2$ and such that B is the G-stabilizer of an edge of T.

2 A TECHNICAL RESULT

Let $S = (B;r_1,r_2)$ be a fixed index-p system, let $\mathcal{G}$ be the grove $(\Gamma;(\Gamma_\alpha,h_\alpha))$ associated with S, and let G be the associated group, acting on $\mathcal{G}$. Let (T,τ) be a G-generating track in $\mathcal{G}$. For $\alpha \in \text{Vert } \Gamma$ and $\beta \in \Delta(\alpha)$, we set

$$Z(\alpha,\beta) = Z(G_{\alpha,\beta}),$$

$$Z(\alpha) = \bigcap_{\beta\in\Delta(\alpha)} Z(\alpha,\beta),$$

and

$$V(\alpha) = \prod_{\beta\in\Delta(\alpha)} (Z(\alpha,\beta) \cap Z(G_{\Delta(\alpha)})).$$

Identify T with $\|\mathbb{Z}\|$, so that for any $\alpha \in \text{Vert } T$ we have $\alpha^\tau = \alpha+2$. Set $P_T(\alpha) = \langle G_{\alpha-1,\alpha}, G_{\alpha,\alpha+1}\rangle$.

We consider the following statements concerning a vertex α of T, letting ε always denote a member of $\{\pm 1\}$.

(L1) For some ε, we have $Z(\alpha+\varepsilon) \neq V(\alpha+\varepsilon) \trianglelefteq P_T(\alpha)$, and there exists $x \in G_{\Delta(\alpha-\varepsilon)} - G_{\Delta(\alpha)}$ such that $[V(\alpha+\varepsilon),x] = 1$.

(L2) For both $\varepsilon = 1$ and -1, we have $V(\alpha)V(\alpha+\varepsilon) \trianglelefteq P_T(\alpha)$ and $V(\alpha+\varepsilon) \neq Z(\alpha+\varepsilon)$. Moreover, there exist elements $x_\varepsilon \in G_{\Delta(\alpha-\varepsilon)} - G_{\Delta(\alpha)}$ such that the following hold.

i) $|V(\alpha)V(\alpha+\varepsilon) : C_{V(\alpha)V(\alpha+\varepsilon)}(x_\varepsilon)| = p$, and

ii) $[V(\alpha+\varepsilon),x_\varepsilon] \subseteq Z(\alpha-\varepsilon,\alpha)$.

(C1) For some ε, we have $|V(\alpha+\varepsilon):Z(\alpha,\alpha+\varepsilon)| = p$, and $[V(\alpha+\varepsilon),G_{\Delta(\alpha)}] \subseteq Z(P_T(\alpha)) \cap Z(P_T(\alpha-\varepsilon))$.

(C2) We have $Z(\alpha) = V(\alpha)$ and, for both $\varepsilon = 1$ and -1, the following hold.

i) $|V(\alpha+\varepsilon):Z(\alpha)| = p^2$, and

ii) $[V(\alpha+\varepsilon),G_{\Delta(\alpha)}] \subseteq Z(\alpha) \cap Z(\alpha-2\varepsilon)$.

(C3) We have $Z(\alpha) = V(\alpha)$ and, for both $\varepsilon = 1$ and -1, the following hold.

i) $|V(\alpha+\varepsilon):Z(\alpha)| = p$, and

ii) there exist elements $x_\varepsilon \in G_{\Delta(\alpha-\varepsilon)} - G_{\Delta(\alpha)}$ such that $[V(\alpha+\varepsilon),x_\varepsilon]$ is a subgroup of order p in $Z(\alpha-2\varepsilon) \cap Z(\alpha-4\varepsilon)$.

We have the following rather technical result.

Theorem A. Let $\mathcal{S} = (B;r_1,r_2)$ be an index-p system, $\mathcal{G}$ its associated grove, G its associated group, and let (T,τ) be a G-generating track in $\mathcal{G}$. Set $M_i = \langle B,B^{r_i}\rangle$, $(i = 1,2)$, regarded as a subgroup of G. Then one of the following holds

a) At least one of the conditions L1, L2, C1, C2, C3, above.

b) $\langle (Z(B_1)Z(B_2))^G\rangle$ is an abelian subgroup of B.

c) There exists a subgroup X of B, such that $|B:X| \leq p^{12}$, and such that $M_i = C_{M_i}(X)X$, for both $i = 1$ and 2.

3 APPLICATIONS

Let L be a group, and let there be given subgroups P_1, P_2, and B of L, and elements c_i of P_i, $(i = 1,2)$, such that the following conditions hold.

1. $L = \langle P_1, P_2 \rangle$.

2. B is a finite p-subgroup of $P_1 \cap P_2$.

3. $P_i = \langle B, B^{r_i} \rangle$, $(i = 1,2)$.

4. $B \cap B^{r_i}$ is maximal in B, and is invariant under r_i,

$(i = 1,2)$. We then say that L is tightly-generated by the index-p system $(B; r_1, r_2)$. Notice that Condition 3 is more restrictive than the statement: $P_i = \langle B, r_i \rangle$. It is this restriction which allows us to derive the following result as a Corollary to Theorem A.

Theorem B. Let L be tightly generated by $(B; r_1, r_2)$, and set $B_i = B \cap B^{r_i}$, $(i = 1,2)$. Then one of the following holds.

a) $C_B(L) \neq 1$.

b) $\langle Z(B_1)Z(B_2))^L \rangle$ is an abelian subgroup of B.

c) $|B| \leq p^{10}$. Moreover, if $p \neq 3$ then $|B| \leq p^7$, and if $p > 3$ then $|B| \leq p^6$.

An example in which $|B| = 3^{10}$ is afforded by Thompson's group F_3, and Aut M_{12} affords an example with $|B| = 2^7$. For all p, $G_2(p)$ affords an example with $|B| = p^6$.

A more interesting application of Theorem A concerns "pushing up" problems involving groups L of the above type.

Theorem C. Let L be tightly generated via $(B; r_1, r_2)$, and set $B_i = B \cap B^{r_i}$, $(i = 1,2)$. Let $\lambda \in$ Aut B. For each i, assume that either $Z(B) \leq P_i$ or that B_i is λ-invariant. Then one of the following holds.

a) $Z(B) \subseteq Z(L)$.

b) There exists a λ-invariant subgroup R of B, containing $Z(B_1)Z(B_2)$, and such that $R \trianglelefteq L$.

c) There exists a subgroup X of B, with $|B:X| \leq p^{12}$, and such that $P_i = C_{P_i}(X)X$ for both $i = 1$ and 2.

As regards part (c): the group $D_4(p) \cdot \mathbb{Z}_3$ affords an example where $X = 1$, $|B| = p^{12}$, and $L \cong p^{3+6} \cdot SL(3,p)$. Other examples of interest occur in Ru (with $L \cong 2^6|G_2(2)$) in $\mathrm{Aut}\,{}^2E_6(2)$ (with $L \cong 3^{3+3} \cdot SL(3,3)$), and in F_1 with $L \cong 5^{3+3} \cdot SL(3,5)$.

We briefly indicate how Theorem C is obtained as a Corollary to Theorem A. There is a "trick" involved in this, which consists of replacing the index-p system S with a system S^* which generates a larger group than L. Namely, if we assume that $Z(B) \ntrianglelefteq L$, then $(B_1)^{\lambda} = B_i$ for some i (say i = 2), and we may form $S^* = (B;r_1,\lambda r_2)$, regarding λr_2 as an automorphism of B_2. Moreover, one may exhibit an infinite group G, containing L, containing also an element λ which induces the given automorphism of B by conjugation, and such that $G = \langle B,r_1,\lambda r_2\rangle$. Theorem A then applies, and the special properties of L and λ then yield Theorem C.

4 FURTHER DIRECTIONS

We list two problems which are suggested by the foregoing discussion.

Problem 1. Let N be a nilpotent, connected Lie group, such that $N/[N,N]$ is simply connected, and let N_1 and N_2 be maximal Lie subgroups of N. Given a pair of automorphisms r_i of N_i, (i = 1,2), generalize the results on index-p systems to results on $(N;r_1,r_2)$.

Problem 2. Extend the notion of index-p system to more general systems, in such a way that all rank-2 BN-pairs over finite fields are "accounted for". For example, and as a first step in this direction, consider the following hypothesis.

> (*) B is a finite group, with $F^*(B) = O_p(B)$ for some prime p. Assume that there exist maximal subgroups B_1 and B_2 such that $B = B_1B_2 = B_iF^*(B)$, (i = 1,2), and let r_1 and r_2 be automorphisms of B_1 and B_2, respectively.

The problem is then to analyze the system $(B;r_1,r_2)$. The condition (*) is probably too broad, however, and may be strengthened by additional hypotheses concerning modules for the groups $\langle B,r_i\rangle$. Thus, assume also

(**) Whenever A is an abelian subgroup of $F^*(B)$, and V is an abelian normal p-subgroup of $\langle B,r_i\rangle = P_i$, such that $|A:C_Z(V/C_V(P_i))| \geq |V/C_V(P_i):C_{V/C_V(P_i)}(A)| \neq 1$, then $B = AB_i$.

AMALGAMS OF FINITE GROUPS AND THE GOLDSCHMIDT-SIMS CONJECTURE

P.S. Fan
Department of Mathematics, University of California,
Berkeley, CA 94720

1 THE GENERAL PROBLEM

An important part of the study of finite groups involves the analysis of group-theoretic configurations satisfying the following criteria:

1) X is a finite group generated by $\{P_1, P_2\}$.

2) No non-identity subgroup of $P_1 \cap P_2$ is normal in both P_1 and P_2.

In [S], Sims observed that a very special case of the above can be treated graph-theoretically. It was not until much later that Goldschmidt introduced in [G] some general techniques and ideas to exploit the geometric nature of the above configuration. Specifically, he produced a complete classification of the amalgams with index (3,3). Please consult [G] for the undefined terms in this article. We introduce below some further terminology. Since any amalgam $P_1 \xleftarrow{\phi_1} B \xrightarrow{\phi_2} P_2$ has a universal completion given by $P_1 \underset{B}{*} P_2$, we shall write $P_1 \supseteq B \subseteq P_2$ for our amalgams, where it is implicitly assumed that $B = P_1 \cap P_2$. A completion X of $P_1 \supseteq B \subseteq P_2$ is said to be proper if the maps of P_i into X for $i = 1,2$, are embeddings, $X = \langle P_1, P_2 \rangle$ and $B = P_1 \cap P_2$ inside X. We observe that every amalgam $P_1 \supseteq B \subseteq P_2$ of finite groups has a proper finite completion; it can for example be given by the symmetric group on the set P_1P_2. We call $P_1 \supseteq B \subseteq P_2$ primitive if $N_{P_i}(K) = B$ whenever $1 \neq K \trianglelefteq P_{3-i}$, and it is called faithful if no non-trivial subgroup of B is normal in both P_1 and P_2. (In the "minimal" case, they are of course equivalent.)

The fundamental problem underlying the recent works on amalgams of finite groups is the problem of Goldschmidt-Sims (G-S):

Classify all primitive amalgams.

It seems one ought to regard the soul of this problem to be the baby G-S problem:

> Produce a "strong" (preferably geometric) bound on the number of primitive amalgams with a given index.

A motivation for concerning ourselves with the G-S problem is that every finite simple group is the proper finite completion of a primitive amalgam. Although it is natural to insist that our amalgams be faithful, it is not quite sufficient. Our primitivity condition is necessitated by concrete examples: Given any pair of integers (n_1,n_2), not both of which are prime, a wreath product construction from an idea of Djokovic [Dj] affords infinitely many faithful amalgams of index (n_1,n_2).

These examples also lead naturally to the consideration of amalgams with prime indices. The known examples are the fifteen Goldschmidt amalgams (see [G]), the ${}^2F_4(2)'$-amalgam and the ${}^2F_4(2)$-amalgam. Indeed, we prove that these are in fact the only "interesting" such amalgams.

<u>Theorem</u>* [F]. Let $\mathcal{A}: P_1 \supseteq B \subseteq P_2$ be a faithful amalgam with index (p_1,p_2) and let B be a p-group, where p_1,p_2 and p are primes and $p_1 \neq p \neq p_2$. Then one of the following holds.

1) $\mathcal{A}$ is a Goldschmidt amalgam.
2) $\mathcal{A}$ is a ${}^2F_4(2)'$-amalgam.
3) $\mathcal{A}$ is a ${}^2F_4(2)$-amalgam.
4) $\mathcal{A}$ is not of char. p-type.

For any rank 2 chev(p)-group X, the amalgam given by two non-conjugate minimal parabolics is called the X-<u>amalgam</u>. We say $P_1 \supseteq B \subseteq P_2$ is of <u>char. p-type</u> if $F^*(P_i) = O_p(P_i)$, for $i = 1,2$. Notice that in any char. p-type amalgam satisfying the above hypothesis, B is a 2-group of order at most 2^{12} and its index is either $(3,3)$ or $(3,5)$. We also remark that according to a theorem of Thompson and Wielandt any non-trivial primitive amalgam $P_1 \supseteq B \supseteq P_2$ is of char. p-type and B is mostly a p-group for some prime p.

---

*This is the main theorem of my doctoral thesis supervised by David Goldschmidt.

2 THE "PROOF"

At this time, we give some sketchy indication as to why our theorem is true. The reader is advised to draw the appropriate picture in following the argument. Set $G = P_1 \underset{B}{*} P_2$. The usual coset geometry of $P_1 \supseteq B \subseteq P_2$ in G affords the following: G is a locally finite and locally transitive group of automorphisms on a tree Γ with valence p_α and p_β, where p_α and p_β are primes and α, β are adjacent vertices in Γ. Further, $G_\alpha \supseteq G_{(\alpha,\beta)} \subseteq G_\beta$ is isomorphic to $P_1 \supseteq B \subseteq P_2$. For the undefined terms below, see §2 of [F]. Given any line T in Γ and $g \in G$, let $T(g)$ denote the fixed point set of g in T. We write $\alpha \in \Gamma$ (resp. $\gamma \subseteq \Gamma$) to mean α (resp. γ) is a vertex (resp. arc) in Γ.

Our game plan has two stages. During the initial stage, we bound $|G_{(\alpha,\beta)}|$ with respect to p_α and p_β. Set N to be the integer for which $G^{[N]}_{(\alpha,\beta)} \neq 1$ and $G^{[N+1]}_{(\alpha,\beta)} = 1$. Take $z \in \Omega_1(Z(G_{(\alpha,\beta)})) \cap G^{[N]}_{(\alpha,\beta)}$. We then observe that it suffices to bound $|T(z)|$, where $(\alpha,\beta) \subseteq T$ and T is a line in Γ.

It is immediate that for any $\delta \in \Gamma$, $G_\delta/G_{\Delta(\delta)}$ is a Frobenius group of degree p_δ. Further, our arguments require a minimal degree of non-degeneracy, which is measured by s. This is defined to be the smallest integer for which there exists a singular arc of length s. It turns out that to say s is small (namely $s \leq 3$) is equivalent to asserting that $G_\alpha \supseteq G_{(\alpha,\beta)} \subseteq G_\beta$ is not of char. p-type. So we might as well assume $s \geq 4$ and $G_{(\alpha,\beta)}$ is a 2-group.

We first achieve a bound on $|T(z)|$ under the condition that (T,τ) is a track and $G_T \neq 1$. This is the only part of our proof that requires any delicate group-theoretic information. We have a certain set $\mathcal{L}$ of 2-locals $N_G(Q)$, where $1 \neq Q \leq G_{(\alpha,\beta)}$, on which we can apply some easy pushing up lemmas to obtain very specific facts about the structure of $N_G(Q)_\alpha$ and $N_G(Q)_\beta$. Our point, of course, is that $L = N_G(\Omega_1(G_T)) \in \mathcal{L}$. For this enables us to conclude that $H = \langle L_\alpha, L^o_\beta \rangle$, where L^o_β is a minimal subgroup of L_β properly containing $L_{(\alpha,\beta)}$, acts vertex transitively on a "generalized" subgraph $\tilde{\Gamma}$ of Γ that happens to be a trivalent tree and such that $H_{\tilde{\Gamma}} \subsetneq G_{T(z)} \subseteq H$. Since these actions are understood from old results, it yields the following facts:

1) There exist small finite subgraphs $\Delta \subseteq \tilde{\Gamma}$ such that $H_\Delta \subseteq H_{\tilde{\Gamma}}$.

2) $H_{\tilde{T}} = G_T$.

From (1), we can obtain that $T(z) \subsetneq \Delta$ and specifically that $|T(z)| \leq 8$.

Part (2) allows us to draw the conclusion that for any track (T,τ) with $(\alpha,\beta) \subseteq T$, if $Z = Z(G_{\Delta(\beta)}) \cap G_T \neq 1$, then $N = N_G(Z) \in L$ and as a consequence, we may then infer that $N_{(\alpha,\beta)} = G_{\Delta(\beta)}$. But this implies $G_{\Delta(\alpha,\beta)} = G_\alpha^{[2]}$ which contradicts $s \geq 4$. Therefore, every non-identity element of $Z(G_{(\alpha,\beta)})$ fixes only a finite subarc of T. To obtain a bound on $|T(z)|$ in general, we analyze this finite subarc $T(z)$ in greater detail. Let us now assume $z \in Z(G_\alpha)$ and set $z = z_\alpha$ and define $z_{\alpha^g} = (z_\alpha)^g$. So for any $\delta \in \Gamma$, an even distance from α, G_δ has a central involution z_δ. Let $r(z_\alpha)$ $(\ell(z_\alpha))$ be the distance between α and the right (left) endpoint of $T(z_\alpha)$. By comparing z_α with $z_\alpha^{<\tau>}$ we are able to arrive at the conclusion that $r(z_\alpha) = \ell(z_\alpha)$ and they are even. In making our analysis of the fixed point configuration of z_δ, $\delta \in T$, it is absolutely crucial to obtain the strongest possible connection between $[x,z_\delta] = 1$ and $x \in G_\delta$.

Our aim of getting a bound on $r(z_\alpha)$ is eventually accomplished by producing a track that has non-trivial fixers. The necessary observation here is that the commutator

$$u = [z_\alpha, z_{\alpha'}] = z_\alpha z_\alpha^{z_{\alpha'}} = z_{\alpha'}^{z_\alpha} z_{\alpha'},$$

where $\alpha' = \alpha + r(z_\alpha) + 2$, appears on both sides of the track. This simple geometric implication of the commutator enables us to attempt to obtain some $x \in G_{(\alpha,\lambda)} - G_{\lambda+1}$ and $y \in G_{(\alpha',\lambda+1)} - G_\lambda$ that centralize u. Whenever $r(z_\alpha)$ is large, such a λ is more likely to and does indeed exist and so u then fixes the track afforded by x and y passing through $(\lambda,\lambda+1)$. In fact with more careful analysis, we can demonstrate that $r(z_\alpha) = \ell(z_\alpha) = 4$. Hence, $N \leq 4$ and $G_\alpha^{[5]} = G_\beta^{[4]} = 1$. Moreover, it is not difficult to show that the set of fixed points of z_α in Γ is precisely $\Delta^{[4]}(\alpha)$. This yields a 1-1 correspondence between $[x,z_\delta] = 1$ and $x \in G_\delta$; in other words, $z_{\delta_1} = z_{\delta_2}$ iff $\delta_1 = \delta_2$. Therefore, since $G_\alpha^{[i]}/G_\alpha^{[i+1]}$, for $i = 1,2$, contains elements generated by z_δ, where δ is near α, they must possess non-central 2-chief factors of G_α. Indeed, it turns out that $G_\alpha^{[3]}$ contains only central 2-chief factors and $G_\alpha^{[3]} = G_\alpha^{[4]} = \langle z_\alpha \rangle$.

Now we can proceed to the second stage of our argument, namely to obtain a very short list of "presentations" such that one of which is admitted by any $G_\alpha \supseteq G_{(\alpha,\beta)} \subseteq G_\beta$. Given any presentation $\mathcal{P}$ for a group P and $B \leq P$, $\mathcal{P}|_B$ consists of generators in $\mathcal{P}$ that lie in B and relations in $\mathcal{P}$ that involve only elements of B. A <u>presentation</u> for $P_1 \supseteq B \subseteq P_2$ is given by a pair of presentations $\mathcal{P}_1$ and $\mathcal{P}_2$ for P_1 and P_2 such that

a) $\mathcal{P}_i|_B$ is a presentation for B, i = 1,2,.

b) The generators in $\mathcal{P}_2|_B$ are known with respect to $\mathcal{P}_1|_B$.

Our first task is then to find a natural "geometric" presentation $\mathcal{P}_\alpha$ of G_α that restricts to a presentation for B. From the above paragraph, it is possible to discover the fixed point set of certain elements in G_α and we can then see the generators that lie in each $G_\alpha^{[i]}/G_\alpha^{[i+1]}$, i = 1,2, and the commutator relations among them. In order to minimize our list of presentations for $G_\alpha \supseteq G_{(\alpha,\beta)} \subseteq G_\beta$, the question we need to answer with precision is how the generators in $\mathcal{P}_\alpha|_{G(\alpha,\beta)}$ must fit into G_β. This is the point underlying the geometry of our presentation for G_α. For since we know where the generators "live" on the graph Γ and their behavior near α, and as β is adjacent to α, it is very conceivable that we can discern how they must behave inside G_β.

3 SOME FURTHER QUESTIONS

In the above discussion, we have tried to justify our theorem without much reference to our hypothesis. This is quite deliberate. Indeed, it reflects our belief that these arguments can be made to work within a broader context. And one of the current goals in the development of this theory of amalgams of finite groups is to obtain results under more general hypothesis. Presently, the works of Goldschmidt [G], Stellmacher [Ste], Stroth [Str 1,2], Delgado [De] and Fan [F] have pretty much completed the rank 2 amalgam problem. (Also relevant is the work of Weiss [W].) This problem asks one to show that those amalgams $P_1 \supseteq B \subseteq P_2$ for which $O^{p'}(P_i/O_p(P_i))$, i = 1,2, is a rank 1 chev(p)-group are in fact essentially isomorphic to the minimal parabolic amalgams in rank 2 chev(p)-groups.

In [T], Timmesfeld proposed a program to classify "weak characteristic 2-type" groups. The motivation is that these groups include the groups of GF(2)-type and over half of the sporadic groups are in this category. Steps (1) and (2) of Timmesfeld's 3 step program can be resolved by a classification of minimal parabolic amalgams $P_1 \supseteq B \subseteq P_2$, which we defined to be 2-constrained amalgams such that for each $i \in \{1,2\}$, there exists $S \in Syl_2(P_i)$ so that $N_{P_i}(S)$ is contained in a unique maximal subgroup of P_i. This is evidently rather difficult, although it ought to be no more than the rank 2 amalgam problem.

REFERENCES

[Dj] Djokovic, D. (1980). A class of finite group amalgams. Proc. of AMS 80, 22-26.

[De] Delgado, A. (1981). Edge-transitive automorphism groups of graphs. Thesis, Berkeley.

[F] Fan, P. Amalgams of prime index. Preprint.

[G] Goldschmidt, D. (1980). Automorphisms of trivalent graphs. Ann. of Math., 111, 377-406.

[Ste] Stellmacher, B. On graphs with edge-transitive automorphism groups. To appear in Illinois J. Math.

[Str 1] Stroth, G. (1983). Graphs with subconstituents containing Sz(q) and $L_2(r)$, J. of Algebra 80, 186-215.

[Str 2] ________. Kantentransitive Graphen und Gruppen von rank 2. To appear.

[W] Weiss, R. (1982). Edge-symmetric graphs. J. of Algebra, 75 no. 1, 261-274.

THE GEOMETRY OF p-TRANSVECTION SUBGROUPS

J.I. Hall*
Department of Mathematics
Michigan State University
East Lansing, MI 48824

1 p-TRANSVECTION SUBGROUPS

Following Aschbacher [2] a p-*transvection subgroup* P of the finite group P is an elementary abelian p-subgroup such that, for each $P_1, P_2 \in P^G$, $\langle P_1, P_2 \rangle$ is isomorphic to one of $P_1 \times P_2$, $L_2(|P|)$, or $SL_2(|P|)$. (Our definition is slightly more restrictive than that of Aschbacher.) The generic examples are the transvection subgroups of $Sp_{2n}(q)$, $U_n(q)$, $O^{\varepsilon}_{2n}(2)$ (as well as the transposition subgroups of S_n), and as such the study of these groups has a very geometric flavor.

Finite groups G generated by a conjugacy class of p-transvection subgroups have been classified under various further restrictions. If $|P| = 2$, then G is precisely a group generated by a class of 3-transpositions in the sense of Fischer [7]. More generally it is easy to see that if $p = 2$, then G is generated by a class of odd transpositions as defined and studied by Aschbacher [1]. Aschbacher and Hall [4] considered the case $|P| = 3$. Finally Aschbacher [2] studied a slightly more general situation for all p at least 5. (Stark's abstract approach to quadratic pairs [13] is also related.) All the classifications provided in these studies involve in one form or another the assumption that G has no normal solvable subgroup. The work presented here is in part concerned with replacing that hypothesis and perhaps avoiding it altogether. One motivation for the attempt is the observation that such a hypothesis, although natural group theoretically, has no elementary geometric interpretation.

An important step in most of these investigations is the determination of subgroups generated by three conjugates of P. More specifically, let $H = \langle P_1, P_2 \rangle$, for $P_1, P_2 \in P^G$, with $H/Z(H) \simeq L_2(q)$ where $q = |P|$. For $P_3 \in P^G$ with P_3 not contained in H, what are the possibilities for $K = \langle P_1, P_2, P_3 \rangle$? A degenerate possibility is

*Partial support provided by the National Science Foundation.

$K = H \times P_3$. Two other possibilities are likely to occur:

a) $K/Z(K) \simeq E_{q^2} \rtimes SL_2(q)$

b) $K/Z(K) \simeq U_3(q)$.

In case a) we write more suggestively $K/Z(K) \simeq Sp_3(q)$. Frequently these three possibilities are the only ones which can occur. This is the case for 3-transposition groups ([7], (1.6)) and also is true in the generic examples given previously. There are however some exceptions. For $|P| = 3$, the only other possibility is that $K/Z(K)$ is a Frobenius group of order 48 ([4], Sec. 3).

2 PARTIAL LINEAR SPACES WHOSE NON-DEGENERATE PLANES ARE SYMPLECTIC

A <u>partial linear space</u> (X,A) is for our purposes a finite set of points X and a collection A of subsets of X, called lines, each line containing at least two points and no two lines intersecting in as many as two points. (X,A) is <u>connected</u> if it is not possible to write X as the disjoint union of two non-empty subsets X_1 and X_2 where no point of X_1 is collinear with any point of X_2.

A <u>subspace</u> (Y,B) of (X,A) is a partial linear space (Y,B) with $Y \subseteq X$ and $B \subseteq A$ such that, whenever a line ℓ of A meets Y in two or more points, then indeed ℓ is a subset of Y and $\ell \in B$. The subspace <u>generated</u> by the subsets $Y' \subseteq X$ and $B' \subseteq A$ is the smallest subspace (Y,B) of (X,A) with $Y' \subseteq Y$ and $B' \subseteq B$. We are particularly interested in the subspaces of (X,A) generated by a line ℓ and a point p not on ℓ. Such a subspace is called a <u>plane</u> of (X,A) and is <u>degenerate</u> if it contains no line other than ℓ and no points other than p and those of ℓ.

If G is a finite group generated by a conjugacy class of p-transvection subgroups P^G, then we may construct a partial linear space (X,A) from G by taking $X = P^G$ and a line of A to be $P^G \cap H$ for some $H = \langle P_1,P_2 \rangle$ with $P_1,P_2 \in P^G$ and $H/Z(H) \simeq L_2(|P|)$. Every line thus contains $q + 1$ points where $q = |P|$. Notice that P_1 and P_2 of P^G are collinear in (X,A) if and only if they do not commute. In particular for a point p of X and a line ℓ of A not containing p, p is collinear with 0, q, or $q + 1$ of the points of ℓ. The first case occurs precisely when the plane π generated by p and ℓ is degenerate.

Suppose $\ell = P^G \cap H$ for $H = \langle P_1, P_2 \rangle$ and $p = P_3$. Then the isomorphism type of the plane π generated by p and ℓ is determined by $K/Z(K)$ where $K = \langle P_1, P_2, P_3 \rangle$ is the subgroup discussed in Section 1. π is degenerate precisely when $K = H \times P_3$. The other two natural structures for K produce nice planar subspaces π of the associated spaces. If $K/Z(K) \simeq U_3(q)$, then π is the classical unital associated with this group, a linear space (i.e., each pair of points is collinear) on $q^3 + 1$ points with each line containing $q + 1$ points. If $K/Z(K) \simeq Sp_3(q)$, then π is the partial linear space of hyperbolic lines in the associated projective symplectic space (whose radical is a point r). This projective space is in fact a projective plane, (S,L), and π can equally well be described as the partial linear space whose point set is $S - \{r\}$ and whose line set is $L - \{\text{all lines through } r\}$.

Our first theorem is a partial converse to the observation that groups generated by a conjugacy class of p-transvection subgroups lead to partial linear spaces whose planar subspaces are nice.

<u>Theorem 1</u>. Let (X,A) be a connected partial linear space containing more than one plane and such that

> (*) every non-degenerate plane is isomorphic to a projective plane with a point and all lines through it removed.

Let G be the automorphism group of (X,A). Then, for each $x \in X$, there is a non-trivial P_x normal in the stabilizer G_x such that $\{P_x | x \in X\}$ is a conjugacy class of p-transvection subgroups of order q. Here $q + 1$ is the number of points on each line of (X,A) and q is a power of the prime p. If π is a non-degenerate plane of (X,A) then, for $K = \langle P_x | x \in \pi \rangle$, $K/Z(K) \simeq Sp_3(q)$.

The proof has two basic parts. First, for each $x \in X$, permutations of X which are candidates for the members of P_x are found. Next, for each such x and each $\ell \in A$, a subspace of (X,A) containing both x and ℓ is found for which the restriction of each candidate is indeed an automorphism. This proves the candidate permutations to be genuine automorphisms of (X,A), and the rest follows easily. If all lines of A have exactly three points (in any case the lines size must be constant by (*) and connectivity), then this program is easy to follow. The non-identity element of P_x must fix all points not collinear with x

and interchange pairs of points on a line with x. The subspace in question is then the plane generated by x and ℓ. (Here (X,A) is in fact a Fischer space in the terminology of Buekenhout [5].) If lines contain $q + 1 > 3$ points, the program is a little more difficult to follow. By a result of Hall [11] we may factor out of (X,A) an appropriate equivalence relation to find a partial linear space (X^*,A^*) which still satisfies (*) and indeed is isomorphic to the space of hyperbolic lines of the non-degenerate projective symplectic space associated with $Sp_{2n}(q)$, for some $n \geq 1$. If $n = 1$, then the conclusion of Theorem 1 follows from a theorem of Thas and DeClerck [14]. If $n > 1$, then the permutation candidates are found by "extending" the transvections of (X^*,A^*) to (X,A). The result of Thas and DeClerck then provides the subspaces within which the candidates can be checked.

3 THE CLASSIFICATION OF CERTAIN GROUPS GENERATED BY p-TRANSVECTION SUBGROUPS

In this section G is a center-free finite group generated by a conjugacy class P^G of p-transvection subgroups of order q. The group theoretic analogue of hypothesis (*) of the previous section is

(**) Let $P_1,P_2,P_3 \in P^G$ with $H = \langle P_1,P_2\rangle \neq \langle P_1,P_2,P_3\rangle = K$ and $H/Z(H) \simeq L_2(q)$. Then either $K = H\times P_3$ or $K/Z(K) \simeq Sp_3(q)$.

Theorem 2. Assuming (**) there is a short exact sequence

$$(***)\ 1 \to V \to G \to G^* \to 1$$

with G^* one of $Sp_{2n}(q)$, (or $PSp_{2n}(q)$ if $V = 1$), $O^{\varepsilon}_{2n}(2)$, or S_m (in these last two cases $q = 2$). V is a direct sum of natural irreducible GF(q)-transvection modules for G^*.

Theorem 3. If $q = 2$ or q is odd, then (***) splits and a complement G^* to V is generated by conjugates of P.

The splitting of (***) for q odd is immediate as $-I \in Sp_{2n}(q)$. For q even, $H^2(G^*,V)$ is in general non-zero (see [6,8,12]); but non-split extensions should not be generated by p-transvection subgroups. We have proven this in the 3-transposition case $q = 2$. This is perhaps the most interesting case as the greatest variety of groups appears. Two different approaches will likely prove Theorem 3 for q even at least 4 as well. Sah [12] has shown that, for q even at least 4, H^2 of $Sp_{2n}(q)$ on its natural dimension $2n$ module has dimension 1, and furthermore Sah has given a generator. (Sah unfortunately excludes the cases $q = 4$, $2 \leq n \leq 5$.) A careful check of Sah's result would presumably give Theorem 3 in all appropriate cases. A geometric proof of Theorem 3 can probably be given using results of Teirlinck [15] provided $q \geq 4$. Teirlinck's theorem should provide a canonical embedding of (X,A) in a (possibly degenerate) projective symplectic space, and an appropriate section of that space then would exhibit the splitting.

4 REMARKS

The next clear step is to complete a proof of Theorem 3 for all q. This would result, among other things, in a complete classification of all partial linear spaces with (*). This problem was first considered by Hale [9,10]. He states a classification in the case $q = 2$ (although a complete proof was never published). As it stands, Theorems 1, 2, and 3 provide a classification for all odd q and $q = 2$. The obvious geometric generalization of the situation considered here is to allow unitals as planes as well. A classification would perhaps be quite difficult. The case $q = 2$ is equivalent to the classification of all finite center-free 3-transposition groups. For $q > 2$, no version of Theorem 1 is known at present, although it seems likely that the only examples would be the expected symplectic and unitary geometries.

We have already mentioned the motivating group theoretic problem, the classification of groups generated by p-transvection subgroups allowing solvable normal subgroups. In Theorems 2 and 3 this is accomplished by adding in the assumption (**), and it remains up in the air as to what restrictions if any must be made. It would be very helpful to know all groups generated by three conjugate p-transvection subgroups. This appears to be known exactly only in the cases $q = 2,3$ [4,7]. With the 3-generated case in hand, complete classification becomes possible (as suggested by Theorems 2 and 3). For example, starting from the results discussed here, we are close to a classification of those center-free groups G generated by a class of 3-transpositions and having $F^*(G)$ not a 3-group.

For other applications and classifications in finite group theory, it would probably be better to weaken our hypotheses. The p-transvection requirement that a p-subgroup Q generated by two conjugates of P be abelian should be replaced by the root subgroup condition that such a Q must be a homomorphic image of an extraspecial group of order q^3. Aschbacher [3] and Timmesfeld [16] have encountered groups generated by root-type subgroups as sections of characteristic 2-type groups. Results, such as those discussed here, concerning groups generated by root-type subgroups allowing normal 2-subgroups might be useful in that context. It is also true that for groups generated by root-type or p-transvection subgroups the allowing of normal solvable subgroups makes induction easier. Thus it is possible that this weaker hypothesis might in the long run lead to easier proofs of some of the known classification theorems.

Theorem 3 for even q focuses on some interesting cohomological questions. What are the cohomology groups not calculated by Sah? Presumably even in those cases the dimension of H^2 equals the known lower bound of 1. In situations where H^2 is non-zero, how much must be known about a specific extension (e.g., splitting when restricted to factors of a Levi complement) to ensure its splitting? Such questions obviously remain interesting outside the present context.

Added in Proof: In the meantime we have a proof (using neither of the suggested methods) that (***) splits always.

REFERENCES

1. Aschbacher, M. (1972). On finite groups generated by odd transpositions I, II, III, IV. Math. Z. 127, 45-56; (1973). J. Alg. 26, 451-459, 460-478, 479-491.
2. Aschbacher, M. (1973). A characterization of the unitary and symplectic groups over finite fields of characteristic at least 5. Pac. J. Math. 47, 5-26.
3. Aschbacher, M. (1981). Weak closure in groups of even characteristic. J. Alg. 70, 561-627.
4. Aschbacher, M. & Hall, M. Jr. (1973). Groups generated by a class of elements of order 3. J. Alg. 24, 591-612.
5. Buekenhout, F. La géométrie des groupes de Fischer. Unpublished.
6. Dempwolff, U. (1974). Extensions of elementary abelian groups of order 2^{2n} by $S_{2n}(2)$ and the degree 2-cohomology of $S_{2n}(2)$. Ill. J. Math. 18, 45-1468.
7. Fischer, B. (1971). Finite groups generated by 3-transpositions. Invent. Math. 13, 232-246.

8. Griess, R.L. Jr. (1973). Automorphisms of extra special groups and nonvanishing degree 2 cohomology. Pac. J. Math. 48, 403-422.
9. Hale, M. Jr. (1976). Locally dual affine geometries. In: Proceedings of the Conference on Finite Groups, Park City 1975, eds., W.R. Scott and F. Gross. New York: Academic Press, 513-518.
10. Hale, M. Jr. (1977). Finite geometries which contain dual affine planes. J. Combin. Th. (A) 22, 83-91.
11. Hall, J.I. (1982). Classifying copolar spaces and graphs. Quart. J. Math. Oxford (2) 33, 421-449.
12. Sah, C.-H. (1977). Cohomology of split group extensions II. J. Alg. 45, 17-68.
13. Stark, B.S. (1977). Another look at Thompson's quadratic pairs. J. Alg. 45, 334-342.
14. Thas, J.A. & DeClerck, F. (1977). Partial geometries satisfying the axiom of Pasch. Simon Stevin 51, 123-137.
15. Teirlinck, L. Combinatorial Structures. Thesis. Free University of Brussels.
16. Timmesfeld, F. (1975). Groups with weakly closed TI-subgroups. Math. Z. 143, 243-278.

FINITE SIMPLE GROUPS VIA p-ADIC GROUPS

W.M. Kantor*
University of Oregon

In this note we will describe a recent approach to the study of the following situation.

(*) G is a finite group generated by a family $\{P_1,\ldots,P_r\}$ of subgroups, $r \geq 3$, such that the following hold for some prime p and all i, j for which $1 \leq i < j \leq r$:

(1) $O^{p'}(P_i/O_p(P_i))$ is a rank 1 Chevalley group of characteristic p;

(2) $B = \bigcap P_i$ projects onto a Borel subgroup of each group in (1); and

(3) $O^{p'}(\langle P_i,P_j\rangle/O_p(\langle P_i,P_j\rangle))$ is either a rank 2 Chevalley group of characteristic p, or the product of the projections of $O^{p'}(P_i)$ and $O^{p'}(P_j)$.

The main theorem concerning (*) is as follows.

<u>Theorem 1</u> (Niles [9]). Assume (*) together with

(i) No Chevalley group in (1) is $A_1(2)$, $A_1(3)$, ${}^2A_2(4)$, ${}^2B_2(2)$ or ${}^2G_2(3)$,

(ii) No group in (3) is $A_2(4)$, and

(iii) Each product arising in (3) is a direct product.

Then $\langle O^{p'}(P_i) \mid 1 \leq i \leq r\rangle$ is a normal subgroup of G having a rank r BN-pair.

Clearly, the results of Tits [15] then determine G modulo $K = \bigcap\{B^g \mid g \in G\}$, if the associated Dynkin diagram is connected.

While the assumptions in (*) are natural, the additional ones in Theorem 1 are unfortunate, and even annoying. However, there are examples showing that the groups P_i in (*) do not necessarily produce BN-pairs unless some kind of additional assumption is made. Before describing these examples, we will need additional notation.

*This research was supported in part by NSF grant MCS 7903130-82.

The diagram of (*) is the obvious analogue of the Dynkin diagram of a group with a BN-pair. Namely, its nodes can be identified with the groups P_i, and two nodes P_i, P_j are joined by 0, 1, 2, 3 or 4 edges when the group in (3) corresponds to a rank 2 BN-pair with Weyl group of order 4, 6, 8, 12 or 16, respectively. The groups P_i are regarded as "minimal parabolic subgroups", while the r groups $M_i = \langle P_j | j \neq i \rangle$, $1 \leq i \leq r$, are "maximal parabolic subgroups".

The following Table contains essentially all of the examples of (*), other than those arising from BN-pairs, presently in print or at least in preprints. (There are even more examples that have been announced, some due to myself but most due to Köhler, Meixner and Wester. Here, "essentially" refers to the omission of some extensions and homomorphic images. Examples of this are groups $\Omega^{\pm}(6,m)$ and $\Omega(5,7)$ in row 8.) The Table only contains the groups M_i; in each case, the groups P_i can be determined using (2). When several M_i are isomorphic, only one of them is listed.

In row 1, the middle group M_i has $M_i/O_5(M_i) = 2(PSL(2,5)PSL(2,5))4 \cong 2A_6 4$. This shows that hypothesis (iii) of Theorem 1 is essential. On the other hand, the remaining examples in the Table show that some of the groups $P_i/O_p(P_i)$ can be $PSL(2,2)$ or $PSL(2,3)$ without $\{P_1,\ldots,P_r\}$ arising from a BN-pair: at least part of hypothesis (i) is essential.

The first 5 columns of the Table are self-explanatory. The last column will be discussed later. The next-to-last column is headed by an "m", which is either definitely or seemingly irrelevant (denoted "-") or stands for all the indicated primes (e.g., all except 2 in the second row). In the latter case, each m produces an example: there is then an infinite family of examples, and for all but the first few values m does not divide any order $|M_i|$. Thus, Niles' theorem fails dramatically when there is a parameter m. It should be noted that m actually does not have to be a prime in any of these cases; for example, in the second row m can be any odd integer > 1, and G then is $P\Omega^+(8,\mathbf{Z}/(m))$, which is far from simple when m is composite.

The references in the Table provide constructions. We next turn to the problem of classification. Here, the main results are due to Timmesfeld. Before stating them, we will add one further assumption to (*):

4) The diagram of $\{P_1,\ldots,P_r\}$ is connected.

TABLE

Diagram	G	p	Ref.	Maximal parabolics	m	Universal 2-cover
	LyS	5	[5]	$G_2(5), 5^{1+4}2A_6 4, 5^3 SL(3,5)$	-	?
	$P\Omega^+(8,m)$	2	[6] [2] if m = 3	$\Omega^+(8,2), SL(2,3)^4 2^3$	≠ 2	$P\Omega^+(8,\mathbb{Q}_2)$
	$\Omega(7,m)$	2	"	$2^6 A_7, Sp(6,2), (SL(2,3)^2 \times A_4)2^2$	≠ 2	$P\Omega(7,\mathbb{Q}_2)$
	$\Omega(7,m)$	2	"	$2^6 A_7, Sp(6,2), P\Omega^-(6,3)2$	≠ 2	??
	A_7	2	[8,2]	$A_6, (A_4 \times 3)2, PSL(3,2)$	-	Itself
	$P\Omega^{\pm}(6,m)$	2	[6] [11,5,2] if m = 3	$2^4 A_6, SL(2,3)^2 2^2$	≠ 2	$P\Omega(\mathbb{Q}_2,f)$
	$G_2(m)$	2	[6] [4] if m = 3	$G_2(2), SL(2,3)^2 2, 2^3 SL(3,2)$	≠ 2	$G_2(\mathbb{Q}_2)$
	$P\Omega^{\pm}(6,m)$ $7^5\Omega(5,7)$ if m = 7	2	[7] [2] if m = 3	A_7	≠ 2	$P\Omega(\mathbb{Q}_2,f)$

TABLE (contd.)

Diagram	G	p	Ref.	Maximal parabolics	m	Universal 2-cover
	PSU(6,m)	3	[7] [5] if m = 2	$P\Omega^-(6,3)3^{1+4}2^{1+4}S_3$	≠ 3	?
	PSU(5,m) $2^{1+8}SU(4,2)$ if m = 2	3	[7]	$\Omega(5,3),3^{1+2}2^{1+4}3$	≠ 3	?
	Suz	2	[11]	$2^4 2^6 3A_6, 2^2 2^8(S_3 \times A_5), PSU(4,2)$	–	?
	$\Omega^+(8,2)$	3	[5]	$(3\times PSp(4,3))2, 3^4 2^3 S_4$	–	?
	$G_2(2)$	2	[2]	$G_2(2), SL(2,3)^2 2$	–	?
	M^c	2	[12]	$2^4 A_7, 2A_8, PSU(4,3)$	–	??
	PSU(3,5)	2	[8,6]	A_7	–	??

Theorem 2 (Timmesfeld [14]). If $r = 3$ then the diagram has a pair of nonadjacent vertices.

Theorem 3 (Timmesfeld [13]). If the diagram has no multiple edges, then either $\{P_1,\ldots,P_r\}$ arises from a BN-pair, or else the diagram is ⨯ or □ and the M_i are exactly as in the Table.

The remainder of this note is concerned with the problem of going a bit further than Theorem 3.

Theorem 4. In Theorem 3, if G is simple then $G \cong P\Omega^+(8,m)$, $\Omega(5,7)$, or $P\Omega^{\pm}(6,m)$ for some odd prime m. (Here, if $m \neq 7$ then $\Omega^+(6,m)$ occurs if and only if $m \equiv 1$, 2, or 4 (mod 7).)

However, we will only outline the proof. On the other hand it will be clear that the ideas involved go far beyond Theorem 4.

Assume (*), and consider the set G/B. The groups P_i determine natural equivalence relations on G/B (namely, $Bg \underset{i}{\equiv} Bh \Longleftrightarrow P_ig = P_ih$). These turn G/B into a chamber system Δ (Tits [16]). In [16] it is shown that there is a universal 2-cover $\tilde{\Delta}$ of this chamber system, arising from a group $\tilde{G}$ generated by a family $\{\tilde{P}_1,\ldots,\tilde{P}_r\}$ of subgroups, such that there is an epimorphism $\sigma:\tilde{G} \to G$ mapping $\tilde{P}_i$ isomorphically onto P_i (with a similar statement for the groups $\langle\tilde{P}_i,\tilde{P}_j\rangle$). Moreover, the map σ has a standard type of universal property. Tits [16] then goes on to show that the chamber system $\tilde{\Delta}$ produced by $\tilde{G}$ and the $\tilde{P}_i$ "is" a building, provided that each subdiagram •══• of the chamber system G/B arises from a rank 3 Chevalley group.

Digression. The fourth, fifth, and last two rows of the Table contain examples in which subdiagrams •══• do not arise from Chevalley groups. The corresponding chamber systems $\tilde{\Delta}$ cannot arise from buildings; and in the fifth row, it is known that $\tilde{\Delta} = \Delta$ (due to Ronan).

Now consider the case where $\tilde{\Delta}$ is a building and the diagram of Δ (and $\tilde{\Delta}$) is an extended Dynkin diagram. In this case, the building $\tilde{\Delta}$ is an affine building. If $r \geq 4$, then Tits [17] has classified all of these: they arise (via [3]) from groups over complete local fields (including the possibility of skew fields). Since $\tilde{G} \leq \mathrm{Aut}\,\tilde{\Delta}$, and the latter group is $\mathrm{Aut}\,G^*$ for a suitable algebraic group G^*, more information can be obtained concerning $\tilde{G}$.

Consider the case  of Theorem 3. Here, all of Tits' results apply, and show that $\tilde{\Delta}$ arises from the affine building associated to a group $D_4(F)$ for a complete local field F. Moreover, since $\sigma:\langle\tilde{P}_i,\tilde{P}_j,\tilde{P}_k\rangle \to \langle P_i,P_j,P_k\rangle$ is a cover, it is an isomorphism. Thus, each M_i is isomorphic to a group $\tilde{M}_i = \langle\tilde{P}_j | j \neq i\rangle \leq \tilde{G} \leq \mathrm{Aut}\, D_4(F)$. Moreover, the residue field of F must be $GF(2)$.

Since we know that $\tilde{M}_i \cong \tilde{\Omega}^+(8,2)$ for 4 values of i, this places restrictions on F. In fact, it is easy to show that F has characteristic 0, and that $\tilde{G} \leq D_4(F)$. It is only slightly harder to use two groups $\tilde{M}_i$ to show that (with respect to a suitable basis) $\tilde{G} \leq D_4(\mathbb{Q}_2)$, and then even that $\tilde{G} = P\Omega(\mathbb{Z}[1/2], \sum_1^8 x_i^2)$. This is, in fact, the situation encountered in [6].

Finally, G must be a finite simple homomorphic image of $\tilde{G}$. In the course of his work on the Congruence Subgroup Problem, Prasad [10] has shown that the only such finite images are $P\Omega^+(8,m)$ with m prime.

The case is similar. In fact, Theorem 3 can be proved in this manner when $r \geq 4$ by first reducing to the case of extended Dynkin diagrams. (The case $r = 3$ is fairly easy.) Moreover, similar results can be proved for the case of all extended Dynkin diagrams of rank $r \geq 4$, provided that all B_3 subdiagrams arise from buildings. In a rather different direction, Aschbacher [1] has classified all situations (*) with $K = 1$ whose diagrams are diagrams of spherical buildings: he showed that only BN-pairs and the A_7 example (row 5) can arise.

There are many open problems, some of which are implicit in the above discussion. Others concern the last column of the Table. That column involves the identification of the universal 2-cover of the example or family of examples. In some cases, this cover is the building $\tilde{\Delta}$ associated with a 2-adic group. The corresponding group $\tilde{G}$ is then flag-transitive on $\tilde{\Delta}$, but is much smaller than $\mathrm{Aut}\,\tilde{\Delta}$ (just as in the  situation discussed earlier). However, usually the universal 2-cover is not known, in which case there is a question mark. Two question marks indicate that the universal 2-cover is not a building (in all other cases it is a building).

REFERENCES

1. Aschbacher, M. Finite geometries of type C_3 with flag transitive groups. (to appear).
2. Aschbacher, M. & Smith, S.D. Tits geometries over GF(2) defined by groups over GF(3). (to appear).
3. Bruhat, F. & Tits, J. (1972). Groupes réductifs sur un corps local. I. Données radicielles valuées. Publ. Math. I.H.E.S. 41, 5-251.
4. Cooperstein, B.N. A finite flag-transitive geometry of extended G_2 type. (to appear).
5. Kantor, W.M. (1981). Some geometries that are almost buildings. Europ. J. Combinatorics 2, 239-247.
6. Kantor, W.M. Some exceptional 2-adic buildings. (to appear in J. Algebra).
7. Kantor, W.M. Some locally finite flag-transitive buildings. (to appear in Europ. J. Combinatorics).
8. Neumaier, A. (unpublished).
9. Niles, R. (1982). BN-pairs and finite groups with parabolic-type subgroups. J. Alg. 75, 484-494.
10. Prasad, G. (unpublished).
11. Ronan, M.A. & Smith, S.D. (1980). 2-local geometries for some sporadic groups. AMS Proc. Symp. Pure Math. 37, 283-289.
12. Ronan, M.A. & Stroth, G. Sylow geometries for the sporadic groups. (to appear).
13. Timmesfeld, F. Tits geometries and parabolic systems in finite groups. (to appear).
14. Timmesfeld, F. Tits geometries and parabolic systems of rank 3. (to appear).
15. Tits, J. (1974). Buildings of spherical type and finite BN-pairs. Springer Lecture Notes 386.
16. Tits, J. (1981). A local approach to buildings. In The Geometric Vein. The Coxeter Festschrift. Springer, New York-Heidelberg-Berlin, pp. 519-547.
17. Tits, J. (unpublished).

EMBEDDINGS OF GROUP GEOMETRIES AND MODULAR REPRESENTATIONS

M.A. Ronan*
Mathematics Department, University of Illinois at Chicago
Chicago, IL 60680

Let G be a finite group having a linear representation on some vector space V. By taking various orbits of G on the set of subspaces of V, one can construct a combinatorial object -- a geometry -- on which G acts as a group of automorphisms. This article, which is mainly an introduction to some joint work with S.D. Smith [3, 4, 5], is concerned with the reverse operation, namely a method of obtaining G-modules V, given some geometry for the group in question.

By a <u>geometry</u> Δ, we shall mean a <u>chamber system</u> [8] of rank n over the indexing set $I = \{1,\ldots,n\}$, and we shall assume the action of a group G of automorphisms which is transitive on the set of chambers. More loosely, the reader may think of a cell complex of dimension $n-1$, built out of simplexes; the vertices of each simplex have distinct <u>types</u> $i \in I$, and every simplex is a face of a simplex of dimension $n-1$ (i.e. having n vertices), called a <u>chamber</u>. Each chamber c has n faces of codimension 1, $\pi_1,\ldots,\pi_n$, where π_i has type $I\backslash\{i\}$; they are called <u>panels</u>. Every chamber system of finite rank is equivalent to such a cell complex (see [2]).

Since we assume our chamber system arises from a group G, its chambers correspond to cosets of some subgroup B, its panels to cosets of subgroups $P_1,\ldots,P_n$ containing B, and all other simplexes to cosets of subgroups generated by the P_i. The idea is to take modules for B and the P_i, and use these to construct modules for G. This is done by setting up a G-equivariant system of coefficients on Δ (which we call a sheaf), and then taking zero-homology. The basic definitions are the following.

Let k be some field. A <u>sheaf</u> F on Δ (see [3, 4]) assigns to each simplex $\sigma \in \Delta$, a k-vector space F_σ. If τ is a face

*Research partially supported by NSF Grant 8201463.

of σ there is a k-linear connecting map $\phi_{\sigma\tau}:F_\sigma \to F_\tau$ such that $\phi_{\rho\sigma}\phi_{\sigma\tau} = \phi_{\rho\tau}$ whenever this composite map is defined (we write all maps and group actions on the right). The F_σ and $\phi_{\sigma\tau}$ are required to be G-equivariant in the following sense. For each $g \in G$ there is a mapping $\tilde{g}:F \to F$ such that $gh = \tilde{g}\tilde{h}$, $F_{\sigma g} = (F_\sigma)\tilde{g}$, and $\tilde{g}\circ\phi_{\sigma g,\tau g} = \phi_{\sigma\tau}\circ\tilde{g}$. In particular F_σ is a kP_σ-module, where P_σ denotes the stabilizer of σ. The homology groups of Δ with coefficients F are kG-modules. We shall be concerned with $H_0(\Delta;F)$, which we abbreviate to $H_0(F)$.

A Universal Construction. If $\sigma \in \Delta$ is a simplex, the chambers containing σ form a chamber system in their own right, called St σ (the star of σ). We will here be concerned with sheaves F such that F_σ is generated by the $(F_c)\phi_{c\sigma}$ where c ranges over all chambers of St σ. Such sheaves we call chamber generated. For the remainder of this paper c is some fixed chamber having stabilizer B in G, and its panels π_i have stabilizers P_i, where $i = 1,\dots,n$.

Theorem 1 [4]. Let F_c be a kB-module, and F_{π_i} a kP_i-module, and suppose there is a kB-module homomorphism $\phi_{c\pi_i}:F_c \to F_{\pi_i}$ for each $i = 1,\dots,n$. Then there exists a sheaf U with connecting maps $\psi_{\sigma\tau}$, such that $U_c = F_c$, $U_{\pi_i} = F_{\pi_i}$ and $\psi_{c\pi_i} = \phi_{c\pi_i}$ for each i, which is universal in the following sense. If $\mathcal{G}$ is any other chamber-generated sheaf having these properties, then $\mathcal{G}$ is a quotient sheaf of U.

Proof. A detailed proof of this theorem is given in [4]. There are two main steps. Firstly one uses G to define terms F_σ at all chambers and panels, since each one is an image under G of one of $c, \pi_1,\dots,\pi_n$. Secondly one uses an inductive construction to define terms at all simplexes σ, as follows. Assume by induction that we have a sheaf on St σ, call it $U|_\sigma$. Then define U_σ to be $H_0(U|_\sigma)$.

Let us now specialize this construction. Suppose the F_c and F_{π_i} are irreducible kB- and kP_i-modules respectively. Then any chamber-generated sheaf with these terms is called panel-irreducible. In the process of constructing U, one can take a quotient of $H_0(U|_\sigma)$ which is irreducible as a kP_σ-module. Doing this (G-equivariantly of course) at each step, we obtain a sheaf in which every term is irreducible, and we call it an irreducible sheaf.

Theorem 2 [3]. Let G be a Chevalley group over $k = \mathbb{F}_q$, and let V be an irreducible kG-module. Then there is an irreducible sheaf F_V on the building for G, such that $H_0(F_V)$ contains a unique maximal kG-submodule with quotient isomorphic to V.

Using Theorem 1 and the discussion above, one can obtain F_V by starting with the appropriate kP_i-modules F_{π_i} (see the discussion on weight symbols below). However there is also a direct way of obtaining F_V, which is not available in general. This is to associate to each parabolic $P = UL$, the fixed point space V^U. By a theorem of S. Smith [7], V^U is an irreducible kP-module.

Weight Symbols. Given a 1-dimensional kB-module λ, and an irreducible kP_i-module F_i, the existence of a kB-module homomorphism $\phi_{c\pi_i}:\lambda \to F_i$, as in Theorem 1, means that F_i is a quotient of the induced module $\lambda|^{P_i}$ (Frobenius reciprocity). If α_i, $i = 1,\ldots,n$, is a symbol denoting a particular irreducible kP_i-module which is a quotient of $\lambda|^{P_i}$ then we say λ admits the weight symbol $(\alpha_1,\ldots,\alpha_n)$. Consider the example of a Chevalley group, for which $L_i (= P_i/O_p(P_i))$, or L_i', is $SL_2(q)$. The irreducible $\mathbb{F}_q$-modules for $SL_2(q)$ are well-known; they are all of the form $V = V_{j_0} \otimes V_{j_1}^{(\sigma)} \otimes \ldots \otimes V_{j_{n-1}}^{(\sigma^{n-1})}$, where V_{j_i} is the unique irreducible $\mathbb{F}_p$-module of dimension (j_i+1) for $SL_2(p)$, tensored with $\mathbb{F}_q$, and (σ^i) denotes the i-th Frobenius twist ($\sigma \in \mathrm{Gal}(\mathbb{F}_q/\mathbb{F}_q)$). Since j_i is one of $0,\ldots,p-1$, V is uniquely specified by giving the integer $\alpha = j_0+j_1p+\ldots+j_{n-1}p^{n-1}$. If F_i is isomorphic to V as an $SL_2(q)$ module, then we set $\alpha_i = \alpha$. For any such weight symbol $(\alpha_1,\ldots,\alpha_n)$ there exists a kB-module λ admitting it, and for this reason one can construct q^n different types of panel irreducible sheaves (there are q choices for each α_i), thus obtaining the q^n different irreducible kG-modules as quotients.

Now suppose G is any finite simple group with Sylow-p normalizer B. Let $P_1,\ldots,P_n$ be subgroups which are minimal with respect to containing B, and such that $\langle P_1,\ldots,P_n\rangle = G$, but no proper subset of the $\{P_i\}$ generates G. This is a minimal parabolic system in the sense of Ronan-Stroth [6], and gives a chamber system Δ in which B stabilizes a chamber c, and P_i stabilizes its panel π_i of type $I\setminus\{i\}$. Now let us assume that $L_i (= P_i/O_p(P_i))$ or L_i' is $SL_2(q)$ for each $i = 1,\ldots,n$, and set $k = \mathbb{F}_q$ (this includes the case of buildings for (untwisted) Chevalley groups). Given a 1-dimensional kB-module λ which admits the weight symbol $(\alpha_1,\dots,\alpha_n)$ (see above), we obtain kP_i-modules $\mathcal{F}_{\pi_i}$, where $\mathcal{F}_{\pi_i}$ affords the $SL_2(q)$-module corresponding to α_i. This enables us to obtain a sheaf $\mathcal{U}(\alpha_1,\dots,\alpha_n)$ as in Theorem 1. By an extension of Frobenius duality to sheaves (see [3; (1.2)]), a kG-module V is a quotient of $H_0(\mathcal{U}(\alpha_1,\dots,\alpha_n))$ if and only if V itself admits the kB-module λ, the kP_i-modules $\mathcal{F}_{\pi_i}$, and the connecting maps $\phi_{c\pi_i}$. In this case let us say that V admits the weight symbol $(\alpha_1,\dots,\alpha_n)$.

Open Question. Given a finite simple group G, and a minimal parabolic system as above, does every irreducible kB-module admit such a weight symbol?

Here is one "sporadic" example which was investigated in collaboration with S.D. Smith [4, 5]. The alternating group A_7 acts on a sporadic C_3 geometry (Neumaier [1]); its diagram is

points lines planes

o————o════o

α_1 α_2 α_3

The panel groups P_i are all of shape $2^2L_2(2)$, so there are eight ($= 2^3$) potential weight symbols $(\alpha_1,\alpha_2,\alpha_3)$. (The trivial $\mathbb{F}_2$B-module admits all eight in the sense above). It turns out that each of the six irreducible $\mathbb{F}_2A_7$-modules admits a weight symbol. The 20-dimensional irreducible admits two weight symbols (0,1,1) and (1,1,1), and the universal sheaf for one of the other weight symbols (1,0,0) has $H_0 = 0$.

In general one would expect different modules to admit the same weight symbol. This appears to be the case for the minimal parabolic system for M_{24} given in [6], where there are eight weight symbols, but thirteen different irreducible $\mathbb{F}_2M_{24}$-modules (see [5]).

Embeddings of Geometries. For any kG-module V, the constant sheaf $\mathcal{K}_V$ on Δ is defined by setting $(\mathcal{K}_V)_\sigma = V$ for all $\sigma \in \Delta$, and letting each $\phi_{\sigma\tau}$ be the identity. A chamber-generated subsheaf of $\mathcal{K}_V$, whose terms generate V, is called an embedding (see [4]).

<u>Theorem 3</u> [4]. Let Δ be the building for a Chevalley group $G(q)$ of rank $n+1$. For each $i \in I = \{1,\dots,n\}$, there is an embedding of Δ in a vector space over $\mathbb{F}_q$ for which the vertices of type i are 1-spaces, and the panels of type $I\setminus\{i\}$ are 2-spaces.

<u>Proof</u>. The proof is given in [4]. The idea is simply to take a sheaf with weight symbol $(0,\dots,0,1,0,\dots,0)$ (the 1 in the ith-position). This is isomorphic to a subsheaf of K_V where V has the same weight symbol (i.e. is a high weight module for the fundamental "restricted" weight λ_i).

This theorem is false in general. For example if Δ is the sporadic C_3 geometry for A_7 (see diagram above), then there is no embedding of this nature for which the points are 1-spaces. This is not surprising because two distinct lines can meet in two distinct points, so the corresponding 2-spaces would coincide, and the embedding would collapse. However, there is a sheaf F for which points, lines, and planes are 1-spaces, 2-spaces, and 3-spaces respectively; it corresponds to the weight symbol $(1,0,0)$. As mentioned above, $H_0(F) = 0$.

<u>Open Problem</u>. Give necessary and sufficient conditions on a geometry Δ and sheaf F such that $H_0(F) \neq 0$.

This is highly relevant to embeddings because if V is a non-zero quotient of $H_0(F)$, then by Frobenius reciprocity [3; (1.2)] F maps onto a subsheaf of K_V, and so some quotient sheaf of F gives an embedding in V.

<u>Universal Embeddings</u>. Suppose E and F are embeddings in kG-modules V and W, and that E and F are isomorphic as sheaves. What can we say about V and W?

<u>Theorem 4</u> [4]. Both V and W are isomorphic to quotients of $H_0(E)$.

This result is straightforward, and allows us to talk of $H_0(E)$ as a <u>universal embedding</u>. Here is an example. Let $G = Sp_{2n}(q) \cong O_{2n+1}(q)$ where q is even, and let Δ be the building for G. Let V denote the 2n-dimensional symplectic module, and W the (2n+1)-dimensional orthogonal module. The totally isotropic subspaces of V give an embedding E of Δ, and the totally singular subspaces of W

an embedding F. As sheaves, E and F are isomorphic. In [3] it is shown that $H_0(E) \cong W$, so this orthogonal embedding is universal.

We conclude with a theorem on minimal weight modules (see [3] for a precise definition of what we mean by this term).

Theorem 5 [3]. Suppose V is a minimal weight module for a Chevalley group G, where $G \ncong Sp_{2n}(q)$ with q even. Then the fixed point sheaf F_V (see above and [3]) is a universal embedding (i.e. $H_0(F_V) \cong V$).

REFERENCES

1. Neumaier, A. Some Sporadic Geometries related to PG(3,2). Archiv. Math. (to appear).
2. Ronan, M.A. (1980). Coverings and Automorphisms of Chamber Systems. Europ. J. Combinatorics 1, 259-269.
3. Ronan, M.A. & Smith, S.D. Sheaves on Buildings and Modular Representations of Chevalley Groups. J. Algebra (to appear).
4. ______. Sheaves on Finite Group Geometries and Modular Representations. Preprint.
5. ______. 2-Local Sheaves and 2-Modular Representations of A_7 and M_{24}. In preparation.
6. Ronan, M.A. & Stroth, G. Minimal Parabolic Geometries for the Sporadic Groups. Europ. J. Combinatorics (to appear).
7. Smith, S.D. (1982). Irreducible Modules and Parabolic Subgroups. J. Algebra 75, 286-289.
8. Tits, J. (1981). A Local Approach to Buildings. In The Geometric Vein (the Coxeter Festschrift), pp. 519-547. Springer-Verlag.

AMALGAMS AND PUSHING UP

P. Rowley
Department of Mathematics, University of Manchester Institute of Science and Technology, Manchester M60 1QD, U.K.

1 INTRODUCTION

In the study of finite simple groups the following kind of question frequently arises:

(1.1) Suppose p is a prime and H is a p-local subgroup of a finite group of G^*. Let $S \in Syl_p H$ and assume that

i) $C_H(O_p(H)) \leq O_p(H)$; and

ii) $N_{G^*}(S) \not\leq H$

Is H contained in a strictly larger p-local subgroup of G^*? (i.e. can H be "pushed up"?)

Observe that if S contains a non-trivial characteristic subgroup which is normal in H, then it is easy to see that H can be pushed up. However, there are many important cases where (1.1) does not have an affirmative answer. Since we may approach (1.1) by attempting to catalogue the counterexamples, the above observation leads us to considering the following problem.

(1.2) Let H be a finite group, p a prime and $S \in Syl_p H$. Assume that

i) $C_H(O_p(H)) \leq O_p(H)$; and

ii) S does not contain any non-trivial characteristic subgroups which are normal in H.

Determine the structure of S and the non-central chief factors of H in $O_p(H)$.

The following theorem, proved by Baumann (for $p = 2$) [1] and Niles [5], successfully solves (1.2) when $H/O_p(H) \cong SL_2(q)$, $q = p^n$.

Theorem (1.3). Suppose H and S are as in (1.2). If $H/O_p(H) \cong SL_2(q)$, $q = p^n$, then either

i) S has nilpotence class 2 and H has one non-central chief factor in $O_p(H)$; or

ii) S has nilpotence class 3, $p = 3$ and H has exactly two non-central chief factors in $O_p(H)$.

(For more information on this configuration see [5].)

The proofs of Baumann and Niles, though quite different, take as their starting point the fact that $Z(S) \not\trianglelefteq H$ and $J_e(S) \not\trianglelefteq H$ and are completely "group theoretic". Here we outline another way of viewing (1.2) which is somewhat more geometric.

Let H and S be as in (1.2), and let K denote that semi-direct product of S with Aut S. There are obvious group monomorphisms ϕ_1, ϕ_2 such that

$$H \xleftarrow{\phi_1} S \xrightarrow{\phi_2} K$$

(in the terminology of Serre [6] we have an amalgam). Set $G = H *_S K$, the amalgamated product of H and K over S and identify H, S and K with their images in G. Now let Γ denote the bipartite graph whose vertices are given by the union of (right) cosets

$$G/H \cup G/K$$

with vertices Hx and Ky joined whenever $Hx \cap Ky \neq \phi$. Making G act up on Γ by right multiplication we see that

(1.4) condition (1.2)(ii) is equivalent to G acting faithfully on Γ.

This alternative view of (1.2) arose in discussions with Bernd Stellmacher at the Rutgers Group Theory Conference and very recently he has succeeded, using this approach, in giving a short proof of (1.3) (see [7]).

It would be desirable to be able to solve (1.2) when $H/O_p(H)$ is a rank 2 Chevalley group of characteristic p. However no general results are presently available for this case. In the following section we will present a result taken from current work on the case $H/O_p(H) \cong SL_3(q)(q = 2^n)$ which shows the "amalgam method" in action.

2 THE AMALGAM METHOD

Let $G = H *_S K$ and Γ be as described in section 1 and suppose $H/O_p(H) \cong SL_3(q)$, $q = 2^n$. Put $\alpha_1 = H$ and $\alpha_2 = K$. So $\alpha_1, \alpha_2 \in \Gamma$ and α_1 and α_2 are adjacent (we write $\alpha \in \Gamma$ to mean α is a vertex of Γ). $d(\, , \,)$ denotes the obvious distance function Γ and, for $\alpha \in \Gamma$, $\Delta(\alpha) = \{\beta | d(\alpha,\beta) = 1\}$ and $Q_\alpha = O_p(G_\alpha)$.

The following properties of Γ are proved in [4].

(2.1) i) Γ is a tree.

ii) G has one orbit on the edges of Γ and two orbits on the vertices of Γ, namely $0 = \{\alpha_1^G\}$ and $X = \{\alpha_2^G\}$.

iii) $G_{\alpha_1} = H$, $G_{\alpha_2} = K$ and $G_{\alpha_1\alpha_2} = S$.

iv) For $\alpha \in \Gamma$, G_α is transitive on $\Delta(\alpha)$.

<u>Remark</u>. Since $S \trianglelefteq K$, (2.1)(iii) and (iv) imply $S = G_{\alpha\alpha_2}$ for all $\alpha \in \Delta(\alpha_2)$.

For $\alpha \in 0$ we define

$$Z_\alpha = \langle \Omega_1(Z(G_{\alpha\beta})) | \beta \in \Delta(\alpha) \rangle \quad \text{and}$$

$$b = \min_{\delta \in \Gamma} \{d(\alpha,\delta) | Z_\alpha \not\leq Q_\delta\}.$$

<u>Definition (2.2)</u>. For $\alpha,\alpha' \in \Gamma$ with $\alpha \in 0$ we say (α,α') is a critical pair if $d(\alpha,\alpha') = b$ and $Z_\alpha \not\leq Q_{\alpha'}$.

Note that, by (1.2)(i) $Z_\alpha \leq Z(Q_\alpha)$ and that b is independent of the choice of α in 0. Also, as a consequence of the above remark, b must be even. So in (2.2) we must have $\alpha' \in 0$ and, since $[Z_\alpha, Z_{\alpha'}] = 1$ forces $Z_\alpha \leq Q_{\alpha'}$, we also have $Z_{\alpha'} \not\leq Q_\alpha$. That b is even is fortunate since our hypothesis gives us the structure of G_α/Q_α when $\alpha \in 0$.

Using a critical pair together with properties of the GF(2)-modules of $SL_3(q)$ it can be shown that

(2.3) Let $\alpha \in 0$. Then $[G_\alpha, Z_\alpha]$ is the direct sum of one or two natural $SL_3(q)$-modules and in the latter case they must both be isomorphic. (The standard 3-dimensional $SL_3(q)$-module over $GF(q)$ and its dual, regarded as GF(2)-modules, we will call natural $SL_3(q)$-modules.)

For $\beta \in X$ and $\alpha \in \Delta(\beta)$ we use $Y_{\beta,\alpha}$ to denote $[G_{\beta\alpha}, Z_\alpha]$. By (2.3) and the structure of natural $SL_3(q)$-modules, $|Y_{\beta,\alpha}| = q^2$ or q^4

and $[G_{\beta\alpha}, Y_{\beta\alpha}] \leq Z(G_{\beta\alpha})$.

One final definition and then we prove the result promised in the first section.

Definition (2.4). (α,β,δ) is a generating triple if (α,β) is a critical pair, $d(\beta,\delta) = 2$ and $G_\beta = \langle Z_\alpha, G_{\beta\delta}\rangle$.

(2.5) Suppose $\alpha,\mu,\varepsilon \in 0$ with $d(\alpha,\varepsilon) = d(\mu,\varepsilon) = 2$, $\mu \neq \alpha$ and $\mu = \alpha^g$ for some $g \in G_\varepsilon$. Also suppose (α,β,δ) and $(\lambda,\mu,\varepsilon)$ are generating triples and (ε,δ) is a critical pair. (Set $\{\delta-1\} = \Delta(\beta) \cap \Delta(\delta)$ and $\{\alpha+1\} = \Delta(\alpha) \cap \Delta(\varepsilon)$.) If $b > 4$ and $d(\beta,\gamma) < b$, then either

i) $Y_{\delta-1,\delta} \trianglelefteq G_\beta$; or

ii) $Z_\varepsilon Y_{\alpha+1,\alpha} \trianglelefteq G_\varepsilon$.

Proof. It helps to draw a picture (note that μ need not be on the path between α and β):

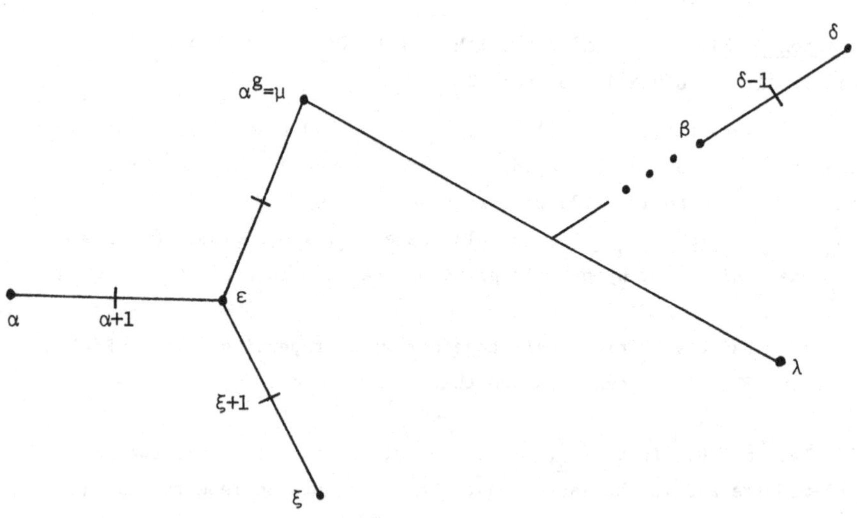

Since (ε,δ) is a critical pair by hypothesis, $Z_\delta Q_\varepsilon/Q_\varepsilon \neq 1$ and so, by the structure of $SL_3(q)$, there exists $h \in G_\varepsilon$ such that

$$G_\varepsilon = \langle Z_\delta, G_{\varepsilon\alpha}h\rangle. \tag{2.5.1}$$

Put $\xi = \alpha^h$ and $\xi+1 = (\alpha+1)^h$. Because (α,β) is a critical pair, $d(\alpha,\beta) = b$, and hence $d(\xi,\beta) = b$. Hence $Z_\xi \leq G_\beta$ and $Z_\beta \leq G_\xi$ and so

$$R = [Y_{\xi+1,\xi}, Z_\beta] \leq [G_\xi, Z_\xi] \cap Z_\beta.$$

Therefore, as $d(\beta,\lambda) < b$, and $b > 2$, we have

$$R \text{ is centralized by } Z_\delta \text{ and } Z_\lambda. \tag{2.5.2}$$

Recalling that $[G_{\xi+1\xi}, Y_{\xi+1,\xi}] \leq Z(G_{\xi+1\xi})$ and that $G_{\xi+1\xi} = G_{\varepsilon\xi}$, (2.5.1) and (2.5.2) give

$$G_\varepsilon = \langle Z_\delta, G_{\varepsilon\xi}\rangle \leq C_{G_\varepsilon}(R),$$

and so $R \leq Z(G_\varepsilon)$. In particular $R \leq Z(G_{\varepsilon\mu})$ and then (2.5.2) and the assumption that $(\lambda,\mu,\varepsilon)$ is a generating triple yield $R \leq Z(G_\mu)$. But then

$$R = R^{g^{-1}h} \leq Z(G_\mu^{g^{-1}h}) = Z(G_\xi).$$

Since $R \leq [G_\xi, Z_\xi]$, (2.3) implies $R = 1$. Thus $[Y_{\xi+1,\xi}, Z_\beta] = 1$ and therefore, as $Y_{\xi+1,\xi} \leq G_\beta$, we obtain $Y_{\xi+1,\xi} \leq Q_\beta$. In particular $Y_{\xi+1,\xi} \leq G_{\beta\delta} \leq G_\delta$.

Now $[Z_\varepsilon, Z_\delta] \neq 1$ and so from (2.3) and properties of natural $SL_3(q)$-modules we deduce that either

A) $\langle [Z_\varepsilon, Z_\delta], [Y_{\xi+1,\xi}, Z_\delta]\rangle = Y_{\delta-1,\delta}$; or

B) $[Y_{\xi+1,\xi}, Z_\delta] \leq [Z_\varepsilon, Z_\delta]$.

If (A) holds, then $Y_{\delta-1,\delta} \leq Z_\varepsilon < Z_\alpha^x | x \in G_\varepsilon\rangle = W$. Because $b > 4$, W is abelian and so Z_α centralizes $Y_{\delta-1,\delta}$. Therefore $Y_{\delta-1,\delta}$ is normalized by $\langle Z_\alpha, G_{\beta\delta}\rangle$ which, since (α,β,δ) is a generating triple, yields (i). In the latter case we have

$$[Y_{\xi+1,\xi}, Z_\delta] \leq [Z_\varepsilon, Z_\xi] \leq Z_\varepsilon$$

and hence $Z_\varepsilon Y_{\xi+1,\xi}$ is normalized by $\langle Z_\delta, G_{\varepsilon\xi}\rangle$. Using (2.5.1) gives $Z_\varepsilon Y_{\alpha+1,\alpha} = (Z_\varepsilon Y_{\xi+1,\xi})^{h^{-1}} \trianglelefteq G_\varepsilon$, which completes the proof of (2.5).

3 OTHER PUSHING UP PROBLEMS

Which other pushing up problems might succumb to the amalgam method? Here we list some possible candidates.

(3.1) The case when $H/O_p(H)$ is a rank 2 Chevalley group has been mentioned earlier; it seems to me there is reason for optimism about $SL_3(q)$, but I have no idea about $Sp_4(q)$.

(3.2) $H/O_p(H) \cong S_n$ or A_n (Goldschmidt, in unpublished work, analyzed this situation when $n = 2m+1$ and $p = 2$ using methods similar to those of [1].)

(3.3) A related type of pushing up problem for $SL_3(q)$, $q = 2^n$, was studied by Campbell [2]. Let H and S be as in (1.2) with $H/O_p(H) \cong SL_3(q)$, and let $S \leq M \leq H$ be such that $M/O_p(M)$ is a maximal parabolic subgroup of $H/O_p(H)$. Then Campbell considers problem (1.2) with (1.2)(ii) replaced by

> (1.2)(ii)' $O_p(M)$ does not possess any non-trivial characteristic subgroups which are normal in H.

By taking an amalgam over $O_p(M)$ the amalgam method can be used here and perhaps analogous results for $Sp_4(q)$ can be obtained.

We finish with a bit of a long shot.

(3.4) Can the amalgam method be pressed into service to give alternative proofs of the "canonical" characteristic subgroup theorems of Glauberman and Niles [3]?

REFERENCES

1. Baumann, B. (1979). Über endliche Gruppen mit einer $L_2(2^n)$ isomorphen Faktorgruppe. Proc. Amer. Math. Soc. 74, 215-222.

2. Campbell, N. (1979). Pushing up in finite groups. Ph.D. Thesis. California Institute of Technology.
3. Glauberman, G. & Niles, R. (1983). A pair of characteristic subgroups for pushing-up in finite groups. Proc. London Math. Soc. (3), 46, 411-453.
4. Goldschmidt, D.M. (1980). Automorphisms of trivalent graphs. Ann. of Math. 111, 377-406.
5. Niles, R. (1979). Pushing up in finite groups. J. Algebra 57, 26-63.
6. Serre, J.-P. (1977). Arbres, Amalgams, SL_2. Asterisque 46, Societé Mathématique de France.
7. Stellmacher, B. Pushing Up. Preprint.

RANK 2 GROUPS

B. Stellmacher
Universität Bielefeld

In this paper we describe a method for the investigation of p-local subgroups. This method was introduced by D. Goldschmidt in [5], and his paper was fundamental for all investigations in this area.

Under a rank 2 group we want to understand a group G together with two proper subgroups M_1 and M_2 which satisfy:

(A_1) G is generated by M_1 and M_2.

(A_2) No non-trivial normal subgroup of G is in $M_1 \cap M_2$.

In the following this notion of a rank 2 group will successively be restricted by additional hypotheses. All stated hypotheses (A_i) will then hold for the remainder of the paper.

Of course, one cannot expect to get substantial information about the structure of M_1 and M_2 (or G) under these very general hypotheses (A_1) and (A_2). Nevertheless, it is possible to describe parts of the method in this generality.

In the following we assume M_1 and M_2 and $B = M_1 \cap M_2$ to be fixed. Let

$$\Gamma = \{M_1x/x \in G\} \cup \{M_2x/x \in G\}$$

be the coset graph of G with respect to M_1 and M_2, where two vertices M_ix and M_jy are adjacent if and only if

$$M_ix \neq M_jy \quad \text{and} \quad M_ix \cap M_jy \neq \emptyset.$$

G operates on Γ by right multiplication.

<u>Lemma 1</u>. Let δ and λ be adjacent in Γ. Then

a) Γ is bipartite and connected.

b) G operates edge- but not vertex-transitively on Γ.

c) $G \leq \mathrm{Aut}(\Gamma)$, i.e. G acts faithfully on Γ.

d) There exists $g \in G$ with $\{G_\delta^g, G_\lambda^g\} = \{M_1, M_2\}$. (Here G_δ denotes the stabilizer of δ in G.)

e) $G_\delta \cap G_\lambda$ is conjugate to B.

f) G_δ is transitive on $\Delta(\delta)$, the set of vertices adjacent to δ.

<u>Lemma 2</u>. Let δ and λ be adjacent in Γ. For $\nu = \delta,\lambda$ let L_ν be a subgroup of G_ν which is transitive on $\Delta(\nu)$. Then the following hold for $\tilde{G} = \langle L_\delta, L_\lambda\rangle$:

a) $\tilde{G}$ is edge-transitive on Γ.

b) $G = \tilde{G}B$.

c) $\tilde{G}$, L_δ and L_λ fulfill the hypotheses (A_1) and (A_2).

<u>Lemma 3</u>. Let δ and λ be adjacent in Γ, and let N be a subgroup of $G_\delta \cap G_\lambda$ such that $N_{G_\nu}(N)$ is transitive on $\Delta(\nu)$ for $\nu = \delta,\lambda$. Then $N = 1$.

The proofs of Lemmas 1, 2 and 3 are elementary (see also [5]). For the next hypotheses let p be a (once and for all) fixed prime.

(A_3) The subgroups M_1 and M_2 are finite and

$$\mathrm{Syl}_p(B) \subseteq \mathrm{Syl}_p(M_i), \quad i = 1,2.$$

(A_4) $C_{M_i}(O_p(M_i)) \leq O_p(M_i), \quad i = 1,2.$

<u>Notation</u>: Let $d(,)$ be the usual distance metric on Γ and $\delta \in \Gamma$:

$$G_\delta^{(1)} = \bigcap_{\lambda\in\Delta(\delta)} (G_\lambda \cap G_\delta),$$

$$Q_\delta = O_p(G_\delta),$$

$$Z_\delta = \langle \Omega_1(Z(T)) \mid T \in \mathrm{Syl}_p(G_\delta)\rangle,$$

$$V_\delta = \langle Z_\lambda / \lambda \in \Delta(\delta)\rangle,$$

$$b = \min_{\delta\in\Gamma} \{d(\delta,\delta') \mid Z_\delta \not\leq G_{\delta'}^{(1)}\}.$$

Let α and α' be two vertices in Γ with $d(\alpha,\alpha') = b$ and $Z_\alpha \not\leq G_{\alpha'}^{(1)}$; and let γ be a fixed path from α to α' of length b. The vertex in γ of distance i from α (resp. α') we denote by $\alpha+i$ (resp. $\alpha'-i$) such that

$$\gamma = (\alpha,\alpha+1,\ldots,\alpha+b) = (\alpha'-b,\ldots,\alpha'-1,\alpha').$$

<u>Lemma 4</u>. The following hold:

a) $Z_\delta \leq Z(Q_\delta) \cap V_\delta$.

b) $Q_\delta \leq G_\delta^{(1)}$, i.e. $b \geq 1$.

c) $Z_{\alpha'} \leq G_\alpha$ and $[Z_\alpha,Z_{\alpha'}] \leq Z_\alpha \cap Z_{\alpha'}$.

d) $Z_\alpha \neq \Omega_1(Z(T))$, $T \in Syl_p(G_\alpha)$.

<u>Proof</u>: The assertions (a) - (c) are direct consequences of (A_3) and (A_4).

Assume that $Z_\alpha = \Omega_1(Z(T))$, $T \in Syl_p(G_\alpha)$. Since Z_α is normal in G_α, we get by (A_3) that $Z_\alpha \leq Z_{\alpha+1}$ and $Z_{\alpha+1} \not\leq G_{\alpha'}^{(1)}$. But this contradicts the minimality of b, since $d(\alpha+1,\alpha') < b$.

The above defined parameter b is closely connected with the structure of M_1 and M_2. Roughly, for $S \in Syl_p(B)$ it describes the weak closure of $\Omega_1(Z(S))$ in M_1 and M_2. We give an easy example.

<u>Example 1</u>. Assume that $p = 2$ and $M_i/O_2(M_i) \cong \Sigma_3$ for $i = 1,2$. This is a situation which was investigated in [5]. Suppose the easiest case, namely $b = 1$. (Further below in Theorem 2 we will see that $b = 1$, 2 or 3 always.)

By Lemma 1 we may assume that

$$\{G_\alpha,G_{\alpha+1}\} = \{M_1,M_2\} \text{ and } G_\alpha \cap G_{\alpha+1} = B.$$

Let $\nu \in \{\alpha,\alpha+1\}$ and $D_\nu \in Syl_3(G_\nu)$. Obviously, $B \in Syl_2(G_\nu)$ and D_ν acts transitively on $\Delta(\nu)$. In addition, we have

$$Z_\alpha \not\leq G_{\alpha+1}^{(1)} = Q_{\alpha+1} \text{ and } [Q_{\alpha+1},Z_\alpha] \leq \Omega_1(Z(B)).$$

Now an elementary argument, together with (A_4), shows:

$$|[Q_\nu, D_\nu]| = 4.$$

On the other hand, Lemma 3 implies

$$C_B(D_\alpha) \cap C_B(D_{\alpha+1}) = 1.$$

It follows that either $M_i \cong \Sigma_4$ or $M_i \cong C_2 \times \Sigma_4$ for $i = 1,2$.

This example shall explain that the investigation of M_1 and M_2 splits into two parts: the determination of b, and then (with the help of b) the determination of the structure of M_1 and M_2.

In this paper we are mainly interested in the first part.

Another subdivision (of the first part) is indicated by Lemma 4. According to Lemma 4(a) and (c), $Z_{\alpha'}$ (resp. Z_α) operates quadratically on Z_α (resp. $Z_{\alpha'}$). In addition, we can choose $\{\delta,\delta'\} = \{\alpha,\alpha'\}$ so that

$$|Z_{\delta'}/C_{Z_{\delta'}}(Z_\delta)| \geq |Z_\delta/C_{Z_\delta}(Z_{\delta'})|.$$

If $[Z_\delta, Z_{\delta'}] \neq 1$, then Z_δ is a so-called failure-of-factorization module for $G_\delta/C_{G_\delta}(Z_\delta)$, a module-type which plays an important role in the investigation of p-local subgroups. Therefore, it seems to be appropriate to treat the cases

$$[Z_\alpha, Z_{\alpha'}] \neq 1 \quad \text{and} \quad [Z_\alpha, Z_{\alpha'}] = 1$$

separately. In both cases we need some information about $C_{G_\delta}(Z_\delta)$. This is given by the next hypothesis.

> (A_5) For $i = 1,2$ and $\bar{M}_i = M_i/O_p(M_i)$ there exists a normal subgroup L_i of M_i containing $O_p(M_i)$ such that
> i) $M_i = L_i N_B(S)$, $S \in Syl_p(B)$.
> ii) $\bar{L}_i/\phi(\bar{L}_i)$ is a minimal normal subgroup of $\bar{M}_i/\phi(\bar{M}_i)$.

By Lemma 1 there exist adjacent vertices α_1 and α_2 in Γ so that $M_i = G_{\alpha_i}$, $i = 1,2$. Every normal subgroup L_i of M_i with $M_i = L_i B$ and $O_p(M_i) \leq L_i$ is transitive on $\Delta(\alpha_i)$. Hence, according to

Lemma 2, there is a normal subgroup $\tilde{G}$ of G for which (A_1) - (A_4) and (A_5)(i) holds with respect to L_1 and L_2. The actual restriction for the structure of G_{α_i} is in (A_5)(ii).

Lemma 5. Let δ and λ be adjacent in Γ and $T \in Syl_p(G_\lambda \cap G_\delta)$, and let N be a normal subgroup of G_δ with $G_\delta \neq N\, N_{G_\delta \cap G_\lambda}(T)$. Then

a) $N \cap Q_\delta \in Syl_p(N)$,

b) $O_p(G_\delta/NQ_\delta) = 1$,

c) $Q_\delta \in Syl_p(G_\delta^{(1)})$,

d) $O_p(G_\delta/C_{G_\delta}(Z_\delta)) = 1$ and $O_p(G_\delta/C_{G_\delta}(V_\delta)Q_\delta) = 1$.

Proof: Set $\bar{G}_\delta = G_\delta/Q_\delta$ and $H = N_{G_\delta \cap G_\lambda}(T)$. By Lemma 1(d) and (A_5) there exists a normal subgroup L_δ in G_δ such that

i) $G_\delta = L_\delta H$, and

ii) $\bar{L}_\delta/\phi(\bar{L}_\delta)$ is a minimal normal subgroup of $\bar{G}_\delta/\phi(\bar{L}_\delta)$,

and by Lemma 1 and (A_3) $T \in Syl_p(G_\delta)$.

It follows that $[\bar{L}_\delta,\bar{N}] \leq \phi(\bar{L}_\delta)$. Set $\bar{T}_o = \bar{T} \cap \bar{N}$. The Frattini-argument implies $\bar{L}_\delta = N_{\bar{L}_\delta}(\bar{T}_o)\phi(\bar{L}_\delta)$ and thus $\bar{L}_\delta\bar{H} = \bar{G}_\delta \leq N_{\bar{G}_\delta}(\bar{T}_o)$. This yields $\bar{T}_o = 1$ and (a).

Let N_o be the preimage of $O_p(G_\delta/NQ_\delta)$ in G_δ. Then $G_\delta \neq N_oH$ and (b) follows from (a).

Assume now that

$$N_1 \in \{G_\delta^{(1)}, C_{G_\delta}(Z_\delta), C_{G_\delta}(V_\delta)Q_\delta\}.$$

By (a), (b) and Lemma 4(a) we may assume that $G_\delta = N_1H$. If $N_1 = G_\delta^{(1)}$, then $G_\delta \leq G_\lambda$ which contradicts (A_1). If $N_1 = C_{G_\delta}(V_\delta)Q_\delta$, then Z_λ is normal in G_δ and G_λ which contradicts Lemma 3. Thus we have $N_1 = C_{G_\delta}(Z_\delta)$ and $Z_\delta = \Omega_1(Z(T))$, and G_δ/N_1 is a p'-group.

Lemma 6. Suppose that $[Z_\alpha,Z_{\alpha'}] \neq 1$ and $\bar{G}_\alpha = G_\alpha/C_{G_\alpha}(Z_\alpha)$. Then $Z_{\alpha'} \not\leq G_\alpha^{(1)}$ and (possibly after reversing the roles of α and α') the following hold:

a) $C_{G_\nu}(Z_\nu)/Q_\nu$ is a p'-group for $\nu = \alpha,\alpha'$.

b) $\bar{Z}_{\alpha'}$ operates quadratically on Z_α.

c) $|\bar{Z}_{\alpha'}| \geq |Z_\alpha/C_{Z_\alpha}(\bar{Z}_{\alpha'})|$.

Proof: By Lemmas 4(a) and 5, $Q_\alpha \in Syl_p(G_\alpha^{(1)})$ and $Z_{\alpha'} \not\leq G_\alpha^{(1)}$; in particular, the assumption on α and α' is symmetric. Therefore, it suffices to prove (a) for $\nu = \alpha$. Then the assertions (b) and (c) follow as above.

Set $N_1 = C_{G_\alpha}(Z_\alpha)$. By Lemma 5 we may assume that $G_\alpha = N_1N_{G_\alpha}(T)$, $T \in Syl_p(G_\alpha)$. It follows that $Z_\alpha = \Omega_1(Z(T))$ which contradicts Lemma 4(d).

Lemma 7. Suppose that $[Z_\alpha, Z_{\alpha'}] = 1$. Then the following hold:

a) $Z_{\alpha'} = \Omega_1(Z(T))$, $T \in Syl_p(G_{\alpha'})$.

b) b is odd.

c) $C_{G_{\alpha'}}(V_{\alpha'})/C_{Q_{\alpha'}}(V_{\alpha'})$ is a p'-group.

d) $V_{\alpha+1} \leq G_{\alpha'}$, $V_{\alpha'} \leq G_{\alpha+1}$ and $[V_{\alpha+1}, V_{\alpha'}] \neq 1$.

e) If $b > 1$, then $V_{\alpha'}$ is elementary abelian, and $V_{\alpha+1}$ operates quadratically on $V_{\alpha'}$.

Proof: Note that $Z_\alpha \leq C_{G_{\alpha'}}(Z_{\alpha'})$ but $Z_\alpha \not\leq Q_{\alpha'}$. Thus, Lemma 5 implies

$$G_{\alpha'} = C_{G_{\alpha'}}(Z_{\alpha'})N_{G_{\alpha'}}(T)$$

for $T \in Syl_p(G_{\alpha'} \cap G_{\alpha'-1})$, and (a) follows. Now, by Lemma 4(d), α is not conjugate to α' (under G) and (b) holds.

If $C_{G_{\alpha'}}(V_{\alpha'})/C_{Q_{\alpha'}}(V_{\alpha'})$ is not a p'-group, then again Lemma 5 yields $G_{\alpha'} = C_{G_{\alpha'}}(V_{\alpha'})(G_{\alpha'} \cap G_{\alpha'-1})$, and $Z_{\alpha'-1}$ is normal in $G_{\alpha'}$, a contradiction to Lemma 3.

Assertion (d) follows directly from (c) and the minimality of b.

Assume that $b > 1$ and $\mu, \lambda \in \Gamma$ with $d(\mu,\lambda) \leq 2$. By (b) and Lemmas 4 and 5, $b \geq 3$ and $Z_\mu \leq Q_\lambda \leq C_{G_\lambda}(Z_\lambda)$. Now (e) follows from (d).

Example 2. We apply the above properties in the proof of the following theorem.

Theorem 1. Suppose that (with the notation of (A_5))

i) $\bar{L}_1/O_{2'}(\bar{L}_1) \cong SL_2(q_1)'$, $p|q_1$,

ii) $\bar{L}_2/O_{2'}(\bar{L}_2) \cong L_2(q_2)'$,

where q_1 and q_2 are powers of (not necessarily the same) odd primes. Then one of the following holds:

a) $G_\alpha \cong M_2$, $[Z_\alpha, Z_{\alpha'}] = 1$ and $b = 1$,

b) $G_\alpha \cong M_1$, $[Z_\alpha, Z_{\alpha'}] \neq 1$ and $b = 2$,

c) $G_\alpha \cong M_1$, and for $\tilde{J} = C_{O_p(M_2)}(Z(J(O_p(M_2))))$ and $L = \langle \tilde{J}^{M_1} \rangle$:

c_1) $\tilde{J} \in Syl_p(L)$,

c_2) $\bar{L}/\phi(\bar{L}) \cong SL_2(q_1)$.

<u>Proof</u>: According to Lemma 1(d) we choose notation so that

$$\{G_\alpha, G_{\alpha+1}\} = \{M_1, M_2\} \quad \text{and} \quad B = G_\alpha \cap G_{\alpha+1}.$$

Pick $S \in Syl_p(B)$. By Lemma 2 we may assume that

$$M_i = L_i S, \quad i = 1,2.$$

Now again by Lemma 1(d) there exists a normal subgroup L_δ in G_δ, $\delta \in \Gamma$, such that

$$G_\delta = L_\delta T, \; T \in Syl_p(G_\delta \cap G_\lambda)$$

and $\lambda \in \Delta(\delta)$, and

$$\{L_\alpha, L_{\alpha+1}\} = \{L_1, L_2\}.$$

We first discuss a situation where case (c) of the Theorem arises. Suppose that $O_p(M_2) \in Syl_p(\langle O_p(M_2)^{M_1} \rangle)$. We use the following notation:

$$Q = O_p(M_2), \; Z = Z(J(Q)), \; L_o = \langle J(Q)^{M_1} \rangle \quad \text{and} \quad Y = L_o \cap Q.$$

By Lemmas 3 and 5 we have $M_1 = L_o S$ and $[Z(Q), O^p(L_o)] \neq 1$. Since $p \neq 2$ we get with a well-known argument (see [4, 2.9]):

$\bar{L}_o/\phi(\bar{L}_o) \simeq L_2(q_1)$,

$Z(O_p(L_o)) = Z(Y)Z(Y^g)$ for suitably chosen $g \in L_o$ and

$C_{O_p(M_1)}(Z(O_p(L_o))) \in Syl_p(C_{M_1}(Z(O_p(L_o))))$.

Since $Z(Y) \leq Z$, it follows that $\tilde{J} \cap \tilde{J}^g \leq C_{M_1}(Z(O_p(L_o))) \cap O_p(M_1)$ and $L_o\tilde{J} = L_oO_p(L_o\tilde{J})$. On the other hand, an argument in [4, 3.3] shows that $Z = C_Z(O^p(L_o))Z(Y)$. Since $L_o = O^p(L_o)J(Q)$ we get $Z = Z(Y)$ and thus

$$\tilde{J} \in Syl_p(L_o\tilde{J}).$$

Now assertion (c) follows.

We now begin to consider the general case. For odd p and q, $SL_2(p)$ is not involved in $L_2(q)$. Furthermore, by Lemma 5, $C_{\bar{M}_i}(\bar{L}_i/\phi(\bar{L}_i))$ is a p'-group. Thus, we get from [6, 3.8.3]:

1) Let V be a $GF(p)$ module for $\bar{M}_2$ and $[V,\bar{L}_2] \neq 1$. Then no non-trivial p-subgroup of $\bar{M}_2$ operates quadratically on V.

Suppose first that $[Z_\alpha, Z_{\alpha'}] \neq 1$, and set $R = [Z_\alpha, Z_{\alpha'}]$ and $P = Z_{\alpha'}Q_\alpha$. By Lemmas 4 and 6, Z_α operates quadratically on $Z_{\alpha'}$, and vice versa, and by (1) we have $G_\alpha = M_1$ and

2) b is even.

In addition, Lemma 6 and well known properties of $SL_2(q_1)$-modules (see [4, 2.3]) imply:

3) $P \in Syl_p(L_\alpha Z_{\alpha'})$ and $Z_\alpha Q_{\alpha'} \in Syl_p(L_{\alpha'} Z_\alpha)$.

4) $L_\alpha Z_{\alpha'}/C_{L_\alpha Z_{\alpha'}}(Z_\alpha) \cong SL_2(q_1)$, and $Z_\alpha/C_{Z_\alpha}(L_\alpha)$ is a natural $SL_2(q_1)$-module.

5) $\Omega_1(Z(P)) = RC_{Z_\alpha}(L_\alpha)$.

Since, by (2), α is conjugate to α', the same statements hold with α and α' reversed.

We now choose $\alpha-1 \in \Delta(\alpha)\setminus\{\alpha+1\}$ such that

6) $L_\alpha Z_{\alpha'} \leq \langle Q_{\alpha-1}, P\rangle$.

Assume that $Z_{\alpha-1} \leq G_{\alpha'}$. By (3) and (4), $Z_{\alpha-1} \leq Z_\alpha C_{G_{\alpha'}}(Z_{\alpha'})$ and by Lemma 4(c)

$$[Z_{\alpha-1}, Z_{\alpha'}] \leq R \leq Z_\alpha.$$

Now (6) implies that $Z_\alpha Z_{\alpha-1}$ is normal in $L_\alpha Z_{\alpha'}$.

If $Q_{\alpha-1} \in Syl_p(C_{G_{\alpha-1}}(Z_{\alpha-1}))$, then $Q_\alpha \cap Q_{\alpha-1} = C_{Q_\alpha}(Z_\alpha Z_{\alpha-1})$ is normal in G_α. Now a short argument using (3) and a property of the Schur-multiplier ([7, V 25.1]) shows that $Q_{\alpha-1} \in Syl_p(\langle Q_{\alpha-1}^{G_\alpha}\rangle)$, and assertion (c) follows.

By Lemma 5, we may assume now that $G_{\alpha-1} = C_{G_{\alpha-1}}(Z_{\alpha-1})(G_{\alpha-1} \cap G_\alpha)$ and $Z_{\alpha-1} = \Omega_1(Z(T))$, $T \in Syl_p(G_{\alpha-1} \cap G_\alpha)$; in particular, $Z_{\alpha-1} \leq Z_\alpha$ and $C_{G_{\alpha-1}}(Z_{\alpha-1})$ is transitive on $\Delta(\alpha-1)$. The application of Lemma 3 and (5) yields

7) $C_{Z_\alpha}(L_\alpha) = 1$ and $Z_{\alpha+1} \leq R$ and $Z_{\alpha'-1} \leq R$.

To prove (b) we assume that $b > 2$ and thus, by (2), $b \geq 4$. We have $V_{\alpha-1} \leq G_{\alpha'-2}$ by the minimality of b. Assume that there exists $\alpha-2 \in \Delta(\alpha-1)$ with $Z_{\alpha-2} \not\leq Q_{\alpha'-2}$. Then the pair $(\alpha-2,\alpha'-2)$ has the same properties as (α,α'), i.e., we can apply Lemma 6 and (7) and get

$$Z_{\alpha-1} \leq [Z_{\alpha-2},Z_{\alpha'-2}] \leq Z_{\alpha'-2}.$$

But now $[Z_{\alpha-1},Z_{\alpha'}] = 1$ since $b > 2$, and (6) implies that $Z_{\alpha-1}$ is normal in $L_\alpha Z_{\alpha'}$ which contradicts Lemma 3.

We have shown that $V_{\alpha-1} \leq Q_{\alpha'-2} \leq G_{\alpha'-1}$. If, in addition, $V_{\alpha-1} \leq G_{\alpha'}$, then as above $[V_{\alpha-1},Z_{\alpha'}] \leq Z_\alpha \leq V_{\alpha-1}$ and (6) and Lemma 3 lead to a contradiction. Thus we have $V_{\alpha-1} \not\leq Q_{\alpha'-1}$.

Set $W_\alpha = \langle V_\lambda / \lambda \in \Delta(\alpha)\rangle$ and $H = [V_{\alpha-1},V_{\alpha'-1}]$. Since $V_{\alpha'-1} \leq G_\alpha$, it follows that $H \leq W_\alpha \cap V_{\alpha'-1}$. If $[H,V_{\alpha-1}] = 1$, we get a contradiction to (1). Hence, $[H,V_{\alpha-1}] \neq 1$ and therefore $[H,Z_{\alpha-2}] \neq 1$ for some $\alpha-2 \in \Delta(\alpha-1)$, and again (1) (applied to $G_{\alpha-1}$) shows that $H \leq Q_{\alpha-1} \leq G_{\alpha-2}$. Now (4), (5) and (7) applied to $G_{\alpha-2}$ (which is conjugate to G_α) imply that $Z_{\alpha-1} \leq [H,Z_{\alpha-2}] \leq V_{\alpha'-1}$. But then again

$$Z_{\alpha-1} \leq V_{\alpha'-1} \leq C_G(Z_{\alpha'}),$$

a contradiction as above.

We may assume now that $[Z_\alpha,Z_{\alpha'}] = 1$. Clearly, Z_α operates quadratically on $Q_{\alpha+1}/\phi(Q_{\alpha+1})$. Therefore, if $b = 1$, we get assertion (a) from (1). Thus we may assume $b > 1$.

It follows from Lemma 6 and (1) that $M_1 = G_{\alpha+1}$. If $[V_{\alpha+1} \cap Q_{\alpha'}, V_{\alpha'}] \neq 1$, then there exists $\lambda \in \Delta(\alpha')$ with $[V_{\alpha+1} \cap Q_{\alpha'}, Z_\lambda] \neq 1$. But by Lemma 7(d), $V_{\alpha+1} \cap Q_{\alpha'}$ operates quadratically on Z_λ which contradicts $G_\lambda \cong M_2$ and (1). With the same argument we have $[V_{\alpha'} \cap Q_{\alpha+1}, V_{\alpha+1}] = 1$. Now as above [5, (2.3)] yields for $V = V_{\alpha+1}/C_{V_{\alpha+1}}(O^p(L_{\alpha+1}))$:

7) $V_{\alpha'} Q_{\alpha+1} \in Syl_p(L_{\alpha+1} V_{\alpha'})$,

8) $L_{\alpha+1} V_{\alpha'} / C_{L_{\alpha+1} V_{\alpha'}}(V) \cong SL_2(q_1)$,

9) V is a natural $SL_2(q_1)$-module.

Set $Z = C_{Z_\alpha}(L_{\alpha+1})$ and $C = C_{G_{\alpha+1}}(Z)$. Note that (9) yields $[Z_\alpha, O_p(C)] \leq Z$. It follows that $O_p(C)$ operates quadratically on Z_α and therefore, by (1), $O_p(C) = Q_\alpha \cap Q_{\alpha+1}$. But now (7) implies that $Q_\alpha \in Syl_p(C)$ and (c) follows, as we have shown above.

In cases (a) and (b) of Theorem 1 one can now deduce the structure of M_1 and M_2 as was indicated in Example 1. Note that the maximal parabolic subgroups of $PSp_4(p^n)$ and $U_4(p^n)$ containing a fixed Sylow p-subgroup, are examples for case (a).

In case (c) any characteristic subgroup of $\tilde{J}$ is normal in M_2. Hence, according to (A_2) no non-trivial characteristic subgroup of $\tilde{J}$ is normal in M_1. Now a pushing-up result of R. Niles [8] leads to a contradiction.

Conversely, Niles' pushing-up result can be proven with the above method (see [12]).

We now begin to discuss a generalization of Goldschmidt's result which was already foreshadowed by Theorem 1.

Goldschmidt investigated subgroups M_i with

(*) $M_i/O_2(M_i) \cong SL_2(2)$ or $SL_2(2)'$ for $i = 1,2$.

The obvious generalization of (*) is to allow all rank 1 groups of Lie type. For this purpose we make the following hypotheses:

(A_6) For $i = 1,2$ one of the following holds:

i) $L_i/O_p(L_i) \cong L_2(p^{n_i})$, $SL_2(p^{n_i})$, $U_3(p^{n_i})$, $SU_3(p^{n_i})$, $Sz(2^{n_i})$ (and $p = 2$), or $Re(3^{n_i})$ (and $p = 3$).

ii) $p = 3$ and $L_i/O_3(L_i) \cong L_2(3)'$, $SL_2(3)'$ or $Re(3)'$ ($\cong L_2(8)$).

iii) $p = 2$ and $L_i/O_2(L_i) \cong L_2(2)'$, $Sz(2)'$ ($\cong C_5$), or $L_i \leq L_i^*$ and $L_i^*/O_2(L_i) \cong U_3(2)$ or $SU_3(2)$.

(A7) For $i = 1,2$ and $S \in Syl_p(B)$ either $S \cap L_i = O_p(L_i)$ or $N_{M_i}(S \cap L_i) = B$.

The characterization of all pairs $\{M_1,M_2\}$ with (A_1^*) - (A_7) is the combined work of several authors: D. Goldschmidt [5], A. Delgado [1], P. Fan [2], K.-H. Schmidt [9], B. Stellmacher [11], and G. Stroth [13, 14]; and P. Rowley announced a contribution at this conference.

In this paper we will discuss a uniform proof given by A. Delgado and B. Stellmacher which uses the above approach and does not depend on the above results.

The major step in the proof is to determine the parameter b as illustrated in Example 2.

Theorem 2. One of the following holds:

a) $[Z_\alpha,Z_{\alpha'}] \neq 1$ and $b = 1$ or 2,

b) $[Z_\alpha,Z_{\alpha'}] = 1$ and $b = 1, 3$ or 5.

With the knowledge of b it is now fairly easy to determine the structure of M_1 and M_2, i.e. the order of M_i, the chief factors of $O_p(M_i)$, the class of $O_p(M_i)$, etc.. We want to refer to this as the type of M_i, not being willing to define it exactly in this paper. We call the pair $\{M_1,M_2\}$ of type $\tilde{X}$, for some finite group $\tilde{X}$, if M_1 and M_2 have the same types as a pair of p-local subgroups in $\tilde{X}$ containing a common Sylow p-subgroup of $\tilde{X}$. In the applications this pair will be unique (up to conjugation) in $\tilde{X}$.

Theorem 3. There exists

$$X \in \{L_3(p^n), PSp_4(p^n), U_4(p^n), U_5(p^n), G_2(p^n), {}^3D_4(p^n), {}^2F_4(2^n), G_2(2)', {}^2F_4(2)', M_{12}, J_2, T\}$$

and $X \leq \tilde{X} \leq Aut(X)$ such that $\{M_1,M_2\}$ is of type $\tilde{X}$, where $p = 2$ if

$X = {}^2F_4(2^n)$, $G_2(2)'$, ${}^2F_4(2)'$, M_{12} or J_2 and $p = 3$ if $X = T$, the Thompson group.

Theorem 3 provides us with a reasonable description of the group theoretic properties of M_1 and M_2 which can be used in the local analysis of finite groups. But it does not yet give the isomorphism type of M_1 and M_2.

We will now give a geometric method to establish isomorphy for the bulk of the above examples. For the exceptional pairs a case by case discussion seems to be necessary.

Let $\Lambda = (\delta_o,\ldots,\delta_n)$ be a path of length n in Γ. We denote its length by $|\Lambda|$. The path Λ is regular, if G_Λ operates transitively on $\Delta(\delta_o)\setminus\{\delta_1\}$ and $\Delta(\delta_n)\setminus\{\delta_{n-1}\}$, otherwise it is singular. We now define

$$r = \max\ \{|\Lambda| / \Lambda \text{ is regular}\},$$

$$s = \min\ \{|\Lambda| / \Lambda \text{ is singular}\}.$$

The following remarkable property singles out the examples which correspond to groups of Lie type. Set

$$\Omega = \{L_3(p^n), PSp_4(p^n), U_4(p^n), U_5(p^n), G_2(p^n), {}^3D_4(p^n), {}^2F_4(2^n)\}.$$

Theorem 4. Let X and $\tilde{X}$ be as in Theorem 3. Then the following statements are equivalent:

a) $r = s - 1$,

b) $\{M_1,M_2\}$ is of type $\tilde{X}$ for $X \in \Omega$.

Define the (free) amalgamated product $\tilde{G} = M_1 *_B M_2$ with respect to the embedding of B into M_1 and M_2. Then elementary properties of amalgamated products imply that $\tilde{G}$ fulfills $(A_1) - (A_7)$ with respect to M_1 and M_2. Let $\tilde{\Gamma}$ be the corresponding coset graph. J.-P. Serre [10] has shown that $\tilde{\Gamma}$ is a tree.

If $r = s-1$ and $\{M_1,M_2\}$ is not of type $G_2(2)$ or ${}^2F_4(2)$, this property together with Theorem 4 allows us to define an equivalence relation $\approx$ on $\tilde{\Gamma}$, which is compatible with the action of $\tilde{G}$, such that

i) $\tilde{\Gamma}/\approx$ is a generalized r-gon,

ii) $\tilde{G}_{\tilde{\Gamma}}$, the kernel of the action of $\tilde{G}$ on $\tilde{\Gamma}/\approx$, has trivial intersection with M_1 and M_2.

iii) There exists a normal subgroup W in $\tilde{G}/\tilde{G}_{\tilde{\Gamma}}$ such that $\tilde{G}/\tilde{G}_{\tilde{\Gamma}} \leq \mathrm{Aut}(W)$ and W is a finite group with a split (B,N)-pair of rank 2. Now a result of P. Fong and G.M. Seitz [3] implies:

<u>Theorem 5</u>. Suppose that $r = s-1$ and $\{M_1, M_2\}$ is of type $\tilde{X}$ and $\tilde{X} \neq G_2(2)$ or ${}^2F_4(2)$. Then M_1 and M_2 are isomorphic to the corresponding subgroups in $\tilde{X}$.

REFERENCES

1. Delgado, A. (1981). On edge-transitive automorphism groups of graphs. Ph.D. Thesis, Berkeley.
2. Fan, P. Amalgams of prime index. Preprint.
3. Fong, P. & Seitz, G. Groups with a (B,N)-pair of rank 2, I, II. Invent. Math. <u>21</u> (1973), 1-57, <u>24</u> (1974), 191-239.
4. Glauberman, G. & Niles, R. (1983). A pair of characteristic subgroups for pushing-up in finite groups. Proc. London Math. Soc. (3), <u>46</u>, 411-453.
5. Goldschmidt, D. (1980). Automorphisms of trivalent graphs. Ann. of Math., <u>111</u>, 377-406.
6. Gorenstein, D. (1968). Finite groups. Harper and Row, New York.
7. Huppert, B. (1967). Endliche Gruppen I. Springer, New York.
8. Niles, R. (1979). Pushing-up in finite groups. J. Algebra <u>57</u>, 26-63.
9. Schmidt, K.-H. (1983). Punkstabilisatoren auf 4-valenten Graphen. Diplomarbeit, Bielefeld.
10. Serre, J.-P. (1977). Arbres, Amalgams, SL_2. Astérisque 46, Soc. Math. de France.
11. Stellmacher, B. On graphs with edge-transitive automorphism groups. Ill. J. of Math. In print.
12. Stellmacher, B. (1983). Pushing-up. Preprint Bielefeld.
13. Stroth, G. Graphs with subconstituents containing $Sz(q)$ and $L_2(r)$. To appear.
14. Stroth, G. Kantentransitive Graphen und Gruppen vom Rang 2. To appear.

PARABOLICS IN FINITE GROUPS

G. Stroth
FU-Berlin, II. Math. Institut, Arnimallee 3
1000 Berlin 33, Federal Republic of Germany

1 MINIMAL PARABOLIC SYSTEMS FOR FINITE GROUPS

Given a finite simple group G of Lie type and characteristic p, the building $\Delta(G)$ may be obtained as a chamber system [13] in the following manner. Let B denote the normalizer of a Sylow p-subgroup of G, and let $P_1,\dots,P_n$ denote the subgroups of G which are minimal with respect to containing B (so called minimal parabolics). Then the chambers of $\Delta(G)$ are the conjugates of B, and two such B^g and B^h are i-adjacent if and only if $gh^{-1} \in P_i$. In this paper a generalization of this construction yielding chamber systems for almost all simple groups is given. Let p be a fixed prime, G any group, B a Sylow p-normalizer, and $P_1,\dots,P_n$ subgroups of G containing B. Then the associated chamber system (chambers B^g, B^h are i-adjacent iff $gh^{-1} \in P_i$) is connected provided that for any two chambers there is a gallery [13] from one to the other. This means that for any $g \in G$ there are elements $x_{i_1},\dots,x_{i_n}$, each $x_{i_j} \in P_{i_j}$, such that $B^{x_{i_1}\cdots x_{i_n}} = B^g$, or in other words, that $\langle P_1,\dots,P_n\rangle = G$. As we wish to consider only connected chamber systems, we shall require that the P_i generate G.

Before giving a definition of a minimal parabolic group in an arbitrary group we state a simple proposition.

Proposition (1.1). Let $G = \langle X \mid B \le X,\ O_p(X) \ne 1\rangle$. Then $G = \langle X \mid B \le X,\ O_p(X) \ne 1,\ B$ lies in exactly one maximal subgroup of $X\rangle$.

This leads now to the following definitions.

Definition (1.2). Let G be a finite group, B the normalizer of a Sylow p-subgroup of G.

a) Let $P > B$. We call P a minimal parabolic subgroup of G if $O_p(G) \ne 1$ and B lies in exactly one maximal subgroup of P.

b) Let $P_1,\ldots,P_n$ be minimal parabolic subgroups of G. Then we call $\{P_1,\ldots,P_n\}$ a minimal parabolic system of rank n providing that $G = \langle P_1,\ldots,P_n\rangle$ and no proper subset of $\{P_1,\ldots,P_n\}$ generates G. We call n the rank of the system.

Given a minimal parabolic system we obtain a chamber system by taking as chambers the conjugates of B, and defining i-adjacency $\underset{i}{\sim}$ by, $B^g \underset{i}{\sim} B^h$ iff $gh^{-1} \in P_i$.

Remarks. a) If G is a finite group of Lie type and characteristic p then there is exactly one such chamber system, namely the building for G.

b) For $p = 2$ every sporadic simple group except J_1 and M_{11} admits a minimal parabolic system of rank at least 2 [9].

c) A group can have more than one minimal parabolic system for a fixed prime p, even the rank is not uniquely determined (see (2.3)).

If G is a group of Lie type and characteristic p, B the normalizer of a Sylow p-subgroup and $P_1,\ldots,P_n$ the minimal parabolic subgroups containing B, then the building may be thought of as a simplicial complex, simplices being precisely all conjugates of the subgroups $P_J = \langle P_j \mid j \in J\rangle$, where $J \subset \{1,\ldots,n\} = I$ and $P_\emptyset = B$, with reverse inclusion indicating that one is a face of another. A simplex is said to have type J if it corresponds to $P_J{}^g$, $g \in G$. The chambers are thus simplices of maximal dimension $n-1$, and two such are i-adjacent if they share a face of type $I - \{i\}$. This leads to the following definition.

Definition (1.3). Let $\{P_1,\ldots,P_n\}$ a minimal parabolic system for G, $I = \{1,\ldots,n\}$, $J \subseteq I$. Set $P_J = \langle P_j \mid j \in J\rangle$, $P_\emptyset = B$, which we call a parabolic subgroup for the system in question. Then we call the system geometric if for all subsets $J, K \subseteq I$ we have

$$P_J \cap P_K = P_{J \cap K}.$$

If a system is geometric, it means that one may define the chamber system as a simplicial complex by starting with the vertices. These are all conjugates of the $P_{I-\{i\}}$, $i \in I$, and the simplices are all sets of vertices whose intersection, as groups, contains a conjugate of B.

Every sporadic group having a chamber system of rank at least two has also a geometric one.

2 EXAMPLES

From now on the prime p in question will be 2. As in [13] the chamber system for the group G coming from a minimal parabolic system generates a geometry whose objects are the conjugates of $P_{I-\{i\}}$, $i \in I$, and two such conjugates are incident if they join a conjugate of B. We describe these geometries by a diagram (see [2]). Each node of the diagram correspond to a maximal parabolic, which is given above that node. The notation $\begin{smallmatrix} B \\ A \end{smallmatrix}$ refers to an extension of the group B by a group A which acts trivially on the residue of the given vertex.

2.1 Geometries involving S_5 and S_3

Let $S_1 = S(22,3,6)$ be the Steiner system for M_{22}. Then we get a geometry whose points are the duads (ab) and lines are the triduads $(ab|cd|ef)$ forming a hexad of the Steiner system. As a diagram we get

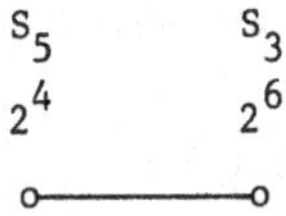

$$\begin{matrix} S_5 & & S_3 \\ 2^4 & & 2^6 \\ \circ & \text{———} & \circ \end{matrix}$$

(if a geometry is of rank 2 the maximal parabolics and the minimal parabolics coincide).

There is a natural extension to $S_2 = S(23,4,7)$ for M_{23}. Take as points the 23 of S_2, as lines all triads and planes one class of projective planes formed from the seven points of a heptad. This gives

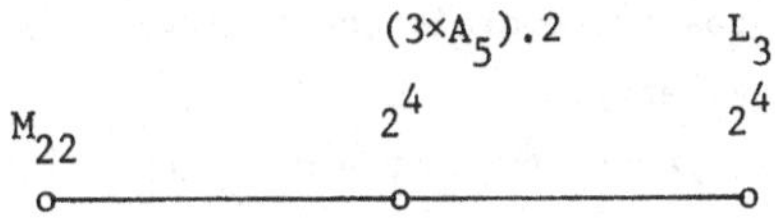

$$\begin{matrix} & & (3\times A_5).2 & & L_3(2) \\ M_{22} & & 2^4 & & 2^4 \\ \circ & \text{———} & \circ & \text{———} & \circ \end{matrix}$$

with the M_{22}-geometry and the projective plane of order 2 as residual.

The M_{22}-geometry seems to be of great importance as we can find this geometry in a great many other geometries. For example

a)

$M_{22}.2$	$S_3 \times S_5$	$L_3(2)$
2^{10}	2^{14}	2^{15}

o————o————o

This is a geometry for .2, the corresponding points, lines and planes are certain 1-spaces, 2-spaces and 3-spaces of the 22-space for .2 (see also (2.2)).

b)

$U_4(2).2$	$S_3 \times S_3$	M_{22}
2^{10}	2^{15}	2^{10}

o————o————o

This is a geometry for M(22). The objects are duads, triduads and 22-ads of Fischer's 3570 transpositions.

c)

	$(3\times U_4(2)).2$	$L_3(2)\times S_3$	M_{23}
2M(22)	2^{11}	2^{15}	2^{11}

o————o————o————o

This is an extension of the geometry (b) and belongs to M(23). The objects are points, triads, heptad-planes (using one class of $L_3(2)$ in A_7 as in the M_{23}-case) and 23-ads of Fischer's 31671 transpositions.

d)

.2	$S_3 \times M_{22}.2$	$L_3(2)\times S_5$	$L_4(2)$
2^{23}	2^{32}	2^{35}	2^{35}

o————o————o————o

This is an extension of the geometry (a) and belongs to F_2.

e)

$3M_{22}.2$	$S_3 \times S_5$	$L_3(2)$
2^{13}	2^{17}	2^{18}

o————o————o

This is a geometry for J_4. It does not contain the M_{22}-geometry but it contains a 3-fold cover of that geometry.

The M_{22}-geometry is a special type of a more general rank 2 geometry. This geometry is of the form

S_5	S_3
A	B
o————	o
1	2

Every object of type 1 is incident with exactly 15 objects of type 2 on which G_1 induces S_5 as a transitive group. Every object of type 2 is incident which exactly 3 objects of type 1 on which G_2 induces the group S_3.

Geometries of this type belong to $\mathrm{Aut}(J_2)$, HiS, Ru, LyS and $U_4(2).2$ as occurring in M(22) above. Furthermore there is such a geometry for A_{11} whose objects of type 1 are the conjugates of $\{(12),(34),(56),(78),(9\ 10)\}$ and objects of type 2 are the conjugates of $\{(1\ 2\ 3\ 4),\ (5\ 6\ 7\ 8)\}$ with the obvious incidence.

2.2 Geometries involving o≈o

Beside the M_{22} geometry described in (2.1) there is another important rank 2 geometry with diagram o≈o occurring in some diagrams for sporadic groups. This diagram indicates a 3-fold cover of the $Sp_4(2)$ quadrangle, its minimal circuits are 5-gons. This geometry occurs as a residual in

	$3Sp_4(2)$	$S_3\times S_3$	$L_3(2)$
a)	2^6	2^8	2^7
	o———	o═══~═══	o

This geometry comes from M_{24}. The points are sextets of the Steiner system S(24,5,8), planes are involutions of type $1^8 2^8$ (Conway's notation) and lines are sets of three mutually commuting involutions of this type belonging to disjoint octads. In terms of the 12-dimensional module over GF(2) obtained from the 24-permutation module after factoring out the Golay code, points lines and planes are certain 1, 2 and 3-spaces.

The same diagram belongs to He, where points, lines and planes are 1, 2 and 3-spaces in the 51-dimensional module over GF(2).

The diagram for M_{24} possesses an extension to

	M_{24}	$S_3\times 3Sp_4(2)$	$L_3(2)\times S_3$	$L_4(2)$
b)	2^{11}	2^{16}	2^{17}	2^{15}
	o———	o———	o═══~═══	o

This geometry belongs to .1 and may be described as follows. Take the Leech lattice Λ, using bars to denote images mod 2. Writing $\Lambda_n = \{\lambda \in \Lambda | \lambda\cdot\lambda = 16n\}$. As in [3] we get $\bar{\Lambda} = \{0\} \cup \bar{\Lambda}_2 \cup \bar{\Lambda}_3 \cup \bar{\Lambda}_4$.

An octad of the Steiner system $S(24,5,8)$ spans an 8-space E of Λ, such that $\overline{E} \cap \overline{\Lambda}_3 = \emptyset$. The form $q(\overline{v}) = v.v/32 \pmod 2$ is well defined on $\overline{E}$. Now $\overline{E}$ contains two types of totally singular 4-spaces, one type is contained in three different $\overline{E}$, whilst those of the other type, which we use as the 4-spaces of the geometry, each lies in a unique $\overline{E}$. The points, lines and planes are simply all 1, 2, 3 - spaces in these 4-spaces.

There is a further extension to a geometry for the monster F_1

c) $.1$ $\quad 2^{25}$ — $S_3{\times}M_{24}$ $\quad 2^{35}$ — $L_3(2){\times}3Sp_4(2)$ $\quad 2^{39}$ — $L_4(2){\times}S_3$ $\quad 2^{39}$ = $L_5(2)$ $\quad 2^{36}$

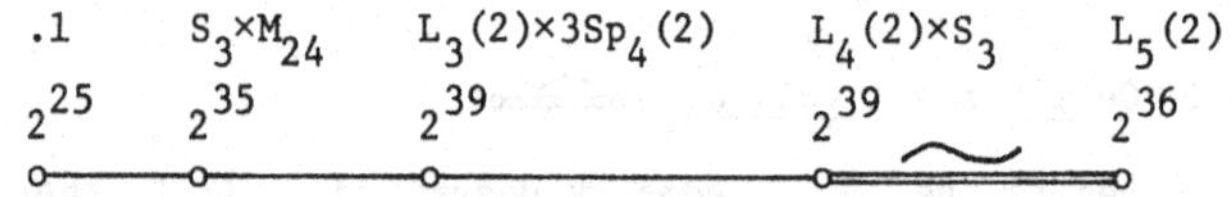

It's tempting to find a module over $GF(2)$ where the objects of this geometry are 1, 2, 3, 4 and 5 - spaces.

2.3 Some geometries for Mc

First of all we will present two geometries for A_7.

a) S_3 $\quad 2^2$ — A_7 — S_3 $\quad 2^2$

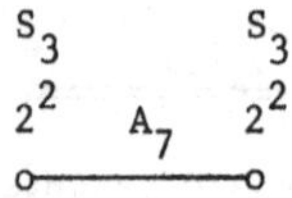

The objects are the biduads (ab)(cd) and triduads (ab)(cd)(ef) of the 7-set (see [5]).

b) A_6 — $(3{\times}A_4).2$ = $L_3(2)$

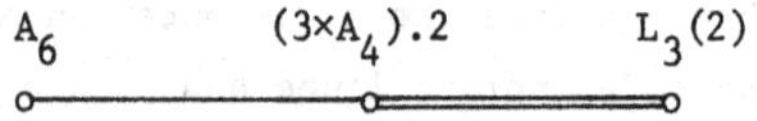

The points are the seven points, lines are all 3-sets and planes are one class of projective planes formed by the seven points.

Using these geometries we get interesting geometries for Mc

c) A_7 $\quad 2^4$ — $S_3{\times}S_3$ $\quad 2^5$ — A_7 — $L_3(2)$ $\quad 2^4$

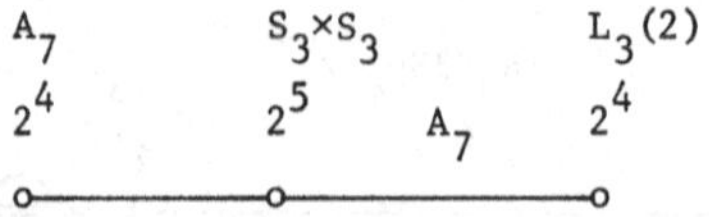

d) 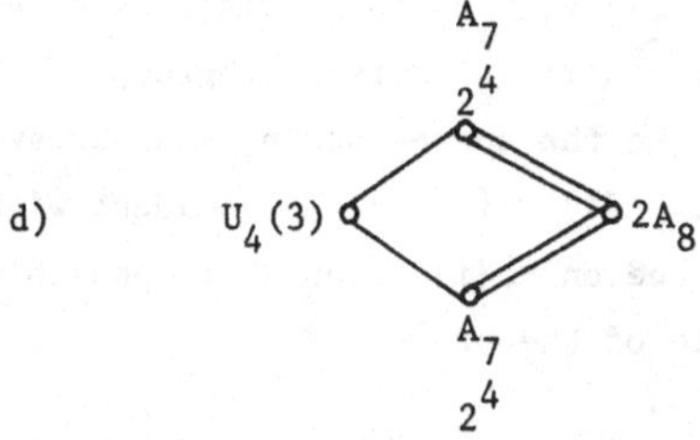

It's worthwhile to mention that the outer automorphism of Mc acts as a diagram automorphism on this geometry.

For more information and further geometries see [9].

2.4 Problems

a) Find a module for F_1 such that the geometry described in 2.2 c) correspond to this module in such a way that the objects of the geometry are a certain class of 5-spaces in this module and all 4, 3, 2 and 1-spaces of these 5-spaces.

b) Determine the apartments of the geometries if there are any.

c) For $p = 2$ determine the geometries for the alternating groups.

d) For $p = 2$ determine the geometries for the finite simple Lie groups of odd characteristic.

3 RESULTS

In this chapter we give some classification results which may be helpful in the revision of the classification of the finite simple groups. Thus as before we have $p = 2$. The idea of the result is as follows. Given the geometry locally, i.e. for each flag Γ the action of G on the residual of Γ (the set of all objects incident with Γ but not in Γ) is given, then determine the structure of the geometry. For applications a little weaker result will suffice. Given $G_\Gamma/O_2(G_\Gamma)$, for each flag Γ, determine G_Γ. It is obvious that a sort of induction enters the proofs. If the rank n of the geometry Δ is 1 this means that we have to determine all minimal parabolics $P/O_2(P)$. This can be done using the results of [1]. For $n = 2$ the combined work of A. Delgado, P. Fan, D. Goldschmidt, B. Stellmacher and G. Stroth gives the following.

<u>Theorem (3.1)</u>. Let $G = \langle P_1, P_2 \rangle$, $P_1 \cap P_2 = B$ the normalizer of a Sylow 2-subgroup of G. Suppose that no nontrivial normal subgroup of G is contained in P_1 or P_2. Let C be the corresponding chamber system and Δ the induced geometry. Set $\Delta(i) = \{x \in \Delta,\ x$ incident with $P_i\}$. If P_i induce Lie groups of rank one on $\Delta(i)$ then it is possible to write down the multiplication table of the P_i's.

<u>Remark</u>. The P_i's are as in Lie groups of rank 2. But also some sporadic examples (M_{12}, J_2, J_3, A_7) occur.

If one of the parabolics P_i does not induce a Lie group of rank one in their natural representation, then there are almost no results. From the examples in chapter 2 we see that the most interesting example is a minimal parabolic P with $P/O_2(P) \cong S_5$ and acting on 15 points. A result due to B. Stellmacher, Huang and G. Stroth 14 says the following.

<u>Theorem (3.2)</u>. Let $G = \langle P_1, P_2 \rangle$, $B = P_1 \cap P_2$ the normalizer of a Sylow 2-subgroup of G, and no nontrivial normal subgroup of G contained in P_1 or P_2. Suppose that P_1 and P_2 are 2-constrained. If $P_1/O_2(P_1) \cong S_5$ and $P_2/O_2(P_2) \cong S_3$ then it is possible to write down the multiplication table of P_1 and P_2.

There are also interesting examples in higher rank containing a minimal parabolic P such that $P/O_2(P) \cong S_5$. As can be seen from chapter 2 there are lots of sporadic examples. But the biggest one is F_2 which is of rank 4. The aim of a joint work of G. Stroth and S.K. Wong is to classify all the minimal parabolics up to $n = 4$, if they are all of the form $P/O_2(P) \cong S_5$ or $P/O_2(P) \cong S_3$. For $n \geq 4$ things should be easier because there are only two examples, $O^-(2n,2)$ and S_m, $m = 2^k.3-1$. If we assume that no rank 2 residual is the A_{11}-geometry described in (2.1), then only $O^-(2n,2)$ occur. But then it is possible to show that if $P_1,\ldots,P_k$ are minimal parabolics and $G = \langle P_1,\ldots,P_k \rangle$, then G contains a subgroup of index two from which we get a new chamber system whose rank two residuals of the corresponding geometry are projective planes of order 2 or generalized quadrangles. This leads to the investigation of the following situation.

<u>Hypothesis (3.3)</u>. Let $G = \langle P_1,\ldots\ldots,P_k \rangle$, $\bigcap_{i=1}^{k} P_i = B$, the normalizer of a Sylow 2-subgroup of G. Suppose $P_i/O_2(P_i) \cong A_5$, $S_3 \times Z_3$, or S_3. Let

C be the corresponding chamber system with induced geometry described by a diagram Δ. Then the diagram is a Coxeter diagram with rank 2 residuals of type o o , o——o , and o═══o . If $P_{ij} = \langle P_i, P_j \rangle$ belongs to o═══o , we assume $O'(P_{ij}/O_2(P_{ij}))$ is isomorphic to $SO^-(6,2)$, $Sp(4,2)$ or $Sp(4,2)'$. If P_{ij} belongs to o——o then $O'(P_{ij}/O_2(P_{ij}))$ is isomorphic to $L_3(2)$. If $P_{ijk} = \langle P_i, P_j, P_k \rangle$ belongs to o——o——o then $O'(P_{ijk}/O_2(P_{ijk}))$ is isomorphic to $L_4(2)$.

Then we get the following theorem

<u>Theorem (3.4)</u>. Suppose Hypothesis (3.3) and $O'(G_\Gamma/O_2(G_\Gamma)) \cong Sp(6,2)$ or $SO^-(8,2)$ for each residual Γ of type o——o═══o . If there is no nontrivial normal subgroup of G contained in one of the P_i's, and there is some residual o═══o , then one of the following is true

i) The geometry is a building of type C_k or F_4 and the group G is isomorphic to $Sp(2k,2)$, $SO^-(2k+2,2)$ or $F_4(2)$.

ii) The diagram is of the form o═══o═══o .

iii) Every rank 2 residual is of the form o═══o or o o, or Γ is of type (o, …, o)>o——o——o——· · ·o——o .

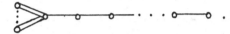

<u>Remark</u>. Because of a quite recent result due to U. Ott [6] and the results in [13] it is possible to replace the assumption on the structure of G_Γ by $G_\Gamma/O_2(G_\Gamma) \not\cong A_7$ or $U_6(2)$, $U_7(2)$.

Geometries of type o═══o═══o can be derived from $U_4(3)$ (see [8]).

The proof of (3.4) is divided in two parts.

i) Assume that (iii) is false. Show that the diagram is o═══o——o ········ o——o , o——o═══o——o or o═══o═══o .

ii) In the former cases show that the geometry is the desired building.

For the sketch of the proof we may assume that all $P_{I-\{i\}}$ are 2-constrained. Then (i) is proved by induction. The first step is to write down the structure of the maximal parabolics for o═══o═══o . We get that they are of the form $P/O_2(P) \cong S_6$ or A_6. Then we prove the following lemma.

<u>Lemma (3.5)</u>. Let G be a group with diagram o═══o═══o , such that all maximal parabolics P are 2-constrained and $P/O_2(P) \cong S_6$ or A_6 if P belongs to o═══o . Then G has no f.f. + 1 module over $GF(2)$.

We say that V is a f.f. + 1 module for G, iff G is faithful and there is an elementary abelian subgroup A of G such that $|V/C_V(A)| \leq 2|A|$.

Lemma (3.5) yields then that the diagram o====o====o is not extendable. We will illustrate the importance of f.f. + 1 modules in the following example.

We assume that our geometry has the following diagram

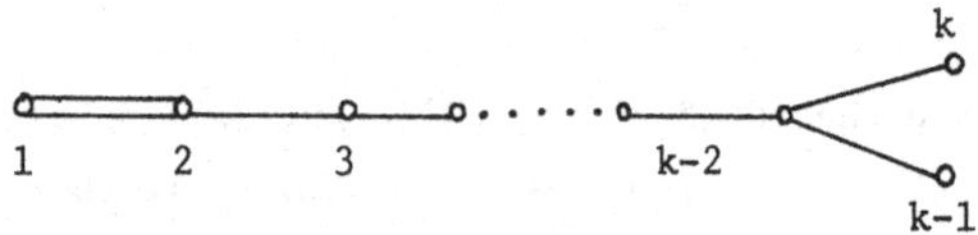

with k big enough ($k \geq 7$ suffices). Suppose that the group belonging to o====o (1 2) is S_6. Set $G_i = P_{I-\{i\}}$, $i = 1,\dots,k$. Then by induction and a result due to Timmesfeld [12] we have $G_1/O_2(G_1) \cong SO^+(2(k-1),2)$ and $G_k/O_2(G_k) \cong G_{k-1}/O_2(G_{k-1}) \cong Sp(2(k-1),2)$. Let $B = N_G(S)$. Then set

$$Z_i = \langle \Omega_1(Z(S))^{G_i} \rangle, \ i = 1,\dots,k.$$

These Z_i play an important part in the proof. If $Z_i \not\leq Z(G_i)$, then an easy argument shows $J(S) \leq O_2(G_i)$ or Z_i is a f.f.-module.

Suppose Z_1 to be a f.f.-module. Then by [4] we have that Z_1 involves only the natural module. The structure of this module yields that neither Z_k nor Z_{k-1} is contained in $Z(G_k)$, $Z(G_{k-1})$ respectively. Thus one of both is a f.f.-module. Again [4] yields that Z_k is the symplectic module. But then $G_k \cap G_1$ centralizes $\Omega_1(Z(S))$, contradicting the action of $SO^+(2(k-1),2)$ on the natural module.

Thus we have that Z_1 is not a f.f.-module. Then we may assume that Z_k is a f.f.-module and by [4] Z_k involves only the natural symplectic module. This implies $Z_1 \not\leq Z(G_1)$.

We have $G_2/O_2(G_2) \cong S_3 \times SO^+(2(k-2),2)$. Let Γ be the graph whose vertices are the cosets of G_1 and G_2 in G and two different cosets are joined iff their intersection is nonempty. Let $\alpha \in \Gamma$ of minimal distance from 1 such that $Z_1 \leq G_\alpha$ but $Z_1 \not\leq O_2(G_\alpha)$ (i.e. Z_1 stabilizes α but not all neighbors of α). Because Z_1 is not a f.f.-module we have that α is in the same class as 2. Then

$|Z_1: Z_1 \cap O_2(G_\alpha)| = 2$. Let β be a neighbor of α such that $Z_1 \not\leq G_\beta$. Then $Z_1 \cap O_2(G_\alpha)$ acts on Z_β. With the same argument we may assume that $Z_\beta \cap O_2(G_2)$ acts on Z_1. If $Z_1 \cap O_2(G_\alpha) \not\leq O_2(G_\beta)$ we get that Z_β and then Z_1 is a f.f. + 1-module. An easy check of the modules for $SO^+(2n,2)$ or [4] yields a contradiction. Thus $[Z_1 \cap O_2(G_\alpha), Z_\beta] = 1$. Then $\langle O_2(G_1), Z_\beta\rangle$ centralizes a subgroup of index 2 in Z_1. As there are no elements in G_1 inducing transvections on Z_1, we have $O_2(G_1) \cap O_2(G_2) \trianglelefteq G_2$ and for some neighbor γ of 2 we have $[Z_\beta, G_2 \cap G_\gamma] \leq O_2(G_1) \cap O_2(G_2)$. This yields $[Z_\beta, Z_1, G_2 \cap G_\gamma] = 1$ and so some $x \in \Omega_1(Z(S))$ is centralized by $G_1 \cap G_2$. Now the structure of the natural module for the symplectic group yields that $G_k = \langle C_{G_k}(x), G_1 \cap G_2 \cap G_k\rangle$. But as $G = \langle G_k, G_2 \cap G_1\rangle$ we get $\langle x\rangle \trianglelefteq G$, a contradiction.

Thus we have shown that k has to be small. With similar arguments we can treat all the diagrams coming up with some small configurations. Then some more delicate arguments are used to investigate modules like $\langle Z_i^{G_j}\rangle$. But the spirit is more or less the same. Here some results due to M. Ronan and S. Smith are helpful [7].

The proof of (ii) is an application of [13, Prop. 9]. The assumptions for the application of this result can be verified using the structure of $G_i/O_2(G_i)$.

3.6 Problems

a. For applications the case (iii) causes no problems because the structure of a Sylow 2-subgroup of G is known. But nevertheless it would be interesting to know whether such chamber systems can occur. It is possible to determine all diagrams. Some examples are the stars

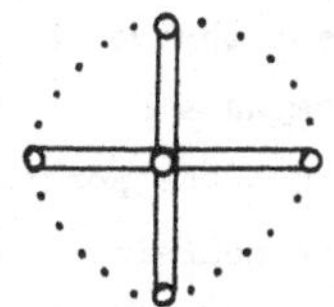

and

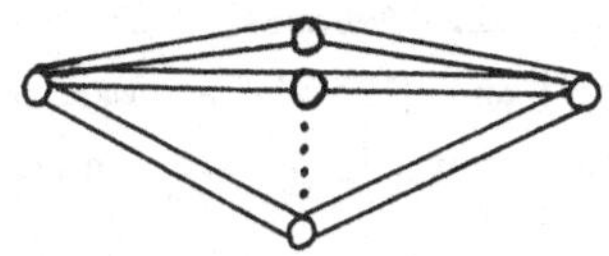

where the second one forces that all parabolics belonging to $\circ\!=\!=\!\circ$ are isomorphic to S_6 or A_6.

b. From the examples it would be interesting to prove the same theorem if $O'(G_\Gamma/O_2(G_\Gamma)) \cong A_7$ for some residual Γ of type o———o═══o. In (2.3) d. we have a nice example. In this example the geometry of type (ii) occurs as a residual too.

c. What is the result if in (3.4) $O'(P_{ij}/O_2(P_{ij})) \cong 3A_6$ or $3S_6$ is added. For $3A_6$ we get the further example Suz: o═══o═══o

For $3S_6$ we have the covering o═══o (with ~) of the $Sp_4(2)$-quadrangle. Because of (2.2) we should then assume $G_\Gamma/O_2(G_\Gamma) \cong M_{24}$ for each residual Γ of type o———o═══o (with ~).

4 CONCLUDING REMARKS

The overall idea in the area of chamber system classifications is the proof by induction. Thus the first problem is to give the exact list of all groups H such that $H \cong P/O_2(P)$, where P is a minimal parabolic in some chamber system of rank 2. In doing this it is helpful to assume that all parabolics are 2-constrained. The non 2-constrained case is related to wreath products and causes only minimal trouble. Assume we want to classify the system $\{P_1,P_2,P_3\}$ of rank 3. Then a good induction would be to know the groups $G_i = P_{\{1,2,3\}-\{i\}}/O_2(P_{\{1,2,3\}-\{i\}})$. But all the classification theorems are more or less of the shape: Given the $P_i/O_2(P_i)$ and perhaps some further condition, then the multiplication table of the maximal parabolics is as follows. In some very good situations where we have buildings it is possible to give the group G generated by the P_i's. In view of the application in the classification of finite simple groups working with a minimal counterexample this seems to be no problem, since we may determine the G_i if we have the parabolics of G_i. But this gives not G_i this gives only $G_i/O(G_i)$. Assume that $P_i/O_2(P_i) \cong L_2(q)$, then there is no reason why we should have $G_1 \cong L_3(q)$. This means that the residual is perhaps not the projective plane but a certain covering of this plane. Thus the possible existence of nontrivial normal subgroups of odd order not contained in the center of G_i gives a lot of trouble. It seems that if $O(P_J'/O_2(P_J')) = 1$ for each J with $|J| = 2$ then it is always trivial. Because of that we made in Chapter 3 and also F. Timmesfeld in [11] and [12] the assumption that $O(O'(G_J)) = 1$ for each $J = \{i,j\}$ such that i, j are connected in the diagram.

From the comments above it follows that it would be helpful to have theorems available avoiding the classification of the G_i. As pointed out in Chapter 3 the classification of minimal parabolic systems is related to question about modules for groups which are near to f.f.-modules. Thus it would be helpful to classify all f.f.+n-modules for small n for a given group using only the minimal parabolics and not the group itself. In other words we need a representation theory using only the parabolics. Some results in this direction can be found in [7] and [10]. For the proof of (3.4) we used such a kind of theory, for example (3.5).

The philosophy in this area is that there are lots of examples up to rank 5 (the highest rank comes from F_1). But if the rank is at least 6 there are only a few series of examples. One of this series are the Lie-groups in characteristic 2, which are more or less classified by J. Tits [13]. Unfortunately there are also series coming from the alternating groups. Thus one of the major problems is to determine all these geometries and hopefully classify them.

REFERENCES

1. Aschbacher, M. (1980). On Finite Groups of Lie Type and Odd Characteristic. J. Alg. 66, 400-424.
2. Buekenhout, F. (1979). Diagrams for Geometries and Groups. J. Comb. Theory A 27, 121-151.
3. Conway, J.H. (1971). Three lectures on Exceptional Groups. In Finite Simple Groups, ed. Higman, Powell, pp. 215-247. Academic Press.
4. Cooperstein, B. (1978). An Enemies List for Factorization Theorems. Comm. Alg. 6, 1239-1288.
5. Goldschmidt, D. (1980). Automorphism of Trivalent Graphs. Annals of Math. 111, 377-406.
6. Ott, U. On Finite Geometries of Type B_3. Preprint.
7. Ronan, M.A. & Smith, S.D. Sheaf Homology on Buildings and Modular Representations on Chevalley Groups. Preprint.
8. Ronan, M.A. & Smith, S.D. (1980). 2-Local Geometries for some Sporadic Groups. AMS Symposia in Pure Math. 37 (Finite Groups) 283-289, American Math. Soc.
9. Ronan, M.A. & Stroth, G. Minimal Parabolic Geometries for the Sporadic Groups. Europ. J., to appear.
10. Smith, S.D. (1982). Irreducible Modules and Parabolic Subgroups. J. Alg. 75, 286-289.
11. Timmesfeld, F. Tits Geometries and Parabolic Systems of Rank 3, preprint.
12. Timmesfeld, F. Tits Geometries and Parabolic Systems in Finite Groups, preprint.
13. Tits, J. (1981). A Local Approach to Buildings. In The Geometric Vein, pp. 519-547. Springer.
14. Huang, Stellmacher, B., Stroth, G. Some parabolic systems of rank two related to sporadic groups, preprint.

6. [illegible]

MINIMAL PARABOLIC SYSTEMS INVOLVING S_5 AND S_3

G. Stroth
Freie Universität Berlin

S.K. Wong
Ohio State University

In [2], M. Ronan and G. Stroth have determined the minimal parabolic systems for all the sporadic simple groups. Looking through their list for the prime 2, we see that half the sporadics have a minimal parabolic system where one minimal parabolic has the form $P/O_2(P) \simeq S_5$ and the rest have $P/O_2(P) \simeq S_3$. The examples are: M_{22}, HiS, Ru, Ly, $\mathrm{Aut}(J_2)$ which are all of rank 2 and the rank 3 examples are M_{23}, $\cdot 2$, J_4, M(22), Mc and Aut(Sz). There are only two rank 4 examples, M(23) and F_2.

The authors have been interested in classifying all the minimal parabolics up to rank 4, if they are all of the form $P/O_2(P) \simeq S_5$ or $P/O_2(P) \simeq S_3$. We have not completed this classification. The aim of this note is to report on a very special case of such a classification.

However, the rank 2 case has been completely solved by B. Stellmacher and G. Stroth. More precisely they have proved

<u>Theorem 1</u>. Let $G = \langle P_1, P_2 \rangle$, $B = P_1 \cap P_2$ the normalizer of a Sylow 2-subgroup of G, and no nontrivial normal subgroup of G is contained in P_1 or P_2. Suppose that P_1 and P_2 are 2-constrained. If $P_1/O_2(P_1) \simeq S_5$ and $P_2/O_2(P_2) \simeq S_3$, then it is possible to write down the multiplication table of P_1 and P_2.

Before stating our result, let us consider the following rank 3 examples. Let G M_{23}, $\cdot 2$ or J_4. Then $G = \langle P_1, P_2, P_3 \rangle$ with $P_1/O_2(P_1) \simeq P_2/O_2(P_2) \simeq S_3$ and $P_3/O_2(P_3) \simeq S_5$. Let $G_i = \langle P_j, P_k \rangle$ with $\{i,j,k\} = \{1,2,3\}$ and $Q_i = O_2(G_i)$. If $G \simeq J_4$ then $G_1/Q_1 \simeq 3 \cdot M_{22} \cdot 2$, $G_2/Q_2 \simeq S_5 \times S_3$ and $G_3/Q_3 \simeq L_3(2)$. If $G \simeq \cdot 2$ then $G_1/Q_1 \simeq \mathrm{Aut}(M_{22})$,

$G_2/Q_2 \simeq S_5 \times S_3$ and $G_3/Q_3 \simeq L_3(2)$. If $G \simeq M_{23}$ then $G_1 \simeq M_{22}$, $G_2/Q_2 \simeq (A_5 \times 3) \cdot 2$ and $G_3/Q_3 \simeq L_3(2)$.

This leads us to investigate the following situation:

<u>Hypothesis 2</u>. Let G be a finite group with $O_2(G) = 1$, $S \in Syl_2(G)$, P_1, P_2, P_3 subgroups of G such that

i) $G = \langle P_1,P_2,P_3\rangle \neq \langle P_i,P_j | i,j\in\{1,2,3\}\rangle$

ii) $P_1 \cap P_2 \cap P_3 = S = N_G(S)$

iii) $P_i/O_2(P_i) \simeq S_3$ or S_5 and for some i, $P_i/O_2(P_i) \simeq S_5$.

iv) P_i are 2-constrained for $i = 1,2,3$.

v) $G_i/Q_i \simeq L_3(2)$, M_{22}, $Aut(M_{22})$, $3 \cdot M_{22} \cdot 2$, $S_3 \times S_3$, $(Z_3 \times Z_3) \cdot 2$, $S_5 \times S_3$, $(A_5 \times 3) \cdot 2$, $S_5 \times S_5$ or $(A_5 \times A_5) \cdot 2$, where $G_i = \langle P_j,P_k\rangle$, $\{i,j,k\} = \{1,2,3\}$ and $Q_i = O_2(G_i)$.

Further, we define the diagram as a graph with vertex set $\{1,2,3\}$ and two distinct vertices i, j are joined iff G_i/Q_i involved $L_3(2)$ or M_{22}.

Then we get the following theorem.

<u>Theorem 3</u>. Assume hypothesis 2 and that the diagram is connected. Then the diagram is of the form •—•—• (1 2 3) with $P_1/Q_2(P_1) \simeq P_2/O_2(P_2) \simeq S_3$, $P_3/O_2(P_3) \simeq S_5$ and G_1 has one of the following possibilities (a) $G_1 \simeq M_{22}$, (b) $G_1 \simeq 2^{1+12} \cdot 3{\cdot}M_{22}{\cdot}2$ or (c) $G_1 \simeq 2^{10}\,Aut(M_{22})$.

The proof of the theorem involves a case by case elimination of many configurations. We will only outline the analysis of the following situation. From now on assume $G_1/Q_1 \simeq M_{22}$, $Aut(M_{22})$ or $3 \cdot M_{22} \cdot 2$, $G_2/Q_2 \simeq S_5 \times S_3$ or $(A_5 \times 3) \cdot 2$ and $G_3/Q_3 \simeq L_3(2)$. If one of the G_i, $i = 1,2,3$ is not 2-constrained, an easy argument shows that $G_1 \simeq M_{22}$ and we have case (a) of the theorem. Hence assume all G_i are 2-constrained. Next we set

$$Z_i = \langle \Omega_1(Z(S))^{G_i}\rangle.$$

We then have to consider the following two cases: $Z_1 \subseteq Z(G_1)$ or

$Z_1 \nsubseteq Z(G_1)$. First let us assume that $Z_1 \subseteq Z(G_1)$. Then we have $J(S) \nsubseteq Q_1$, $Q_1 \nsubseteq Q_2$ and $G_2/Q_2 \simeq S_5 \times S_3$. Let $E = \langle Z_2^{G_1} \rangle$. We want to show E is non-abelian. Hence assume E is abelian. Then Z_2 is a fours-group and Z_1 is of order 2 and $[E,Q_1] \subseteq Z_1$. Let Γ be the graph whose vertices are the cosets of G_1 and G_2 in G, and two different cosets are joined iff their intersection is non-empty. We identify G_1 with 1 and G_2 with 2. If $G_\alpha = G_i^g$ and $V_i \triangleleft G_i$ then let $V_\alpha = V_i^g$.

Let $\alpha \in \Gamma$ of minimal distance from 1 such that $E \subseteq G_\alpha$ but $E \nsubseteq Q_\alpha$. If $G_\alpha = G_1^g$, then we have $E_\alpha \subseteq G_1$. Since there are no transvections in M_{22}, it follows that $E_\alpha \nsubseteq Q_1$. Let $x \in E \backslash E \cap Q_\alpha$, then $|E_\alpha : C_{E_\alpha}(x)| \leq 2 \cdot |E_\alpha : E_\alpha \quad Q_1| \leq 2^6$. Then by a result of M. Aschbacher [1], E_α has only one irreducible module and this is of dimension 10 or 12. Based on the structure of this module, we arrive at a contradiction. Now we have $G_\alpha = G_2^g$. Since $E \nsubseteq Q_\alpha$, there exists a neighbor β of α such that $E \nsubseteq G_\beta$. Now $E_{\alpha-1} = \langle Z_\alpha^{G_{\alpha-1}} \rangle$ and by choice of α, we have $Z_\alpha \subseteq Q_1$. Hence $[Z_\alpha, E] = Z_1 \subseteq Z_\alpha \subseteq E_\beta = \langle Z_\alpha^{G_\beta} \rangle$. So $Z_\alpha = Z_1 Z_\beta$. Let u be a neighbor of 1 such that the distance of α from u is one less than the distance of α from 1. Then $E_\beta \subseteq G_u$ and $[E_\beta, Z_1] = 1$. Hence $E_\beta \subseteq Q_u \cdot Q_1$, otherwise $Z_1 \triangleleft G_u$, which is a contradiction. This shows that $E_\beta \subseteq G_1$. If $E_\beta \subseteq Q_1$, then $[E_\beta, E] = Z_1$ and so E normalizes E_β. Hence $E_\beta \triangleleft \langle G_\alpha, G_\beta \rangle = G$, which is impossible. Now if $E \cap Q_\alpha \subseteq Q_\beta$, then $[E \cap Q_\alpha, E_\beta] \subseteq Z_\beta \cap E = 1$. This is impossible as M_{22} contains no transvections. Let $e \in E \backslash E \cap Q_\beta$. Then $|E_\beta : C_{E_\beta}(e)| \leq 2^6$. Again [1] yields that E_β has only one irreducible module which is of dimension 10 or 12. A check of the modules yields similarly a contradiction. With this we have $E = \langle Z_2^{G_1} \rangle$ is nonabelian. We show next that Z_2 is a fours-group and since $[Q_1, E] \subseteq Z_1$ we have $E' = \phi(E) = Z_1$. Hence $E = Z(E) * F$, where F is an extra-special 2-group. It can be shown that M_{22} acts trivially on Q_1/E. Since the action of M_{22} on $Q_1/C_{Q_1}(Z(E))$ is dual to that on $Z(E)/Z_1$, we have M_{22} is trivial on $Z(E)$. This gives $Z(E)$ is of order 2 and E is an extra-special 2-group. Then $Q_1 = EC_{Q_1}(E)$ and further analysis gives $Q_1 = E$. From here it is an easy argument to show that $|Q_1| = 2^{13}$ and we obtain case (b) of the theorem. The analysis of the case $Z_1 \nsubseteq Z(G_1)$ is similar. In this case we show that $Q_1 = Z_1$ is elementary abelian of order 2^{10} and this gives case (c).

We conclude this note by referring the reader to section 4 of the article of G. Stroth in these proceedings for a discussion of some of the ideas that may be relevant in treating the more general problem.

REFERENCES

1. Aschbacher, M. (1982). GF(2)-Representations of finite groups. American Journal of Math., Vol. 104, 683-771.
2. Ronan, M.A. & Stroth, G. Minimal Parabolic Geometries for the Sporadic Groups. Europ. J. To appear.

TITS GEOMETRIES AND REVISIONISM OF THE CLASSIFICATION OF FINITE SIMPLE GROUPS OF CHAR. 2-TYPE

F.G. Timmesfeld

1 INTRODUCTION

In [12] Tits defined a <u>chamber system</u> $\mathcal{C} = (C, (\mathfrak{p}_i)_{i \in I})$ over some index I to be a set C of <u>"chambers"</u> together with partitions $\mathfrak{p}_i$, $i \in I$ of C. If one has such a chamber system $\mathcal{C}$ then $\mathfrak{p}_J$ is the join of all partitions $\mathfrak{p}_j$, $j \in J$ for each $J \subseteq I$. If $c \in C$ and $J \subseteq I$, then <u>$\Delta_J(c)$</u> is the element of $\mathfrak{p}_J$ containing c. Notice that this set-up is very inductive, since the $\Delta_J(c)$ are again chamber systems over J with partitions $\mathfrak{p}_j$, $j \in J$ restricted to $\Delta_J(c)$. We say $\mathcal{C}$ is <u>connected</u> if and only if $\Delta_I(c) = C$ for some $c \in C$.

Now the consideration of such chamber systems with transitive automorphism group G (i.e. G acts transitively on C and respects all partitions $\mathfrak{p}_i$) is equivalent to the following group theoretic situation: G is a group, B a subgroup of G and X_i, $i \in I$ are subgroups of G containing B. Namely in this situation one can construct the chamber system <u>$\mathcal{C}(G/B, (G/X_i)_{i \in I})$</u>, which has a chamber set the cosets Bg of B, two cosets Bg, Bh being i-adjacent if and only if $X_i g = X_i h$. Moreover, it is easy to see that $\mathcal{C}(G/B, (G/X_i)_{i \in I})$ is connected if and only if $G = \langle X_i | i \in I \rangle$.

Also if G is a transitive automorphism group of $\mathcal{C} = (C, (\mathfrak{p}_i)_{i \in I})$, $c \in C$, $B = G_c$ and the X_i are the stabilizers of $\Delta_i(c)$ for each $i \in I$, then the map $c^g \to Bg$ is a canonic isomorphism of $\mathcal{C}$ onto $\mathcal{C}(G/B, (G/X_i)_{i \in I})$.

In [9] I suggested the following 3-step program for an alternative classification of the finite simple groups of (weak) characteristic 2-type. If G is such a group and $S \in Syl_2(G)$ let $G = N(S)$ and P_i, $i \in I$ the <u>minimal parabolic</u> subgroups of G containing B, where P_i is a minimal parabolic if it satisfies:

1) $F^*(P_i) = O_2(P_i)$
2) B is contained in an unique maximal subgroup of G.

Then it is easy to see that either $G \neq C(G,S)$ (characteristic core as defined by M. Aschbacher [1]) or $G = \langle P_i | i \in I \rangle$. So by the C(G,S)-theorem one may assume that the latter is the case. Hence choosing I minimal (i.e. $G \neq \langle P_j | j \in J \rangle$ for each proper subset J of I) one has the following program:

1) Restrict the structure of $O^{2'}(P_i/O_2(P_i))$.

2) Determine $O^{2'}\langle P_i,P_j\rangle/O_2\langle P_i,P_j\rangle$ using Goldschmidt's graph approach.

3) Determine the groups G of rank $|I| > 2$ using the theory of chamber systems $C(G/B,(G/P_i)_{i \in I})$.

As I remarked in [9] step (1) is the great problem in this program. It is not enough to determine "abstractly" the structure of a minimal parabolic subgroup, since it might be very complicated. (For example a wreath product of some rank 1 Chevalley-group with an arbitrary 2-group acting regularly!) One has to restrict the structure of such a P_i inside a simple group of (weak) characteristic 2-type. At the moment no methods seem to be known for doing this. Still I think it is worthwhile to carry on with step (2) and (3) of this program under the assumption that the P_i's have a restricted structure. (Essentially each $O^{2'}(P_i/O_2(P_i))$ is a rank 1 Chevalley-group in char.2.) Namely first of all it leads to a better understanding of the groups of (weak) characteristic 2-type and secondly the case of small $|I|$ is useful for the revisionism of the classification of groups with small $e(G)$, since in this case the structure of the P_i's is necessarily restricted.

As can be seen from these proceedings, step (2) and (3) are already in good shape under the hypothesis that $O^{2'}(P_i/O_2(P_i))$ is essentially a rank 1 Chevalley-group in char.2. Delgado and Stellmacher have now a uniform theorem determining the structure of $O^{2'}(P_i/O_2\langle P_i,P_j\rangle)$ for each pair $i,j \in I$, which does not use the earlier results of Goldschmidt, Stellmacher and Stroth. Using known local classifications this determines $O^{2'}(\langle P_i,P_j\rangle)/O_2(\langle P_i,P_j\rangle)$, since we assume that $S \in Syl_2(\langle P_i,P_j\rangle)$. The generic and more difficult case is when the $\bar{P}_{i,j} = O^{2'}\langle P_i,P_j\rangle/O_2(\langle P_i,P_j\rangle)$ are "essentially" all rank 2 Chevalley-groups in characteristic 2. (Neglecting the possibility of non-central cores in $\bar{P}_{i,j}$!) Namely my work on the rank 3 case is based on FF-modules (failure of factorization) [3], and it is always an easy case if for some pair i,j the group $\bar{P}_{i,j}$ either has no FF-module or exactly one such module and the "offending subgroups" of $\bar{P}_{i,j}$ for this module are uniquely determined up to conjugacy. So on this basis the treatment of step (3) if some $\bar{P}_{i,j}$ is not "essentially" a Chevalley-group (for example $\bar{P}_{i,j} \simeq A_7$ or M_{11}!) should be easier.

In this note I will describe, mostly without proof, the results obtained so far in the case where all $\bar{P}_{i,j}$ are "essentially" Chevalley-groups.

2 NOTATION AND CONNECTIONS

(2.1) A weak parabolic system $\mathfrak{P}$ in characteristic p of some group G, is a system of subgroups P_i, $i \in I$ satisfying:

1) $G = \langle P_i | i \in I\rangle \neq \langle P_j | j \in J\rangle$ for each proper $J \subset I$.

2) There exists a finite p-subgroup $S \leq \bigcap_{i\in I} P_i$ such that $S \in Syl_p(\langle P_i, P_j\rangle)$ for each pair $i,j \in I$.

3) $\bar{P}_i = O^{p'}(P_i/O_p(P_i))$ is a rank 1 Chevalley-group in char.p for each $i \in I$.

4) $\bar{P}_{i,j} = O^{p'}\langle P_i,P_j\rangle/O_p\langle P_i,P_j\rangle$ is a rank 2 Chevalley-group in char.p for all $i,j \in I$.

(In (4) we allow all kinds of degeneracies such as $(\mathbb{Z}_3 \times \mathbb{Z}_3)\mathbb{Z}_2 \subseteq \Omega^+(4,2)$, A_6, $U_3(3)$, ${}^2F_4(2)'$ in char.2 or $(Q_8 \times Q_8)\mathbb{Z}_3 \subseteq SL_2(3)*SL_2(3)$ in char.3.) The diagram $\underline{\Delta(\mathfrak{P})}$ has vertex set I with edges of strength n_{ij} connecting i and j; where the Weyl-group of $\bar{P}_{i,j}$ is defined by the relations $w_i^2 = 1 = w_j^2 = (w_iw_j)^{n_{ij}}$. (In cases like A_6, where $\bar{P}_{i,j}$ is not a Chevalley-group in the strict sense one has to take the Weyl group of $\Sigma_6 \simeq Sp(4,2)$ and so on!) $\mathfrak{P}$ is called a parabolic system, if $S \in Syl_p(G)$. (Notice that in general G is not necessarily finite, although by (2) and (3) all the P_i's are finite!) If for some $i,j \in I$ we have $\bar{P}_{i,j} \simeq \hat{A}_6$ or $\hat{\Sigma}_6$, the perfect extension of A_6 or Σ_6 by $\mathbb{Z}_3$, and if $P_i/O_2\langle P_i,P_j\rangle \simeq \Sigma_4$ resp. $\Sigma_4 \times \mathbb{Z}_2$, then we write $\overset{i}{\text{o}}\!\overset{\sim}{=\!=}\!\overset{j}{\text{o}}$.
In this case $\mathfrak{P}$ is called a semi-parabolic system! (If $\bar{P}_{i,j} \simeq \hat{A}_6$ and $P_i/O_2\langle P_i,P_j\rangle \simeq \Sigma_4 \times \mathbb{Z}_3 \simeq P_j/O_2\langle P_i,P_j\rangle$ then $\overset{i}{\text{o}}\!=\!=\!\overset{j}{\text{o}}$ and $\mathfrak{P}$ remains a weak parabolic system. The case $\bar{P}_{i,j} \simeq \hat{\Sigma}_6$ and $P_i/O_2\langle P_i,P_j\rangle \simeq \Sigma_4 \times \Sigma_3$ is not considered by condition (3)!)

(2.2) Let now $C = (C,(P_i)_{i\in I})$ be a connected chamber system with transitive automorphism group G. Let $B = G_c$ for a fixed $c \in C$ and call B a Borel-subgroup of G. We call C of type CM (i.e. classical type M), if for each $J \subseteq I$ with $|J| = 2$ one of the following holds:

1) $\Delta_J(c) \simeq C(G_o/B_o, G_o/X_1, G_o/X_2)$, where G_o is a finite rank 2 Chevalley-group, B_o a Borel-subgroup of G_o and X_1, X_2 the parabolic subgroups of G_o containing B_o.

2) $\Delta_J(c) \simeq C(X/X_1 \cap X_2, X/X_1, X/X_2)$, where $X = X_1X_2$ and X_1, X_2 are subgroups of X.

(In case (2) the chamber system $\Delta_J(c)$ is called a generalized digon. X must be a product of X_1 and X_2, but not necessarily a direct product!)

We call C of type SM if for each, $J \subseteq I$ with $|J| = 2$ either (1) or (2) holds for $\Delta_J(c)$ or

3) $\Delta_J(c) \simeq C(\hat{\Sigma}_6/S, \hat{\Sigma}_6/X_1, \hat{\Sigma}_6/X_2)$; where $S \in Syl_2(\hat{\Sigma}_6)$ and X_1, X_2 are two subgroups of $\hat{\Sigma}_6$ containing S isomorphic to $\Sigma_4 \times \mathbf{Z}_2$ with $\hat{\Sigma}_6 = \langle X_1, X_2 \rangle$.

Now the diagram $\Delta(C)$ is defined in the same way as for weak parabolic (or semi-parabolic) systems. So the problem is in some sense to classify all chamber systems of type CM (or even SM) with transitive automorphism group G and finite Borel-subgroup B and also these automorphism groups.

This is slightly more general than the similar problem for weak parabolic systems, since as shown in [9, (4.2)] one obtains from each weak parabolic system a chamber system with the same diagram Δ, transitive automorphism group and finite Borel-subgroup by slightly altering the groups P_i, $i \in I$. Also, as shown in [9, (4.4)], the classification of weak parabolic systems reduces easily to the classification of those with connected diagrams. So we will state theorems in §3 only for connected diagrams.

(2.3) Now, since we want to apply geometric classifications, and since even chamber systems with diagram $\Delta = A_3$ are not necessarily buildings (i.e. chamber systems of buildings!) we need a connection with diagram geometries.

A geometry Γ over an index set I, is a triple $\Gamma = (V, *, \tau)$; where V is a set of "vertices" or "objects", τ the "type map" a surjective map from V onto I and $*$ an incidence relation on V, satisfying $v * w$ and $\tau(v) = \tau(w)$ if and only if $v = w$. A flag F of V is a set of pairwise incident vertices of V. The residue res. F of a flag F is the set

$$V_F := \{v \in V-F \mid v * f \text{ for each } f \in F\}$$

with incidence relation and type map restricted to res.F. It is clear that res.F is again a geometry over $\tau(\text{res.F})$. Such a geometry Γ is called <u>residually connected</u> if it satisfies:

1) The graph $(V,*)$ is connected and the same property holds for $(V_F,*)$ for each flag F of corank at least 2. (I.e. $|I-\tau(F)| \geq 2$.)

2) $V_F \neq \emptyset$ for each flag of corank 1.

The canonical examples of such geometries are the <u>group geometries</u> $\Gamma = \Gamma(G,(G_i)_{i \in I})$; where G is a group and the G_i are (not necessarily distinct) subgroups of G. The vertices of type i are the cosets G_ig, $g \in G$ with $G_ig * G_jh$ if and only if $G_ig \cap G_jh \neq \emptyset$. Γ is connected if and only if $G = \langle G_i \mid i \in I\rangle$. Moreover, each geometry Γ with automorphism group G acting transitively on the pairs of incident vertices of each type which has a flag F of type I is isomorphic to such a $\Gamma(G,(G_i)_{i \in I})$.

A residually connected geometry Γ is of <u>type M</u>, if for each flag F of corank 2 and type $\{i,j\}$ res.F is a generalized n_{ij}-gon. It is of <u>type CM</u> (classical type M) if whenever $n_{ij} > 2$ then res.F is isomorphic to the generalized n_{ij}-gon of some finite rank 2 Chevalley-group. Similarly we define type SM. The diagram $\Delta(\Gamma)$ is now defined in the same way as for chamber systems. We have the following:

<u>Theorem (2.4)</u>. Let Γ be a residually connected geometry. Then the following holds:

1) Suppose Γ is of type A_n or D_n, $n \geq 3$. Then Γ is a building of that type.

2) Suppose Γ is of classical type C_n, $n \geq 3$ and admits a flag-transitive automorphism group G. Then either Γ is a building of type C_n or $G \simeq A_7$ and Γ is isomorphic to the unique geometry of type C_3 obtained from A_7.

3) If Γ is of classical type A_n, C_n or D_n and G is a flag-transitive automorphism group of Γ, then either G is an extension of a Chevalley-group of such a type by diagonal and field automorphisms or $G \simeq A_7$ and Γ is either of type A_3 or C_3.

Proof. For type A_n (1) is well-known (see [12]). For type D_n see [9, (3.3)]. (2) was proved by M. Aschbacher [2] using the work of Tits [12]. (3) is a consequence of (1), (2) and a theorem of Seitz [8]*, which determines the groups acting flag-transitively on a finite building. (Γ is finite since it is of spherical type and the rank 2 residues are finite!)

(2.5) If Γ is a residually connected geometry, then one can define a chamber system $C(\Gamma)$ in the following way: The chambers are the flags of Γ of type I, two such chambers being i-adjacent if and only if they coincide in a flag of type $I-\{i\}$. It is easy to see that $C(\Gamma)$ is connected and that the diagrams $\Delta(\Gamma)$ and $\Delta(C(\Gamma))$ coincide.

On the other hand, if C is a connected chamber system we may define a geometry $\Gamma(C)$, whose vertices of type i are the "cells" $\Delta_{I-\{i\}}(c)$, $c \in C$ and two such vertices $\Delta_{I-\{i\}}(c)$, $\Delta_{I-\{j\}}(d)$ are incident if and only if $\Delta_{I-\{i\}}(c) \cap \Delta_{I-\{j\}}(d) \neq \emptyset$. Now unfortunately one has in general neither $\Gamma \simeq \Gamma(C(\Gamma))$ nor $C \simeq C(\Gamma(C))$. Tits has shown that the first holds for residually connected geometries Γ, but the second does not hold in such generality.

So if C is a connected chamber system let $\mathcal{P}_\emptyset = \{\{c\} | c \in C\}$. We say that C satisfies the fundamental condition if for every $J,J' \subseteq I$ and $c \in C$ we have

$$\Delta_J(c) \cap \Delta_{J'}(c) = \Delta_{J \cap J'}(c).$$

We say the diagram $\Delta(C)$ is a string, if the index set I can be ordered in a way such that $\Delta_{\{i,i+1\}}(c)$ is a generalized digon for $i = 1,\dots,n-1$ and $c \in C$. (This is defined for even more general diagrams then those considered in this paper.) Now by [6] we have:

Proposition (2.6). Let C be a connected chamber system over I having a string diagram Δ and satisfying the fundamental condition. Then the following holds:

1) There exists a canonic isomorphism $C \to C(\Gamma(C))$.

2) Each flag of $\Gamma(C)$ is contained in a flag of type I.

3) $\Gamma(C)$ belongs to the diagram Δ and is residually connected.

---

*In his published version Seitz overlooked two groups, $2^4 \cdot F_{20}$ and $L_3(4) \cdot 2$ acting flag-transitively on the PSp(4,3) resp. PSU(4,3) generalized quadrangle.

4) If G is a group of automorphisms acting transitively on $\mathcal{C}$, then $\mathcal{C}$ is flag-transitive on $\Gamma(\mathcal{C})$.

(2.7) A <u>2-covering</u> of a chamber system $\tilde{\mathcal{C}}$ over I onto a chamber system $\mathcal{C}$ over I is a surjective morphism $\pi:\tilde{\mathcal{C}} \to \mathcal{C}$ satisfying $\pi|_{\Delta_J(c)}$ is an isomorphism for each $J \subseteq I$ with $|J| = 2$ and each $c \in \tilde{\mathcal{C}}$. Abusing notation we also call $\tilde{\mathcal{C}}$ a 2-covering of $\mathcal{C}$. The connected chamber system $\mathcal{C}$ is <u>simply connected</u>, if each 2-covering $\tilde{\mathcal{C}} \to \mathcal{C}$ with connected chamber system $\tilde{\mathcal{C}}$ is an isomorphism. Finally for a connected chamber system the 2-covering $\pi:\tilde{\mathcal{C}} \to \mathcal{C}$ is called <u>universal</u> if $\tilde{\mathcal{C}}$ is connected and if for each 2-covering $\varphi:\mathcal{D} \to \mathcal{C}$ with connected $\mathcal{D}$ and $\tilde{c} \in \tilde{\mathcal{C}}$, $d \in \mathcal{D}$ with $\pi(\tilde{c}) = \varphi(d)$, there exists a 2-covering $\sigma:\tilde{\mathcal{C}} \to \mathcal{D}$ such that $\pi = \varphi\sigma$ and $\sigma(\tilde{c}) = d$. Tits has shown in [12, (5.1)] that each connected chamber system possesses such a universal 2-covering and that it is uniquely determined up to isomorphism.

Moreover, since Tits has shown in [12] that a building is uniquely determined by its chamber system we call a chamber system a building, if it is the chamber system of a building. Now we may phrase the main theorem of Tits [12] for our case of chamber systems of type CM in the following way:

<u>Theorem (2.8)</u>. Suppose $\mathcal{C}$ is a chamber system of type CM over I and $|I| \geq 3$. Then the following holds:

1) Let $\pi:\tilde{\mathcal{C}} \to \mathcal{C}$ be a universal 2-covering. Then $\tilde{\mathcal{C}}$ is a building if and only if for each $J \subseteq I$ for which the diagram $\Delta|_J$ is of type A_3 or C_3 and for each $c \in \mathcal{C}$, $\Delta_J(c)$ is a building.

2) Let $G \leq \mathrm{Aut}(\mathcal{C})$, $A = \{\alpha \in \mathrm{Aut}(\tilde{\mathcal{C}}) \mid \alpha\pi = \pi\}$ and $B = \{\alpha \in \mathrm{Aut}(\tilde{\mathcal{C}}) \mid \alpha\pi \in \pi G\}$. Then $A \trianglelefteq B$, $B/A \simeq G$ and A acts regularly on $\pi^{-1}(c)$ for each $c \in \mathcal{C}$.

3) If G is transitive on $\mathcal{C}$, then B is transitive on $\tilde{\mathcal{C}}$.

As a corollary of (2.4) and (2.8) we obtain in the next section a complete classification of weak parabolic systems and of chamber systems of type CM with transitive automorphism groups and spherical diagrams.

3 <u>SOME THEOREMS</u>

<u>Theorem (3.1)</u>. Suppose $\mathcal{C}$ is a chamber system of type CM of rank ≥ 3

with transitive automorphism group G and spherical diagram $\Delta(C)$. Then either C is a finite building of type Δ or C is the chamber system of type C_3 obtained from A_7. Moreover, either $G \simeq A_7$ or G is an extension of a Chevalley-group with Dynkin-diagram Δ by diagonal and field automorphisms.

Proof. We split the proof in 3 steps. First we show:

1) If rank $C = 3$, then (3.1) holds.

To prove (1) we may by [9, (3.12)] assume $\Delta(C) = \overset{1}{\circ}\!\!-\!\!\!-\!\!\overset{2}{\circ}\!\!=\!\!=\!\!\overset{3}{\circ}$. Fix $c \in C$ and let for $J \subseteq \{1,2,3\}$ X_J be the stabilizer of $\Delta_J(c)$ and K_J the kernel of the action of X_J on $\Delta_J(c)$. Then by a theorem of Seitz [8], either $X_{1,2}/K_{1,2}$ and $X_{2,3}/K_{2,3}$ is an extension of a Chevalley-group with Dynkin-diagram A_2 or C_2 by diagonal and field automorphisms or one of the following holds:

a) $\Delta_{1,2}(c)$ is defined over $GF(2)$ or $GF(8)$ and $X_{1,2}/K_{1,2}$ is a Frobenius group of order 21 or 73.9.

b) $\Delta_{2,3}(c)$ is defined over $GF(3)$ and $X_{2,3}/K_{2,3}$ is isomorphic to $2^4 \cdot F_{20}$, $L_3(4)\cdot 2$ or $2^4 \cdot A_5$.

Hence in any case one of the two groups contains a rank 2 Chevalley-group acting naturally on $\Delta_{1,2}(c)$ or $\Delta_{2,3}(c)$, so that in any case X_2/K_2 contains a rank 1 Chevalley-group acting naturally on $\Delta_2(c)$. But by the structure of the groups in (a) and (b) this is only possible if $X_{2,3}/K_{2,3} \simeq L_3(4)\cdot 2$ or $2^4 \cdot A_5$ and $\Delta_{2,3}(c)$ is the $U_4(3)$ resp. $Sp(4,3)$ generalized quadrangle.

We will show that this is impossible.

Assume first $X_{2,3}/K_{2,3} \simeq L_3(4)\cdot 2$. Then $\Delta_{1,2}(c)$ is a projective plane over $GF(3)$ or $GF(9)$. In the second case $K_2 = K_{2,3}$ and by the structure of $L_3(9)$ we have $SL_2(9) \subsetsim X_2/K_2$. But it is easy to see that $SL_2(9) \nsubseteq L_3(4)$. In the first case X_2/K_2 is the stabilizer of a point in the $U_4(3)$ generalized quadrangle. So $|X_{2,3}:X_2| = 2^3 \cdot 5 \cdot 7$ and $|X_2/K_{2,3}| = 2^4 \cdot 3^2$. Since $X_2/K_{2,3}$ contains the normalizer of a 3-Sylow subgroup of $L_3(4)$, as it contains a "Borel-subgroup", it easily follows that $X_2/K_{2,3} \simeq (\mathbb{Z}_3 \times \mathbb{Z}_3)Q_8 \times \mathbb{Z}_2$. But then $X_2/K_{2,3}$ contains no section isomorphic to $PSL_2(3)$, a contradiction to the action of $L_3(3)$ on its projective plane.

Next assume $X_{2,3}/K_{2,3} = \bar{X}_{2,3} \simeq 2^4 \cdot A_5$. Then, since $PSp(4,3) \simeq PSU(4,2)$, $O_2(\bar{X}_{2,3})$ is the "orthogonal" $A_5 \simeq \Omega^-(4,2)$-module. Counting the flags of the $Sp(4,3)$ generalized quadrangle it follows that a Borel-subgroup $\bar{B}$ of $\bar{X}_{2,3}$ has order 6. Since $X_{2,3} = \langle X_2, X_3 \rangle$ it follows that $\bar{B} \simeq \mathbb{Z}_6$. It is now easy to see that $X_2/K_{2,3}$ is isomorphic to one of the following groups: $\mathbb{Z}_2 \times A_4$, $SL_2(3)$ or $(\mathbb{Z}_2 \times \mathbb{Z}_2 \times \mathbb{Z}_3)\langle t \rangle$, where $t^2 = 1 \neq t$, t inverts $\mathbb{Z}_3$ and interchanges 2 involutions in $\mathbb{Z}_2 \times \mathbb{Z}_2$. But by the structure of $L_3(3)$ we have $X_2/K_2 \simeq PGL_2(3) \simeq \Sigma_4$, a contradiction since none of these groups has a section isomorphic to Σ_4.

This finally shows that X_J/K_J is an extension of a Chevalley-group by diagonal and field automorphisms for $J = \{1,2\}$ and $\{2,3\}$ if C is such a chamber system of type C_3. But then X_i is maximal in X_J for $i \in J$, J as before. Now [6, §4] shows that the hypothesis of (2.6) is satisfied and so $\Gamma(C)$ is a geometry with diagram $\Delta = C_3$, G acts flag-transitively on $\Gamma(C)$ and $X_{1,2}$ resp. $X_{2,3}$ are stabilizers of vertices of type 3 resp. 1 in $\Gamma(C)$. Hence (1) is a consequence of (2.4) (2).

(Aschbacher didn't treat the additional cases $2^4 \cdot F_{20}$ and $L_3(4) \cdot 2$ of Seitz's theorem. But since we have shown that $X_{1,2}/K_{1,2}$ and $X_{2,3}/K_{2,3}$ contain Chevalley-groups of the respective types and are stabilizers of vertices in $\Gamma(C)$, the proof is still complete!)

2) There does not exist any extension C of the A_7 chamber system of type C_3 with transitive automorphism group and linear diagram $\Delta(C)$.

Proving (2) by induction on rank C we may assume rank $C = 4$. Use the same notation as in (1). Then without loss $\Delta_{1,2,3}(c)$ is the chamber system of type C_3 obtained from A_7. So the structure of A_7 implies $X_{1,3}/K_{1,3} \simeq (\mathbb{Z}_3 \times \mathbb{Z}_3)\mathbb{Z}_2$.

By assumption $\Delta_{3,4}(c)$ is a classical generalized n-gon in characteristic 2. Since again by the structure of A_7 we have $X_3/K_3 \simeq \Sigma_3$, the theorem of Seitz shows that $\bar{X}_{3,4} = X_{3,4}/K_{3,4} \trianglerighteq \bar{Y}$, where $\bar{Y}$ is a rank 2 Chevalley-group over $GF(2)$ or $\bar{Y} \simeq A_6$, $U_3(3)$ or ${}^2F_4(2)'$. Now $X_{1,3,4} = X_1 \cdot X_{3,4}$ and thus $|X_{1,3,4} : X_{3,4}| = 3$, since $|X_1 : B| = 3$. Let $H = \langle X_1, Y \rangle$ and $H^* = H/K_{1,3,4}$. Since $X_1/K_1 \simeq \Sigma_3$, it is easy to see that $H = X_1 \cdot Y$ and $H^* \simeq \Sigma_3 \times \bar{Y}$ or $(\mathbb{Z}_3 \times \bar{Y})\langle t \rangle$, where $t \in B$ inverts $\mathbb{Z}_3$. Hence $X_1(X_3 \cap Y)K_{1,3}/K_{1,3} \simeq \Sigma_3 \times \Sigma_3$, since the stabilizer of $\Delta_3(c)$ in $\bar{Y}$ induces a Σ_3 on $\Delta_3(c)$. This is a contradiction to $X_{1,3}/K_{1,3} \simeq (\mathbb{Z}_3 \times \mathbb{Z}_3)\mathbb{Z}_2$.

We are now in a position to prove (3.1). By (1) and (2) we may assume rank $C \geqq 4$ and if $\Delta_J(c)$ is a cell of type A_3 or C_3, then

$\Delta_J(c)$ is a building of that type. Let $\pi: \tilde{C} \to C$ be a universal 2-covering of C. Then by (2.8)(1) and [11] $\tilde{C}$ is a finite building of type $\Delta(C)$ and by (2.8)(2) and (3) G is lifted to a group B, which acts transitively on the chambers of $\tilde{C}$. Since rank $\tilde{C} \geqq 4$ the theorem of Seitz [8] implies that $F^*(B)$ is a Chevalley-group of type Δ defined over some finite field. Hence $F^*(B)$ is already transitive on $\tilde{C}$. Thus the normal subgroup A of B which acts transitively on each $\pi^{-1}(c)$, $c \in C$ and satisfies $B/A \simeq G$ must be trivial. But then $G \simeq B$ and $C \simeq \tilde{C}$ which proves (3.1).

Similar as in [9, (4.3)] we obtain a corollary for weak parabolic systems of spherical type.

Theorem (3.2). Let $\mathfrak{P} = \{P_1,\ldots,P_n\}$ be a weak parabolic system of the group G of characteristic p, rank $n \geqq 3$ and spherical diagram $\Delta(\mathfrak{P})$. Let $G_o = \langle O^{p'}(P_i) | i = 1,\ldots,n\rangle$. Then $G_o \trianglelefteq G$ and either $O^{p'}(G_o/O_p(G_o))$ is a quasisimple Chevalley group of type $\Delta(\mathfrak{P})$ defined over some finite field of characteristic p or $\Delta(\mathfrak{P}) = A_3$ or C_3, $p = 2$ and $O^{2'}(G_o/O_2(G_o)) \simeq A_7$.

The proof, reducing (3.2) to (3.1) is exactly the same as the proof of [9, (4.3)], which reduced the same statement for diagrams $\Delta(\mathfrak{P})$ with only single bonds to the similar statement for chamber systems.

(3.3) Now, if one believes that all chamber systems of type CM with connected diagram and transitive automorphism groups are essentially of spherical type, then the problem is by (3.1) to exclude one-node extensions of spherical diagrams. Unfortunately there seem to exist a great number of exceptions in small rank and over small field, [4], [5], which makes this task more complicated. But still this has been carried through in [9] for the case where $\Delta(\mathfrak{P})$ has only "single or no" bonds. Moreover, most of the one-node extensions of spherical diagrams are so called "affine" diagrams and Tits has classified (unpublished) the affine buildings of rank $\geqq 4$. If one knows these buildings and their automorphism groups very well it might be possible in view of (2.8) to show that there do not exist any finite projections with transitive automorphism groups except in special cases. (Also the proofs in [9] show that the higher rank cases are easier!)

This leaves us with the rank 3 case, which is also the more important one for the revisionism of the classification of the finite simple groups of characteristic 2-type with $e(G) = 3$ (or ≤ 3!). I have been working on this rank 3 case for some time and hope to get a complete local classification in the near future. I will describe some of these results in the rest of this section.

Let C be a chamber system of type SM with transitive automorphism group G and rank 3. Fix $c \in C$ and let G_i be the stabilizer of $\Delta_{I-i}(c)$, $i = 1,2,3$ and K_i the kernel of the action of G_i on $\Delta_{I-1}(c)$. Then by the theorem of Seitz either G_i/K_i contains a rank 2 Chevalley-group in characteristic p or $p = 2$ and $G_i/K_i \simeq F_{21}$ or $F_{73.9}$. Now it has been shown in [10, §9] that this can be reduced to the following situation:

1) $G = \langle G_i, G_j \rangle$ for each pair $i \neq j \leq 3$.

2) $O^{p'}(G_i/O_p(G_i))$ is essentially a quasisimple rank 2 Chevalley-group in char.p for each $i \in I$ for which $\Delta_{I-i}(c)$ is a generalized n_{ij}-gon with $n_{ij} > 2$.

3) There exists a p-subgroup $S \leq \bigcap_{i=1}^{3} G_i$ such that $S \in Syl_p(G_i)$ for each $i \in I$ and $B = N_{G_i}(S) = N_{G_j}(S)$ for all $i,j \in I$.

Let now $Z = \Omega_1(Z(S))$ and $V_i = \langle Z^{G_i} \rangle$. Then after splitting of some trivial cases where $O_p(G_i) \leq Z(G_i)$ for some i (which occur for example with $\Delta(C)$ $\overset{1}{\circ}=\overset{2}{\circ}=\overset{3}{\circ}$ and $G_1 \simeq \Sigma_6 \simeq G_3$ as shown by Köhler, Meixner and Wester in Giessen).

We obtain

4) $V_i \leq O_p(G_i)$ $i = 1,2,3$.

Moreover, $G_i \cap G_i$ is a parabolic of both G_i and G_j by the action of G_i and G_j on their chamber system $\Delta_{I-i}(c)$ resp. $\Delta_{I-j}(c)$.

Now, forgetting about the third group for some time, consider just the following:

(3.4) a) $G = \langle G_1, G_2 \rangle$, $\bar{G}_i = O^{2'}(G_i/O_2(G_i))$ are finite quasisimple rank 2 Chevalley-groups in char.2 on $\hat{A}_6$, $\hat{\Sigma}_6$. (Allowing all exceptional cases like $U_3(3)$, ${}^2F_4(2)'$ and A_6!)

b) $G_1 \cap G_2$ is a maximal parabolic subgroup of both G_1 and G_2 or $|G_i : G_1 \cap G_2| = 3^2 \cdot 5$ in case $\bar{G}_i \simeq \hat{A}_6$ or $\hat{\Sigma}_6$.

c) Let $S \in Syl_2(G_1 \cap G_2)$ and $Z = \Omega_1(Z(S))$. Then $Z \ntrianglelefteq G_1$ and $Z \ntrianglelefteq G_2$.

d) There does not exist any $N \trianglelefteq G$ with $N \leq G_1 \cap G_2$.

In this situation I obtain, based on the graph approach due to D. Goldschmidt, a complete list of the possibilities for G_1 and G_2.

The proof depends on the hypothesis, that the $\bar{G}_i$ are defined in characteristic 2 only so far, that we use properties of certain GF(2)-representations of these groups. If similar properties also hold in odd characteristic, the proof will carry over.

As an immediate consequence of this result one obtains:

Theorem (3.5). Suppose C is a rank 3 chamber system of type SM in characteristic 2 with transitive automorphism group G. Then one of the following holds:

1) $\Delta(C)$ is a string diagram.

2) $\Delta(C) =$ o<o>o. All rank 2 cells are projective planes over GF(2) or GF(8) and their stabilizers in G are just Frobenius groups of order 21 resp. 73.9.

This theorem slightly generalizes the main theorem of [10] without using it. Namely if $\Delta(C)$ is not a string one gets by what has been described in (3.3) either possibility (2) or $F^*(\bar{G}_i)$ is quasisimple for $i = 1,2,3$. But in the latter case one finds a pair $1 \leq i \neq j \leq 3$ such that for G_i, G_j the hypothesis of (3.4) holds. Hence one obtains by (3.4) a description of the possible structure of G_i and G_j. Using now the third group G_k, $\{i,j,k\} = \{1,2,3\}$, one obtains a contradiction. (Essentially $O_2(G_i)$, $O_2(G_j)$ are either trivial or natural modules for $\bar{G}_i$ and $\bar{G}_j$. This determines the structure of the common 2-Sylow subgroup $S \leq \bigcap_{i=1}^{3} G_i$.)

Moreover, also based on (3.4), I obtained in the meantime a complete local classification of such chamber systems with string diagrams, except for the cases o——o═o and o——o$\overset{\sim}{=}$o. I hope to finish the "local" characterization of rank 3 chamber systems of type SM in the near future.

4 PROBLEMS

1) In [7] Niles has shown that, if G is a finite group with weak parabolic systems $\mathcal{P} = \{P_1,\dots,P_n\}$ and all the P_i are defined over fields of order ≥ 5, then $G_o = \langle O^{p'}(P_i) \mid i = 1,\dots,n\rangle \trianglelefteq G$ and

G_o has a BN-pair. By [11] this shows that $G_o/O_p(G_o)$ is a finite Chevalley-group.

Now there are many counterexamples to this theorem known if some P_i is defined over GF(2) or GF(3), but none if all the P_i's are defined over fields of order $\geqq 4$. So it would be desirable to extend the theorem of Niles to this case. It would make the representation arguments involved in the treatment of the remaining cases easier, if one could assume that the prime field is a splitting field for the $\bar{P}_i$.

2) Try to obtain a classification of affine chamber systems C of type CM and rank $\geqq 4$ with transitive automorphism group, using the Tits classification of affine buildings of rank $\geqq 4$ and (2.8). (C is affine if the Cartan-matrix corresponding to $\Delta(C)$ is positive semi-definite.) This would be at least half the way of classifying all chamber systems of type CM with transitive automorphism group and rank $\geqq 4$, although not all such chamber systems are either spherical or contain an affine subdiagram. (For example

```
        o====o
Δ(C) =  |    |  !)
        o====o
```

3) Of course, as mentioned in the introduction, the deepest and most interesting problem in this area is the restriction of the structure of the minimal parabolic subgroups in a simple group of (weak) characteristic 2-type. Any success in this problem will change completely the existing classification of finite simple groups.

REFERENCES

1. Aschbacher, M. (1980). Groups of char.2-type. Proc. Sympos. Pure Math. 37. Providence Rhode Island: Amer. Math. Soc.
2. Aschbacher, M. Finite geometries of type C_3 with flag transitive groups. To appear in Geo. Ded.
3. Cooperstein, B. S and F-pairs for groups of Lie-type in char.two. Proc. Sympos. Pure Math. 37, Providence Rhode Island: Amer. Math. Soc.
4. Kantor, W. Some exceptional 2-adic buildings. To appear in J. of Alg.
5. Meixner, Th., Wester, M. Some 2-adic building arising from Kantor's construction of the $\tilde{D}_4$-building. Preprint.
6. Meixner, Th. & Timmesfeld, F. A note on chamber systems with string diagrams. Geometria Dedicata 15 (1983) 115-123.
7. Niles, R. (1982). Finite groups with parabolic type subgroups must have a BN-pair. Journal Alg. 75, 484-494.
8. Seitz, G. (1973). Flag-transitive subgroups of Chevalley groups. Ann. of Math. (2) 97, 27-56.
9. Timmesfeld, F. (1983). Tits Geometries and Parabolic Systems in Finitely Generated Groups. I, MZ 184, 377-396, II, MZ 184, 449-487.

10. Timmesfeld, F. Tits geometries and parabolic systems of rank 3. Preprint.
11. Tits, J. (1974). Buildings of spherical type and finite BN-pairs. Lecture Notes in Math. 386. New York-Heidelberg-Berlin: Springer-Verlag.
12. Tits, J. (1981). A local approach to Buildings. In The geometric Vein (Coxeter Fetschrift) pp. 519-547. Springer-Verlag, New York-Heidelberg-Berlin.

FISCHER'S CLASSIFICATION OF GROUPS GENERATED BY 3-TRANSPOSITIONS

R. Weiss*
Department of Mathematics, Tufts University, Medford, MA 02155

Let G be a group. A subset D of G is called a set of 3-transpositions if $|x| = 2$ for all $x \in D$ and $|xy| = 3$ whenever $x,y \in D$ do not commute. In 1971 [5] Fischer classified all finite groups G with $G' = G''$ and $O_2(G) = O_3(G) = Z(G) = 1$ which are generated by a conjugacy class D of 3-transpositions. In the course of carrying out this classification, he discovered the three sporadic simple groups $M(22)$, $M(23)$ and $M(24)$ (it is actually $M(24)'$ which is simple), the construction of which occupies a good portion of [5]. In a forthcoming book, Timmesfeld will be surveying the enormous influence of this work on the classification of finite simple groups. In this paper, we describe some recent efforts to simplify Fischer's classification by developing further certain underlying graph theoretical aspects of the problem.

We begin with some definitions. Let Γ be an arbitrary undirected graph. We denote by Γ_x the set of neighbors of a vertex x. Vertices x and y will be called 2-equivalent if $\{x\} \cup \Gamma_x = \{y\} \cup \Gamma_y$ and 3-equivalent if $\Gamma_x = \Gamma_y$. The graph Γ will be called p-<u>reduced</u> (for $p = 2$ or 3) if all its p-equivalency classes are of size one. Γ will be called <u>reduced</u> if it is both 2- and 3-reduced.

Let D be a set of 3-transpositions which generate a finite group G and consists of a union of conjugacy classes of G. For each $x \in D$, let $D_x = \{y \in D: |xy| = 2\}$ and $A_x = \{y \in D: |xy| = 3\}$. For each subset X of D, we denote by $[X]$, respectively (X), the graph with vertex set X where two elements of X are joined by an edge whenever they commute, respectively do not commute.

The first step in the classification is to assemble a list of some of the ways in which group theoretical properties of D are reflected in properties of the graph $[D]$ or its complement (D):

*Research supported by NSF Grant MCS 8102011.

Lemma 1. D consists of a single conjugacy class if and only if (D) is connected.

Lemma 2. Suppose $D \cap Z(G) = \emptyset$. If $\bar{D}$ denotes the image of D in $G/Z(G)$, then $[D] \cong [\bar{D}]$ and $Z(G/Z(G)) = 1$. ([5, (2.1.1)] or [12, (2.4)]) (Note that whenever $N \trianglelefteq G$ and $N \cap D = \emptyset$, the image of D in G/N is a set of 3-transpositions.)

Lemma 3. Let $x,y \in D$. Then x is p-equivalent to y in $[D]$ (for $p = 2$ or 3) if and only if $xy \in O_p(G)$. ([5, (2.13-4)])

Lemma 4. Suppose $[D]$ is reduced and (D) connected. Then $[D]$ is connected if and only if $(G/Z(G))'$ is simple. (This is essentially [5, (3.2.7)]; also [11, (1.5)].)

Lemma 5. If $[D]$ is reduced and (D) connected, then (D_d) is connected for each $d \in D$. (Essentially [5, (3.2.2]; see below.)

Lemma 6. If $[D]$ is reduced and (D) connected, then $[A_d]$ is connected which implies that $C_G(d)$ acts transitively on A_d (for each $d \in D$). (Essentially [5, (3.3.4)]; see [11, (1.4)].)

Lemma 7. If $[D]$ and $[D_d]$ (for $d \in D$) are reduced and (D) connected, then $[D_d]$ is connected. ([11, (1.7)])

Lemma 1 is obvious since $x^{yx} = (xy)^2x = y$ whenever $x,y \in D$ do not commute. Lemmas 2 and 7 and Lemma 3 for $p = 2$ are easy to prove. Lemma 3 for $p = 3$ follows easily from the fact that if $[D]$ has no edges, then $xy \in O_3(G)$ for all $x,y \in D$. This fact was proved in [4]; it also follows from Glauberman's Z*-Theorem or, as was pointed out in [7], it may be proved in an elementary fashion using [8]. The essential elements of the proofs of Lemmas 5 and 6 were beautifully unified and revised in [1] and [2]. See [11] for a thorough discussion of Lemmas 4-7.

We now proceed to state Fischer's result. To do so, we require a few more definitions: For each n we denote by S_n the set of transpositions in the symmetric group Σ_n, by Sp_{2n} the set of symplectic transvections of a vector space of dimension n over $GF(2)$ endowed with a nonsingular symmetric bilinear form, by $O^{\varepsilon}_{2n}(2)$ the subset of Sp_{2n}

leaving invariant a quadratic form of type ε belonging to the bilinear form, and by H_n the set of hermitian transvections of the projective space associated with a vector space of dimension n over $GF(4)$ endowed with a nonsingular hermitian form. Let V be a vector space of dimension n over $GF(3)$ endowed with a nonsingular symmetric bilinear form of type ε. For n even, let $O_n^{\varepsilon}(3)$ denote the set of orthogonal reflections of the projective space $PG(V)$ corresponding to all $v \in V$ such that $(v,v) = 1$. For n odd, let $O_n^{\varepsilon}(3) = D_d$ where $D = O_{n+1}^{\varepsilon}(3)$ and $d \in D$; then $\langle O_n^+(3)\rangle \cong \langle O_n^-(3)\rangle$ and $[O_n^{\varepsilon}(3)]$ is independent of the choice of d up to isomorphism but $[O_n^+(3)] \ncong [O_n^-(3)]$.

Here, now, is Fischer's result:

<u>Theorem</u>. Suppose the group G is finite, that $Z(G) = 1$ and that G is generated by a conjugacy class D of 3-transpositions. Suppose that $[D]$ and (D) are connected and $[D]$ reduced. Let $d \in D$. Then:

i) If $\hat{G}$ is a second group generated by a conjugacy class of 3-transpositions $\hat{D}$ with $Z(\hat{G}) = 1$ and if $\phi:[D] \to [\hat{D}]$ is an isomorphism, then ϕ extends to an isomorphism from G to $\hat{G}$.

ii) If $[D_d]$ is not 3-reduced, then $[D] \cong [S_5]$.

iii) If $[D_d]$ is not 2-reduced, then $[D] \cong [Sp_{2n}]$ for some $n \geq 2$ or $[D] \cong [H_n]$ for some $n \geq 4$.

iv) If $[D_d]$ is reduced, then $[D] \cong [S_n]$ for some $n \geq 7$, $[D] \cong [O_{2n}^{\varepsilon}(2)]$ for $\varepsilon = +1$ or -1 and some $n \geq 4$, $[D] \cong [O_n^+(3)]$ for some $n \geq 5$, $[D] \cong [O_n^-(3)]$ for some $n \geq 6$ or

a) $[D_d] \cong [H_6]$, $|D| = 3510$ and $[D]$ is uniquely determined,

b) $[D_d]$ is as in (a), $|D| = 31571$ and $[D]$ is uniquely determined or

c) $[D_d]$ is as in (b), $D = 306936$ and $[D]$ is uniquely determined.

Cases (ii) and (iii) are proved in [5, (6.1), (8.2) and (9.2.11)]; see alternatively [3] and [10]. It turns out in the course of these proofs ([10, Step 14]) that $[D]$ must be a <u>triple-graph</u>; this means that for each noncommuting pair $x,y \in D$, $C_D(D_x \cap D_y) = \{x,y,x^y\}$. It is easily seen [12, (2.7)] that assertion (i) of the Theorem holds whenever $[D]$ is a triple-graph.

Part (iv) is the generic case of the classification. By Lemmas 1, 2, 5 and 7 above, D_d (or, rather, its image in $\langle D_d\rangle/Z(\langle D_d\rangle)$)

fulfills the hypotheses of the Theorem and so $[D_d]$ may be assumed to be known. Since it is easily seen that $[D]$ is a triple-graph if $[D_d]$ is, we may assume that they both are, in fact, triple-graphs (and (i) holds in general). We are thus in a position to apply the following lemma, which is at the heart of the classification ([12, (1.1)]; compare [5, (10.8)]):

<u>Lemma 8</u>. Let $[D]$ and $[\hat{D}]$ be triple-graphs. Let $d,x \in D$ and $i,j \in \hat{D}$ be noncommuting pairs. Suppose there exists an isomorphism $\phi:[D_d] \to [\hat{D}_i]$ such that

I) $\phi(u)^{\phi(v)} = \phi(u^v)$ for all $u,v \in D_d$ and

II) $\phi(D_d \cap D_x) = \hat{D}_i \cap \hat{D}_j$.

Then ϕ extends to an isomorphism from $[D]$ to $[\hat{D}]$.

In each application of Lemma 8, $[D_d]$ is a triple-graph; condition (I) holds by part (i) of the Theorem. Thus, the classification is reduced to determining, for each $[D_d]$, the subset $D_d \cap D_x$ for $x \in A_d$ up to the action of $\mathrm{aut}([D_d])$. Choose $x \in A_d$ and let $R = D_d \cap D_x$. It follows easily from Lemma 6 ([12, (4.4)]) that $N_G(R)$ acts transitively on R and that $[R \cap D_e] \cong [D_d \cap D_g \cap D_h]$ for each $e \in R$ and each noncommuting pair $g,h \in D_d$. These observations provide enough information to determine R in every case. The R's so determined correspond each time to the various classical groups in the statement of the Theorem except when $[D_d] \cong [\mathcal{H}_6]$. In this case, it turns out that $[R] \cong [\mathcal{O}_6^-(3)]$ and R is determined up to the action of $\mathrm{aut}([D_d])$ (although not up to the action of $\langle D_d \rangle$). If such a D exists, then $[D]$ is uniquely determined and must itself be substituted inductively for $[D_d]$, etc. In this manner, the uniqueness portion of the Theorem may be completed without addressing the question of whether the sets D described in (a)-(c) (and the corresponding groups G called M(22), M(23) and M(24)) actually exist (see [12] for the exact statement).

One might justify omitting entirely the construction of M(22), M(23) and M(24) with the observation that M(24)' exists as a section of F_1 which was constructed in [6]; see [6, (13.3)]. While it is true that the identification of F_1 (and M(24)' as a section of F_1) rests on [5] (see [9]), it in fact requires only the uniqueness part of the classification. At the same time, an "easy" construction of these groups is certainly desirable. The proof of Lemma 8 can be modified to give a succinct description of $[D]$ in terms of $[D_d]$ and the subgraph $[R]$.

The author is currently working on the problem of showing how the vertex-transitivity of the graph [D] so described (and with it, the existence of G) may be deduced from some simple list of geometric properties of the pair [R] and [D]. If this attempt succeeds, it is the author's hope that the existence of M(22), M(23) and M(24) might be deduced in a fairly simple and uniform manner. (In [5], we point out, the graph for M(22) is constructed with reference to the pair $[R] \cong [O_7^+(3)]$ and $[D_e \cap R] \cong [Sp_6]$ for $e \in D_d - R$ of M(23), not to the pair $[D_d] \cong [H_6]$ and $[R] \cong [O_6^-(3)]$ of M(22).)

In conclusion, we emphasize that the outline of Fischer's original classification remains in every essential way unaltered. With our more graph theoretical standpoint, however, we have managed to simplify or avoid some of the technical difficulties encountered in [5]. For example, in the case $[D_d] \cong [O_n^\varepsilon(3)]$ where, it turns out, $O_3(\langle R\rangle) \neq 1$, the uniqueness of R up to the action of $\langle D_d\rangle$ is established in a few lines in [12, (6.3)] (without even the trouble of determining the isomorphism type of [R]) whereas Fischer with some effort must first determine the structure and uniqueness of the subgroup $\langle R\rangle$ in $\langle D_d\rangle$ (see [5, (13.2.5-6)]). It is perhaps of interest as well to observe that in our version of the classification, no mention is made of the Mathieu groups in connection with the sporadic groups M(22), M(23) and M(24), and the results [5, (2.2.6-9)] on solvable groups are not required.

REFERENCES

1. Aschbacher, M. (1976). A homomorphism theorem for finite groups. Proc. Amer. Math. Soc. 54, 468-470.
2. Blass, A. (1978). Graphs with unique maximal clumpings. J. Graph Th. 2, 19-24.
3. Danielson, S., Guterman, M. & Weiss, R. (1983). On Fischer's characterizations of Σ_5 and Σ_n. Comm. Algebra 11, 1501-1510.
4. Fischer, B. (1964). Distributive Quasigruppen endlicher Ordnung. Math. Z. 83, 257-303.
5. Fischer, B. (1971). Finite groups generated by 3-transpositions. Invent. Math. 13, 232-246; and University of Warwick Lecture Notes (unpublished).
6. Griess, R. (1982). The friendly giant. Invent. Math. 69, 1-102.
7. Hall, J. (1980). On the order of Hall triple systems. J. Comb. Theory A 29, 261-262.
8. Hall, M. Jr. (1962). Automorphisms of Steiner triple systems. In Proceedings, Symposium Pure Math., Vol. VI, pp. 47-66. Amer. Math. Soc., Providence.

9. Parrott, D. (1981). Characterizations of the Fischer groups I, II, III. Trans. Amer. Math. Soc. 265, 303-347.
10. Weiss, R. (1983). On Fischer's characterizations of $Sp_{2n}(2)$ and $U_n(2)$. Comm. Algebra 11, 2527-2554.
11. Weiss, R. 3-Transpositions in infinite groups (submitted).
12. Weiss, R. A uniqueness lemma for groups generated by 3-transpositions (submitted).

SOME RECENT TRENDS IN MODULAR REPRESENTATION THEORY

D. Benson
Yale University, Department of Mathematics,
New Haven, Connecticut 06520

Introduction

The study of modular representation theory was in some sense started by L.E. Dickson [11] in 1902. However, it was not until R. Brauer [7] started investigating the subject that it really got off the ground. In the years between 1935 and his death in 1977, he almost single-handedly constructed the corpus of what is now regarded as the classical modular representation theory. Brauer's main motivation in studying modular representations was to obtain number theoretic restrictions on the possible behavior of ordinary character tables, and thereby find restrictions upon the structure of finite groups. His work has been indispensable in the classification of the finite simple groups. For a definitive account of modular representation theory from the Brauer viewpoint (as well as some more modern material) see Feit [13].

It was really J.A. Green who first systematically attacked the study of modular representation theory from the point of view of looking at the set of indecomposable modules, starting with his paper [14]. Green's results formed an indispensable tool in the treatment by Thompson, and then more fully by E.C. Dade [10], of blocks with cyclic defect groups. Since then, many other people have become interested in the study of the modules for their own sake.

I wish to discuss here the developments of the last five years or so in this area. I shall not discuss the progress made in the theory of blocks and representations of particular classes of groups, where a lot of progress has also been made recently. I shall concentrate on three main topics, namely almost split sequences, representation rings, and algebraic varieties associated with the cohomology of modules.

Almost split sequences

In the course of studying the representation theory of Artin algebras (of which modular group algebras are an example), M. Auslander and I. Reiten [2] developed the notion of an almost split sequence (also called an Auslander-Reiten sequence).

Definition. An almost split sequence is a short exact sequence of modules

$$0 \to A \to B \overset{\sigma}{\to} C \to 0$$

with the following properties.

i) The sequence does not split.

ii) A and C are indecomposable.

iii) Given any map $\rho: D \to C$, either ρ is a split epimorphism (i.e. C is isomorphic to a direct summand of D with projection ρ) or there is a map $\phi: D \to B$ with $\rho = \phi\sigma$.

(Thus for example σ splits on every proper submodule of C.)

Theorem (M. Auslander and I. Reiten, [2]). If Λ is an Artin algebra and C is a non-projective indecomposable Λ-module, then there exists an almost split sequence terminating in C. This sequence is unique up to isomorphism of short exact sequences.

The existence of almost split sequences has also been established for lattices over an order, so that p-adic group algebras are covered as well as modular group algebras. However, the construction is quite different, so that for a modular group algebra we have $A \cong \Omega^2 C$ while for a p-adic group algebra $A \cong \Omega C$ (Ω denotes the 'Heller operator', of taking the kernel of the projective cover; we shall also need the operator $\mho$ which sends a module to the cokernel of the injective hull).

The almost split sequences fit together to form the Auslander-Reiten quiver as follows.

If U and V are indecomposable modules, we write $\mathrm{Rad}(U,V)$ for the space of non-invertible maps from U to V (if $U \not\cong V$ then $\mathrm{Rad}(U,V)$ is just $\mathrm{Hom}(U,V)$) and $\mathrm{Rad}^2(U,V)$ for the subspace spanned by elements of the form $\alpha\beta$ for $\alpha \in \mathrm{Rad}(U,W)$ and $\beta \in \mathrm{Rad}(W,V)$. Then $\mathrm{Irr}(U,V) = \mathrm{Rad}(U,V)/\mathrm{Rad}^2(U,V)$ is an $\mathrm{End}_{kG}(U)$ - $\mathrm{End}_{kG}(V)$ bimodule, possibly zero. Let (a_{UV}, a'_{UV}) be the lengths as left $\mathrm{End}_{kG}(U)$-module and right $End_{kG}(V)$-module respectively. Note that if k is algebraically closed then $a_{UV} = a'_{UV} = \dim_k Irr(U,V)$. The Auslander-Reiten quiver is the directed graph whose vertices are the indecomposable modules, and with a directed edge from U to V labelled (a_{UV}, a'_{UV}) whenever $Rad(U,V) \neq Rad^2(U,V)$.

Theorem (Auslander and Reiten). The Auslander-Reiten quiver of an Artin algebra is a locally finite graph. If V is not projective, then there is a directed edge from U to V if and only if U is a direct summand of the middle term of the almost split sequence terminating in V. In this case a'_{UV} is its multiplicity as a direct summand. If V is projective, there is a directed edge from U to V if and only if U is a direct summand of Rad V, and its multiplicity then a'_{UV}. Dually, if U is not injective, then there is a directed edge from U to V if and only if V is a direct summand of the middle term of the almost split sequence commencing with U. In this case a_{UV} is its multiplicity as a direct summand. If U is injective, there is a directed edge from U to V if and only if V is a direct summand of $U/Soc\ U$, and its multiplicity is then a_{UV}.

Using techniques developed by the people who study Artin algebras, and an invariant related to the complexity of a module (see later in this article), P. Webb [17] showed that the possible shape of a connected component of the Auslander-Reiten quiver of a group algebra is quite restricted. One corollary of Webb's work is the following theorem.

Theorem. Suppose P is a (non-simple) projective indecomposable module for a group algebra. Then $Rad\ P/Soc\ P$ is a direct sum of at most four indecomposable modules.

For the group A_4 over $GF(2)$ there is a projective indecomposable module P such that $Rad\ P/Soc\ P$ has three direct summands, but I know of no examples with four.

Representation rings

The representation ring (or Green ring) of a finite group is a complex vector space $A_k(G)$ whose basis elements are in one-one correspondence with the indecomposable kG-modules. Multiplication in $A(G)$ is given by tensor products, and is well defined by the Krull-Schmidt theorem.

Green first investigated this ring in the case of a cyclic group of order p, and showed that $A(C_p)$ is semisimple. It is now known that $A(G)$ is semisimple whenever kG has finite representation type (i.e. the Sylow p-subgroups of G are cyclic, where $p = \text{char } k$), as well as a few other cases in characteristic two, while it is not semisimple in general (Zemanek, [19,20]).

In 1980, Richard Parker and I started our investigation of representation rings, the outcome of which will appear in [4]. We defined bilinear forms $(\, , \,)$ and $\langle \, , \, \rangle$ on $A(G)$ as follows.

$$(V,W) = \dim_k \operatorname{Hom}_{kG}(V,W)$$

$$\langle V,W\rangle = \dim_k (V,W)_1^G$$

where $(V,W)_1^G$ is the space of homomorphisms from V to W which factor through a projective module.

The second form has the advantage of being symmetric, and the forms are related as follows. Let $u = P_k - \mho(k)$ and $v = P_k - \Omega(k)$ as elements of $A(G)$, where P_k is the projective cover of the trivial module k. Then $u^* = v$ and $uv = 1$, and

$$(V,W) = \langle v \cdot V,W\rangle = \langle V,u \cdot W\rangle$$

$$\langle V,W\rangle = (u \cdot V,W) = (V,v \cdot W)$$

$$(\text{hence } (V,W) = (W,v^2 \cdot V)).$$

Now if V is an indecomposable module, define

$$\tau(V) = \begin{cases} \operatorname{Soc} V & \text{if } V \text{ is projective} \\ X_{\mho V} - \Omega V - \mho V & \text{otherwise} \end{cases}$$

where $0 \to \Omega V \to X_{\mho V} \to \mho V \to 0$ is the almost split sequence terminating in $\mho V$.

$$v \cdot \tau(V) = \begin{cases} V - \operatorname{Rad} V & \text{if } V \text{ is projective} \\ V + \Omega^2 V - X_V & \text{otherwise} \end{cases}$$

where $0 \to \Omega^2 V \to X_V \to V \to 0$ is the almost split sequence terminating in V.

For general $x = \Sigma a_i V_i \in A(G)$, define $\tau(x) = \Sigma \bar{a}_i \tau(V_i)$.

<u>Theorem</u> (Benson and Parker [4]). If V and W are indecomposable modules then

$$\langle V, \tau(W)\rangle = (V, v\cdot\tau(W)) = \begin{cases} d_V & \text{if } V \cong W \\ 0 & \text{otherwise} \end{cases}$$

where $d_V = \dim_k(\mathrm{End}_{kG}(V)/J(\mathrm{End}_{kG}(V)))$

(= 1 if k is algebraically closed)

Thus for general $x \in A(G)$, $\langle x, \tau(x)\rangle \geq 0$ and $(x, v\cdot\tau(x)) \geq 0$ with equality if and only if $x = 0$.

<u>Corollary</u>. Suppose V and W are kG-modules with the property that for any kG-module U, $\dim_k \mathrm{Hom}_{kG}(V,U) = \dim_k \mathrm{Hom}_{kG}(W,U)$. Then $V \cong W$ (and similarly if $\dim_k \mathrm{Hom}_{kG}(U,V) = \dim_k \mathrm{Hom}_{kG}(U,W)$).

We also investigated certain direct sum decompositions of $A(G)$ associated with subgroups of G, and used these to investigate the notions of <u>species</u>, <u>vertex</u> and <u>origin</u>, which I shall not define here. As an example of such a decomposition, let $i_{H,G}$ and $r_{G,H}$ denote the induction and restriction maps.

<u>Theorem</u> (Benson and Parker [4]). Let $H \leq G$. Then

$$A(G) = \mathrm{Ker}(r_{G,H}) \oplus \mathrm{Im}(i_{H,G})$$

as a direct sum of ideals, and

$$A(H) = \mathrm{Im}(r_{G,H}) \oplus \mathrm{Ker}(i_{H,G})$$

as a direct sum of a subring and a subspace.

<u>Corollary</u>. Let $H \leq G$.

i) Suppose V_1 and V_2 are kG-modules such that $V_1{\downarrow_H}{\uparrow^G} \cong V_2{\downarrow_H}{\uparrow^G}$. Then $V_1{\downarrow_H} \cong V_2{\downarrow_H}$.

ii) Suppose W_1 and W_2 are kH-modules such that $W_1{\uparrow^G}{\downarrow_H} \cong W_2{\uparrow^G}{\downarrow_H}$. Then $W_1{\uparrow^G} \cong W_2{\uparrow^G}$.

We then went on to apply the nondegeneracy of the inner products, and the notions of species, vertex and origin, to investigate finite dimensional summands of A(G) satisfying certain natural properties. We showed how to set up, for each such summand, a pair of dual tables resembling Brauer's modular irreducible and projective indecomposable character tables. Our analogues for the centralizer orders are certain real algebraic numbers which need be neither positive nor rational! P. Webb [18] has also investigated some numerical properties of these tables.

In [5], I went on to investigate a certain special lambda-ring structure on A(G), and to relate this to the concepts introduced in [4].

Complexity and cohomology varieties

This subject started off with some work of Quillen [15,16] describing the structure of the set of prime ideals of the (even) equivariant cohomology ring of a compact Lie group $H_G^{ev}(X)$ with coefficients in a permutation representation X. His main results, when interpreted for finite groups, give a description of Spec $H^{ev}(G,\mathbb{Z}/p\mathbb{Z})$ in terms of the elementary abelian p-subgroups and their normalizers (the Quillen stratification theorem). In particular he showed that dim Spec $H^{ev}(G,\mathbb{Z}/p\mathbb{Z})$ is equal to the maximal rank of an elementary abelian p-subgroup of G.

The next move was made by Chouinard [9], who showed that an arbitrary module V in characteristic p is projective if and only if $V{\downarrow_E}$ is projective for all elementary abelian p-subgroups E.

Since then, the subject has progressed quite far, and the following is a summary of the present state of the subject.

Definitions

Let k be a field of characteristic p, and V a kG-module. Let $\alpha_1,\dots,\alpha_r$ be the degrees of homogeneous generators for $H^*(G,k)$ (which is known to be finitely generated). If

$$\cdots \to P_1 \to P_0 \to V \to 0$$

is a minimal projective resolution of V then the Poincaré series $\eta_V(t) = \Sigma t^i \dim(P_i)$ is a rational function of the form $p(t)/\pi(1-t^{\alpha_i})$, with p a polynomial with integer coefficients. The order of the pole of $\eta_V(t)$ at $t = 1$ is called the <u>complexity</u> of V, written $cx_G(V)$, and measures the rate of growth of $\dim(P_i)$.

<u>Remark.</u> The invariant used by Webb in his analysis of the Auslander-Reiten quiver was the value of the analytic function $(\pi\alpha_i)(1-t)^{cx_G(V)}\eta_V(t)$ at $t = 1$. (I believe this is related to the cardinality of the generic fibre of the natural map $\mathrm{Proj}(Z(\mathrm{Ext}_G^{ev}(V,V))) \to \mathrm{Proj}(\mathrm{Ext}_G^{ev}(k,k))$, but I have not had time to check this yet.)

<u>Theorem</u> (Properties of complexity). Let $H \leq G$, let V be a kG-module and W a kH-module.

i) $cx_G(V) = cx_G(V^*) = cx_G(V \otimes V^*) = cx_G(\Omega V)$

ii) $cx_G(V) \geq cx_H(V{\downarrow_H})$

iii) $cx_G(W{\uparrow^G}) = cx_H(W)$

iv) If $0 \to V_1 \to V_2 \to V_3 \to 0$ is a short exact sequence of kG-modules then $cx_G(V_i) \leq \max(cx_G(V_j), cx_G(V_k))$, $\{i,j,k\} = \{1,2,3\}$. In particular, the two largest complexities are equal.

v) $cx_G(V \oplus V') = \max(cx_G(V), cx_G(V'))$

vi) If D is a vertex of V then $cx_G(V) = cx_D(V{\downarrow_D})$

vii) $cx_G(V) + cx_G(V') - p\text{-rank}(G) \leq cx_G(V \otimes V') \leq \min(cx_G(V), cx_G(V'))$

viii) $cx_G(V) \leq cx_G(k) = p$-rank of G, where k denotes the trivial kG-module (c.f. (xii)).

ix) $cx_G(V) = 0$ if and only if V is projective.

x) (Eisenbud [12]) $cx_G(V) = 1$ if and only if V is periodic.

xi) If $|G:H| = p^n \cdot r$ with $(p,r) = 1$ then

$$cx_H(V{\downarrow_H}) \leq cx_G(V) \leq cx_H(V{\downarrow_H}) + n$$

xii) (Alperin, Evens [1]) $cx_G(V) = \max_E cx_E(V{\downarrow_E})$ as E ranges over the elementary abelian p-subgroups of G.

xiii) If V and V' are in the same connected component of the Auslander-Reiten quiver of kG-modules and neither V nor V' is projective then $cx_G(V) = cx_G(V')$.

Note that (xii) is a generalization of Chouinard's result.

<u>Definitions</u>. Suppose k is algebraically closed. Denote by X_G the affine variety $\mathrm{Max}(H^{ev}(G,k))$ of maximal ideals of the even cohomology ring, with the Zariski topology. Then X_G is a union of lines through the origin, so we may form a projective variety $\bar{X}_G = \mathrm{Proj}(H^{ev}(G,k))$ of one smaller dimension.

Denote by $\mathrm{Ann}_G(V)$ the ideal of $H^{ev}(G,k)$ consisting of those elements annihilating $H^*(G,V)$. The <u>support</u> of a module V, written $X_G(V)$, is the set of all maximal ideals $\mathcal{M} \in X_G$ which contain $\mathrm{Ann}_G(V \otimes S)$ for some module S. Denote by $I_G(V)$ the ideal of $H^{ev}(G,k)$ consisting of those elements x such that for all modules S, there exists a positive integer j with $H^*(G, V\otimes S)\cdot x^j = 0$ (cup product). Then $X_G(V) = \mathrm{Max}(H^{ev}(G,k)/I_G(V))$ is a homogeneous subvariety of X_G, and $\bar{X}_G(V) = \mathrm{Max}(H^{ev}(G,k)/I_G(V))$ is a projective (closed) subvariety of $\bar{X}_G$.

For H a subgroup of G, denote by $t_{H,G}$ the map from X_H to X_G induced by $\mathrm{res}_{G,H}:H^{ev}(G,k) \to H^{ev}(H,k)$.

<u>Theorem</u> (Properties of cohomology varieties). Let $H \leq G$, let V be a kG-module and W a kH-module.

i) $\dim(X_G(V)) = cx_G(V)$

ii) $X_G(V) = X_G(V^*) = X_G(V\otimes V^*) = X_G(\Omega V)$

iii) $X_H(V{\downarrow_H}) = t_{H,G}^{-1}(X_G(V))$

iv) $X_G(W{\uparrow^G}) = t_{H,G}(X_H(W))$

v) If $0 \to V_1 \to V_2 \to V_3 \to 0$ is a short exact sequence of kG-modules then $X_G(V_i) \subseteq X_G(V_j) \cup X_G(V_k)$, $\{i,j,k\} = \{1,2,3\}$

vi) $X_G(V\oplus V') = X_G(V) \cup X_G(V')$

vii) (Avrunin, Scott [3]) $X_G(V \otimes V') = X_G(V) \cap X_G(V')$

viii) $X_G(V) = \{0\}$ if and only if V is projective

ix) $X_G(V) = \bigcup_E t_{E,G}(X_E(V{\downarrow_E}))$ as E ranges over the elementary abelian p-subgroups of G.

x) If V and V' are indecomposable modules in the same connected component of the Auslander-Reiten quiver of kG-modules and neither V nor V' is projective then $X_G(V) = X_G(V')$.

xi) Given a closed homogeneous subvariety $X \subseteq X_G$ there is a module V with $X_G(V) = X$.

xii) If $X_G(V) \cap X_G(W) = \{0\}$ then $\mathrm{Ext}^i_G(V,W) = 0$ for all $i > 0$.

Definitions. If E is an elementary abelian p-group of rank r, then X_E is a vector space of dimension r. We define

$$X^+_E = X_E \setminus \bigcup_{E'<E} t_{E',E}(X_{E'})$$

$$X_{G,E} = t_{E,G}(X_E) \qquad X^+_{G,E} = t_{E,G}(X^+_E)$$

$$X^+_E(V) = X_E(V) \setminus \bigcup_{E'<E} t_{E',E}(X_{E'}(V))$$

$$X_{G,E}(V) = t_{E,G}(X_E(V)) \qquad X^+_{G,E}(V) = t_{E,G}(X^+_E(V))$$

Thus X^+_E is the space X_E with all the hyperplanes defined over $\mathbb{Z}/p\mathbb{Z}$ removed.

Theorem (Quillen stratification for modules, Avrunin, Scott [3]). $X_G(V)$ is a disjoint union of the locally closed subvarieties $X^+_{G,E}(V)$ as E runs over a set of representatives of conjugacy classes of elementary abelian p-subgroups of G. The group $W_G(E) = N_G(E)/C_G(E)$ acts freely on $X^+_E(V)$, and $t_{E,G}$ induces a finite homeomorphism

$$X^+_E(V)/W_G(E) \to X^+_{G,E}(V)$$

(i.e. homeomorphism in the Zariski topology; Quillen calls this map an 'inseparable isogeny').

The natural map

$$\varinjlim_E X_E(V) \to X_G(V)$$

is a bijective finite morphism.

Finally, the following recent result of Carlson seems to be very important, since it is practically our only tool apart from Mackey decomposition for showing that a module decomposes as a direct sum.

Theorem (Carlson, [8]). If $X_G(V) \subseteq X_1 \cup X_2$, where X_1 and X_2 are closed homogeneous subvarieties of X_G with $X_1 \cap X_2 = \{0\}$, then we may write $V = V_1 \oplus V_2$ with $X_G(V_1) \subseteq X_1$ and $X_G(V_2) \subseteq X_2$. In particular if V is indecomposable then $\bar{X}_G(V)$ is topologically connected.

We may express these results in terms of $A(G)$ as follows. If X is a subset of $\bar{X}_G$, denote by $A(G,X)$ the linear span in $A(G)$ of the modules V for which $\bar{X}_G(V) \subseteq X$.

Theorem (properties of $A(G,X)$). Let $H \leq G$, X a subset of $\bar{X}_G$ and X' a subset of $\bar{X}_H$.

i) $A(G,X)$ is an ideal in $A(G)$

ii) $A(H,t^{-1}_{H,G}(X)) \supseteq r_{G,H}(A(G,X))$

iii) $A(G,t_{H,G}(X')) \supseteq i_{H,G}(A(H,X'))$

iv) $A(G,X)$ is closed under Ω, under taking dual modules, under forming extensions of modules, and under taking direct summands of modules.

v) $A(G,\phi)$ is the linear span of the projective modules

vi) $A(G,X_1 \cap X_2) = A(G,X_1) \cap A(G,X_2)$

vii) $A(G,X_1 \cup X_2) \supseteq A(G,X_1) + A(G,X_2)$ with equality if $X_1 \cap X_2 = \phi$.

viii) If $X_1 \subsetneqq X_2$ then $A(G,X_1) \subsetneqq A(G,X_2)$

ix) The indecomposable modules in $A(G,X)$ form a union of connected components of the Auslander-Reiten quiver of kG-modules.

x) The bilinear forms (,) and < , > are non-singular on $A(G,X)$.

xi) Let ψ^n and λ^n denote the operations on $A(G)$ introduced in [5]. Then $\psi^n(A(G,X)) \subseteq A(G,X)$ for n coprime to p, while $\psi^p(A(G,X)) \subseteq A(G,X^{(p)})$, where $^{(p)}$ is the Frobenius map on varieties. Thus if $X = X^{(p)}$, $\lambda^n(A(G,X)) \subseteq A(G,X)$ for all n.

Remark. Most of the topics discussed here are investigated at greater length in my forthcoming book [6].

REFERENCES

1. Alperin, J.L. & Evens, L. (1981). Representations, resolutions, and Quillen's dimension theorem. J. Pure Appl. Algebra 22, 1-9.
2. Auslander, M. & Reiten, I. (1975). Representation theory of Artin algebras, III. Comm. in Alg. 3, 239-294.
3. Avrunin, G.S. & Scott, L.L. (1982). Quillen Stratification for Modules. Invent. Math. 66, 277-286.

4. Benson, D.J. & Parker, R.A. The Green Ring of a Finite Group. J. Algebra 87 (1984), 290-331.
5. Benson, D.J. Lambda and Psi Operations on Representation Rings. J. Algebra 87 (1984), 360-367.
6. Benson, D.J. Modular Representation Theory via Representation Ring. To appear in SLN.
7. Brauer, R. (1980). Collected Papers, Vols. I-III. MIT Press, ed. P. Fong and W. Wong.
8. Carlson, J. The Variety of an Indecomposable Module is Connected. To appear.
9. Chouinard, L.G. (1976). Projectivity and Relative Projectivity over Group Rings. J. Pure Appl. Algebra 7, 287-302.
10. Dade, E.C. (1966). Blocks with cyclic defect groups. Ann. of Math. (2) 84, 20-48.
11. Dickson, L.E. (1902). On the group defined for any given field by the multiplication table of any given finite group. Trans. AMS 3, 285-301.
12. Eisenbud, D. (1980). Homological algebra on a complete intersection with an application to group representations. Trans. AMS 260, 35-64.
13. Feit, W. (1982). The Representation Theory of Finite Groups. North Holland.
14. Green, J.A. (1959). On the indecomposable representations of a finite group. Math. Z. 70, 430-445.
15. Quillen, D. (1971). The spectrum of an equivariant cohomology ring I. Ann. of Math. 94, 549-572.
16. Quillen, D. (1971). The spectrum of an equivariant cohomology ring II. Ann. of Math. 94, 573-602.
17. Webb, P.J. (1982). The Auslander-Reiten quivers of a finite group. Math. Z. 179, 97-121.
18. Webb, P.J. On the Orthogonality Coefficients for Character Tables of the Green Ring of a Finite Group. Preprint.
19. Zemanek, J.R. (1971). Nilpotent Elements in Representation Rings. J. Alg. 19, 453-469.
20. Zemanek, J.R. (1973). Nilpotent Elements in Representation Rings over Fields of Characteristic 2. J. Alg. 25, 534-553.

BLOCKS OF REPRESENTATIONS OF CLASSICAL GROUPS

P. Fong
Department of Mathematics, University of Illinois at Chicago
Chicago, Illinois 60680

Any theory of the representations of finite Lie groups over fields of finite characteristic r unequal to the defining characteristic p of the group must, on first impression, appear to be somewhat unnatural. Perhaps this accounted for the relative neglect of the problem up to 1979, the year of the Santa Cruz Conference. However, the impression is a misleading one. In fact, the field is particularly ripe for investigation as recent work by Bhama Srinivasan and myself have shown. The Deligne-Lusztig theory, the Brauer theory, and the character theory of Weyl groups fit nicely and naturally together, pointing to a theory similar to the Nakayama theory for the symmetric groups.

The Deligne-Lusztig theory [1], which deals with the characteristic 0 representations of the finite Lie groups over the ℓ-adic field $\bar{Q}_\ell$, $\ell \neq p$, is highly developed. It is thus reasonable to study characteristic r representations by Brauer's approach of decomposing characteristic 0 representations modulo r. In other words, we study the block theory with respect to r for these groups.

The results are simplest to state when $G = GL(n,q)$, q being a power of p. The unipotent irreducible characters of G, in the case of the linear groups, are the constituents of $(1_B)^G$, where B is a Borel subgroup of G. Since the centralizer algebra of $(1_B)^G$ is a Hecke algebra of type S_n, the symmetric group of degree n, the unipotent characters of G are in 1-1 correspondence with the irreducible characters of S_n. We express this correspondence as $\chi_\mu \longleftrightarrow \phi_\mu$, where $\mu \vdash n$ is a partition of n and ϕ_μ is the irreducible character of S_n corresponding to μ.

To describe the remaining characters of G, we need the following notation. Let

$$F = \{\Delta \in F_q[X]: \Delta \text{ is monic, irreducible, } \Delta \neq X\}.$$

The degree of Δ in $F_q[X]$ will be denoted by d_Δ. Given a semisimple element t in G and Δ in $\mathcal{F}$, let $m_\Delta(t)$ be the multiplicity of Δ as an elementary divisor of t. The primary decomposition of t will be denoted by $t = \Pi t_\Delta$, where t_Δ is the primary component of t corresponding to Δ. Here and elsewhere, products are over the elements of $\mathcal{F}$. In particular, $m_\Delta(t) = m_\Delta(t_\Delta)$, and $C_G(t) = \Pi C_G(t)_\Delta$ with

$$C_G(t)_\Delta \simeq GL(m_\Delta(t), q^{d_\Delta}).$$

Given a partition μ_Δ of $m_\Delta(t)$ and hence a unipotent character χ_{μ_Δ} of $C_G(t)_\Delta$ for each Δ in $\mathcal{F}$, we shall write $\mu = \Pi\mu_\Delta$ for the function mapping $\Delta \longmapsto \mu_\Delta$, and

$$\chi_\mu = \Pi\chi_{\mu_\Delta} \tag{1}$$

for the corresponding unipotent character of $C_G(t)$. This makes sense since μ_Δ is the empty partition $\{-\}$ for almost all Δ.

The irreducible characters of G are then in 1-1 correspondence

$$\chi_{t,\mu} \longleftrightarrow (t,\mu) \tag{2}$$

with G-conjugacy classes of pairs (t,μ), where t is a semisimple element of G, $\mu = \Pi\mu_\Delta$, and $\mu_\Delta \vdash m_\Delta(t)$ for each Δ in $\mathcal{F}$. The correspondence (2) is given by

$$\chi_{t,\mu} = \varepsilon_G\varepsilon_L R_L^G(\hat{t}\chi_\mu), \tag{3}$$

where ε_G, ε_L are signs depending on G, L respectively, R_L^G is the Deligne-Lusztig operator, $L = C_G(t)$, $\hat{t}$ is the linear character of L dual to t, and χ_μ is the unipotent character (1) of L. The duality between t and $\hat{t}$ depends on fixing an isomorphism from $\bar{F}_q^\times$ into $\bar{Q}_\ell^\times$, where $\bar{F}_q$ is an algebraic closure of F_q. This description of the characters of G is essentially Green's Theorem [4] stated in the language of Deligne and Lusztig.

Now let r be an odd prime different from p. For each Δ in $\mathcal{F}$, let e_Δ be the multiplicative order of q^{d_Δ} modulo r. We assume from the representation theory of the symmetric groups [6] the notions due to Nakayama of an e-hook and the e-core of a partition.

Theorem 1. ([3]) Let $G = GL(n,q)$. Then the following hold:

a) The r-blocks of G are in 1-1 correspondence

$$B_{s,\lambda} \longleftrightarrow (s,\lambda)$$

with G-conjugacy classes of pairs (s,λ), where s is a semisimple r'-element of G, $\lambda = \Pi\lambda_\Delta$, and λ_Δ is the e_Δ-core of a partition of $m_\Delta(s)$.

b) The representative s in $B_{s,\lambda}$ may be chosen in $C_G(R)$, where R is a defect group of $B_{s,\lambda}$.

c) An irreducible character $\chi_{t,\mu}$ belongs to $B_{s,\lambda}$ if and only if t is G-conjugate to sy for some y in R, and μ_Δ has e_Δ-core λ_Δ for each Δ in $\mathcal{F}$.

This is the analogue of the Nakayama conjecture for the symmetric groups. A prophetic note is contained in L. Scott's article in the Proceedings of the Santa Cruz Conference [7]. "Little has been done with representations in fields of characteristic $\ell \neq p$, though Alperin has suggested that this theory should parallel the modular theory of the symmetric groups."

Theorem 2. ([2]) Let $B_{s,\lambda}$ be a block of $GL(n,q)$ with cyclic defect group R. Then the following hold:

a) There is a unique elementary divisor Γ of s such that

$$m_\Delta(s) = \begin{cases} |\lambda_\Gamma| + e_\Gamma & \text{for } \Delta = \Gamma \\ |\lambda_\Delta| & \text{for } \Delta \neq \Gamma. \end{cases}$$

b) The non-exceptional characters in $B_{s,\lambda}$ are e_Γ in number and have the form χ_{s,μ_i}, $i = 1,2,\ldots,e_\Gamma$.

c) The exceptional characters in $B_{s,\lambda}$ have the form $\chi_{sy,\mu}$, where y is in $R^\#$.

d) The tree of $B_{s,\lambda}$ is the open polygon

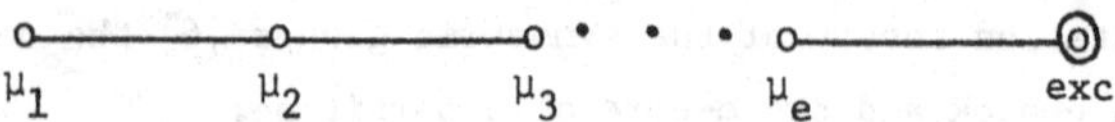

where the μ_i are arranged so that the unique e_Γ-hook in $(\mu_i)_\Gamma$ has leg length $i-1$. (The μ_i in the diagram refer to the second component in (s,μ_i).)

Note: The assumption $r > 2$ in the preceding theorems is unnecessary by recent work of Broué.

Analogues of these theorems hold for the unitary groups. Let $G = U(n,q)$, where the underlying vector space for G is over the field F_{q^2}. As in the linear case, the unipotent characters χ_μ of $U(n,q)$ are in 1-1 correspondence with the irreducible characters of S_n and hence with partitions μ of n. This uses the fact that $U(n,q)$ is a twisted form of $GL(n,q)$. A set F serving as elementary divisors for the semisimple elements of G can also be defined. If Λ is a monic polynomial in $F_{q^2}[X]$ with non-zero roots $\alpha_1,\alpha_2,\ldots$ in $\bar{F}_{q^2}$, let $\Lambda^\#$ be the monic polynomial in $F_{q^2}[X]$ with roots $\alpha_1^{-1},\alpha_2^{-1},\ldots$. Let

$$F_1 = \{\Lambda: \Lambda \neq X,\ \Lambda = \Lambda^\#,\ \Lambda \text{ is monic, irreducible}\}$$
$$F_2 = \{\Lambda\Lambda^\#: \Lambda \neq \Lambda^\#,\ \Lambda \text{ is monic, irreducible}\},$$

and let $F = F_1 \cup F_2$. Given a semisimple element t in G and Δ in F, it makes sense to talk of the multiplicity $m_\Delta(t)$ of Δ as an elementary divisor of t. Then $C_G(t)$ decomposes as $\Pi C_G(t)_\Delta$, where

$$C_G(t)_\Delta \simeq \begin{cases} U(m_\Delta(t),q^{d_\Delta}) & \text{for } \Delta \text{ in } F_1 \\ GL(m_\Delta(t),q^{\frac{1}{2}d_\Delta}) & \text{for } \Delta \text{ in } F_2. \end{cases}$$

The unipotent characters of $C_G(t)$ are then of the form (1), where $\mu_\Delta \vdash m_\Delta(t)$ for Δ in F. The analogue of Green's Theorem holds, namely, the irreducible characters of G are in 1-1 correspondence (2) with G-conjugacy classes of pairs (t,μ), and the correspondence is given by (3) as before.

Now let r be an odd prime different from p. For each Δ in F, we define e_Δ to be the smallest positive integer m such that

$$|U(m,q^d)| \equiv 0 \pmod r \quad \text{if} \quad \Delta \in F_1$$

$$|GL(m,q^{\frac{1}{2}d})| \equiv 0 \pmod r \quad \text{if} \quad \Delta \in F_2.$$

With this as the definition of e_Δ, the analogues of Theorem 1, and Theorem 2, (a), (b), (c) hold. The proof is in [3]. The tree of a cyclic block $B_{s,\lambda}$ is still an open polygon, but the exceptional vertex need not be at one end. The explicit labelling of the vertices leads to some nice combinatorics. Let $\kappa_i = (\mu_i)_\Gamma$, where Γ is the unique elementary divisor of s for which $m_\Gamma(s) \neq |\lambda_\Gamma|$. Let Ξ_i be the 2-core of κ_i, and let σ_i, τ_i be the partitions in the 2-quotient of κ_i (see [5]). Here Ξ_i is a partition of the form $\{1,2,\ldots,d_i\}$, and σ_i, τ_i are partitions derived from the configuration of even hooks in κ_i. The pair (σ_i,τ_i) is ordered so that σ_i corresponds to the even hooks in κ_i having residues of the same parity as d_i.

<u>Theorem 3</u>. Let $B_{s,\lambda}$ be a cyclic block of $U(n,q)$. With the preceding notation, the following hold:

a) Suppose $e_\Gamma = 2f$ is even. Then the κ_i have the same 2-core, and can be arranged so that $\sigma_1 = \sigma_2 = \ldots = \sigma_f$ and $\tau_{f+1} = \tau_{f+2} = \ldots = \tau_{2f}$ have no f-hooks, τ_i has a unique f-hook of leg length $i-1$ for $1 \leq i \leq f$, and σ_i has a unique f-hook of leg length $2f-i$ for $f+1 \leq i \leq 2f$. The tree of $B_{s,\lambda}$ is

$$\underset{\mu_1}{\circ}\!\!-\!\!-\!\!\underset{\mu_2}{\circ} \;\cdots\; \underset{\mu_f}{\circ}\!\!-\!\!-\!\!\underset{\text{exc}}{\circledcirc}\!\!-\!\!-\!\!\underset{\mu_{f+1}}{\circ} \;\cdots\; \underset{\mu_{2f-1}}{\circ}\!\!-\!\!-\!\!\underset{\mu_{2f}}{\circ}$$

b) Suppose e_Γ is odd. Then there are at most two distinct 2-cores $\Xi \neq \Xi'$ arising from the κ_i. The tree of $B_{s,\lambda}$ is

$$\underset{\mu_{i_1}}{\circ}\!\!-\!\!-\!\!\underset{\mu_{i_2}}{\circ} \;\cdots\; \underset{\mu_{i_\alpha}}{\circ}\!\!-\!\!-\!\!\underset{\text{exc}}{\circledcirc}\!\!-\!\!-\!\!\underset{\mu_{j_\beta}}{\circ} \;\cdots\; \underset{\mu_{j_2}}{\circ}\!\!-\!\!-\!\!\underset{\mu_{j_1}}{\circ}$$

Here $\kappa_{i_1},\kappa_{i_2},\ldots,\kappa_{i_\alpha}$ have 2-core Ξ, and the leg lengths of the e_Γ-hooks in these κ_i increase with increasing index $i_1 < i_2 < \ldots < i_\alpha$. Similarly, $\kappa_{j_1},\kappa_{j_2},\ldots,\kappa_{j_\beta}$ have 2-core Ξ', and the leg lengths of the e_Γ-hooks in these κ_j increase with increasing index $j_1 < j_2 < \ldots < j_\beta$.

Examples. We consider cyclic blocks of the form $B_{1,\lambda}$. The non-exceptional characters are then unipotent characters χ_μ, where μ has a unique e_Γ-hook with e_Γ-core λ.

a) Let $n = 7$, and let $\mu_1 = \{7\}$, $\mu_2 = \{4,3\}$, $\mu_3 = \{3^2 1\}$, $\mu_4 = \{2^2 1^3\}$, $\mu_5 = \{2,1^5\}$ be the 5 partitions of 7 with 5-core $\lambda = \{2\}$. Suppose $G = U(7,q)$ and r is an odd prime dividing q^5+1, but no $q^{2i}-1$ for $1 \leq i \leq 4$. Then $B_{1,\lambda}$ is a cyclic block, $e_\Gamma = 5$, and the tree has the form

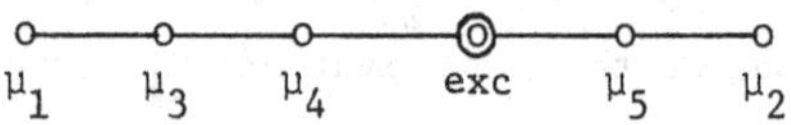

Here μ_1, μ_3, μ_4 have 2-core $\{1\}$; μ_2, μ_5 have 2-core $\{2,1\}$.

b) Let $n = 7$, and let $\mu_1 = \{7\}$, $\mu_2 = \{5,2\}$, $\mu_3 = \{4,2,1\}$, $\mu_4 = \{3,2,1^2\}$, $\mu_5 = \{2^2 1^3\}$, $\mu_6 = \{1^7\}$ be the 6 partitions of 7 with 6-core $\lambda = \{1\}$. Suppose $G = U(7,q)$ and r is an odd prime dividing q^3-1, but not q^2-1 nor $q-1$. Then $B_{1,\lambda}$ is a cyclic block of G, $e_\Gamma = 6$, and the tree has the form

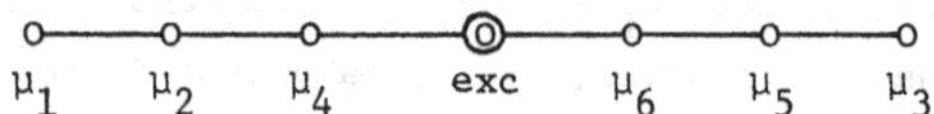

The μ_i have 2-core $\{1\}$; μ_1, μ_2, μ_4 have respectively 2-quotients $(\{3\},\{0\})$, $(\{2,1\},\{0\})$, $(\{1^3\},\{0\})$; and μ_3, μ_5, μ_6 have respectively 2-quotients $(\{0\},\{3\})$, $(\{0\},\{2,1\})$, $(\{0\},\{1^3\})$.

The difference in the structures of the trees in $GL(n,q)$ and in $U(n,q)$ is partly explained by the fact that the unipotent characters of $GL(n,q)$ are constituents of $(1_B)^G$, and hence form a single family in Harish-Chandra's classification, whereas the unipotent characters of $U(n,q)$ form several such families. The 2-core Ξ of a partition μ characterizes the Harish-Chandra family containing χ_μ. The appearance of the Harish-Chandra families in the structure of the trees is a feature of trees in the other classical groups as well.

Work in progress indicates that corresponding results will hold in the remaining classical groups. A description of the situation for these groups would require too much notation for the purpose of these proceedings. In short, a theory for the representations over characteristic r is developing and should soon be at the point reached by the modular theory of symmetric groups when the Nakayama conjecture was established. One obvious continuation of this work is to parallel the developments that that theory undertook. One problem of importance is finding the irreducible modular representations. In the case of GL(n,q), a start has been made by R. Dipper, who showed that under certain restrictions, the decomposition matrix of a block is essentially an upper triangular matrix. Another obvious pursuit is the extension of this theory to the exceptional Lie groups.

REFERENCES

1. Deligne, P. & Lusztig, G. (1976). Representations of reductive groups over finite fields. Annals Math. 103, 103-161.
2. Fong, P. & Srinivasan, B. (1980). Blocks with cyclic defect groups in GL(n,q). Bull. AMS 3, 1041-1044.
3. Fong, P. & Srinivasan, B. (1982). The blocks of finite general linear and unitary groups. Inv. Math. 69, 109-153.
4. Green, J.A. (1955). The characters of the finite general linear groups. Trans. AMS. 80, 402-447.
5. Frame, J.S., Robinson, G. de B. & Thrall, R.M. (1954). The hook graphs of the symmetric groups. Can. J. Math. 6, 316-324.
6. James, G.D. (1978). The representation theory of the symmetric groups. Lecture Notes in Math., No. 682, Springer.
7. Scott, L. (1980). Representations in characteristic p. Proc. Symp. in Pure Math. 37, 319-331.

MODULAR HECKE ALGEBRAS

P. Landrock*
Institute for Advanced Study, Princeton, NJ 08540
Aarhus University, 8000 Aarhus C, Denmark

Let G be a finite group and K any field. By the Hecke algebra $\mathcal{H}_K(H)$ associated with a subgroup H of G, we understand the endomorphism algebra of the right K[G]-module K[H\G] spanned by the right cosets of H in G with right translation as the action.

There are intimate relations between the algebraic structure of $\mathcal{H}_K(H)$ and the structure of K[H\G], and a number of classical as well as new results in modular representations can be obtained and appreciated through the properties of certain Hecke algebras. This will be demonstrated below in the ambitious hope to attract attention to the theory. It is our conviction there is still much to be discovered in this area.

The key to our subject is that K[H\G] has a natural basis which is independent of the field. This in fact was known to Schur. More precisely, if $\mathcal{H}_{\mathbb{Z}}(H)$ is the endomorphism ring of $\mathbb{Z}$[H\G], then $\mathcal{H}_K(H) = \mathcal{H}_{\mathbb{Z}}(H) \otimes_{\mathbb{Z}} K$. This is easy to see and will be demonstrated below. One important aspect of this is that it is possible to develop block theory for the triple $(\mathcal{H}_F, \mathcal{H}_R, \mathcal{H}_S)$ (we suppress H), where (F,R,S) is a p-modular system. This just means $R \subseteq \mathbb{C}$ is a complete discrete valuation ring with maximal ideal (π), while S is the quotient field of R. This block theory of $(\mathcal{H}_F, \mathcal{H}_R, \mathcal{H}_S)$ was developed by L. Scott (1973) and has since found a number of applications. Recently a completely new aspect has emerged, when D. Burry (1982) found a marvelous way of introducing lower defect groups through permutation modules. Hopefully this is just the beginning of a new approach to some of the unsolved problems in modular representation theory.

Sometimes it is even possible to find a natural embedding of $\mathcal{H}_K$ into K[G]. One obvious advantage of this is that idempotents of $\mathcal{H}_K$ then are idempotents of K[G]. We will explain below what we mean by natural.

---

*Supported in part by NSF Grant MCS-8108814 (A01) and The Danish Natural Science Research Council.

There are at least three different ways to think of transitive permutation modules over any p.i.d. O. First some notation: For $H \leq G$, $H\backslash G$ resp. G/H resp. $H\backslash G/H$ denote an arbitrary set of representatives of the right resp. left resp. double cosets of H in G. If $X \subseteq G$, we set $\bar{X} = \sum_{x \in X} x$.

The permutation module $O[H\backslash G]$ is

i) The free O-module with basis $\{Hx \mid x \in H\backslash G\}$ and action by right translation.

ii) The trivial O[H]-module O induced from H to G, i.e. $O\uparrow_H^G = \bigoplus_{x \in H\backslash G} O \otimes x$, where $O \otimes xg = O \otimes y$ if $xg \in Hy$, $y \in H\backslash G$.

iii) The right ideal of O[G] generated by $\bar{H}$.

If we use the second interpretation, the Nakayama Relations (Frobenius Reciprocity) and Mackey Decomposition yield that $\text{rank}_O \mathcal{H}_O = |H\backslash G/H|$. (See Landrock (1983), Lemma II.12.3). This proves that $\mathcal{H}_F = \mathcal{H}_R/\mathcal{H}_R\pi$. In the following, we will use the third interpretation. It is then a straightforward observation that as a basis of $\mathcal{H}_O$, we may use

$$\{A_g \mid g \in H\backslash G/H,\ A_g(\bar{H}) = \overline{HgH}\} \tag{1}$$

Indeed, as $O[H\backslash G] = \bar{H}O[G]$, it suffices to describe $A_g(\bar{H})$. Notice that

$$\overline{HgH} = \bar{H}\overline{gH \cap H^g\backslash H} = \overline{H/H \cap H^{g^{-1}}}\, g\bar{H} \tag{2}$$

The first equality shows that $A_g(\bar{H}) \in \bar{H}O[G]$, the second that A_g commutes with G in the action on $O[H\backslash G]$. (For details, see p. 177-178 of Landrock (1983).) Moreover, (2) shows that the action of A_g is just multiplication from the left by $a_g = \overline{H/H \cap H^{g^{-1}}}\, g$.

As mentioned earlier, we are especially interested whenever this gives rise to a natural embedding of $\mathcal{H}_K$ into K[G]. By natural, we mean an embedding

$$\mathcal{H}_K \xrightarrow{\ \Phi\ } K[G] \tag{3}$$

such that $\Phi(A_g)\bar{H} = a_g\bar{H}$. This is trivially possible if $|H|$ and charK are mutually prime, since then $\bar{H}$ may be replaced by $\varepsilon_H = \frac{1}{|H|}\bar{H}$, which is an idempotent, and $\varepsilon_H K[G]$ is a direct summand of K[G]. But in general, it seems difficult to give a useful criterion for an affirmative solution to this cohomological problem. We will return to this toward the end of our discussion.

The first and foremost direct connection between $K[H\backslash G]$ and $\mathcal{H}_K$ is that a direct sum decomposition

$$K[H\backslash G] \cong \oplus M_i \tag{4}$$

corresponds to an idempotent decomposition $1 = \Sigma e_i$ of 1 in $\mathcal{H}_K$, where e_i is the projection of $K[H\backslash G]$ onto M_i with kernel $\bigoplus_{j\neq i} M_j$. Moreover, e_i is primitive if and only if M_i is indecomposable.

Now, as we have chosen R to be complete, all idempotents of $\mathcal{H}_F$ are liftable to $\mathcal{H}_R$. Hence each M_i lifts naturally to a direct summand $\hat{M}_i$ of $R[H\backslash G]$. Thus we may define the character of M_i as the character of $\hat{M}_i \otimes_R S$. On the other hand, by Nakayama Relations,

$$\mathrm{Hom}_{K[G]}(K, K\!\uparrow_H^G) \cong \mathrm{Hom}_{K[H]}(K,K) \cong \mathrm{Hom}_{K[G]}(K\!\uparrow_H^G, K) \cong K \tag{5}$$

Thus $K[H\backslash G]$ has exactly one factor module and one submodule isomorphic to the trivial $K[G]$-module, independently of which field we have. In particular, exactly one indecomposable direct summand M_o of $F[H\backslash G]$ has a character, determined as indicated above, which picks up the trivial character of $S[G]$. Thus an elementary classical argument shows that F is a submodule and a factor module of M_o. (For details, see Landrock (1983), Lemma II.12.7.) This unique direct summand $M_o = M_o(H)$ of $F[H\backslash G]$ is called the Scott-module in Burry (1982). It also easily follows that $M_o(H_1) = M_o(H_2)$ if and only if the Sylow p-subgroups P_i of H_i are conjugate in G and that P_i is a vertex of $M_o(H_i)$. Indeed, as the trivial $F[H]$-module is a direct summand of $F[P_i\backslash H_i]$, we deduce that $M_o(H_i) = M_o(P_i)$.

We now let G act on $K[G]$ by conjugation, $a\cdot g = g^{-1}ag$, which makes $K[G]$ a right $K[G]$-module. Let $x_1,\ldots,x_k$ be representatives of the conjugacy classes in G, and set $C_i = C_G(x_i)$. We next observe that under conjugate action,

$$K[G] = \bigoplus_{i=1}^{k} K[C_i\backslash G]. \tag{6}$$

Again we restrict ourselves to the modular case $K = F$. Then $F[G] = \bigoplus_{i=1}^{r} B_i$ as an algebra, where the B_i's are the blocks of $F[G]$, i.e. the 2-sided ideals which are indecomposable as direct summands of $F[G]$. Thus Krull-Schmidt asserts that each B_i is isomorphic under conjugate action to a direct sum of indecomposable direct summands of the $K[C_i\backslash G]$'s (i.e. of so-called trivial source modules). Now Burry's work allows us to introduce

Definition. Let $P \leq G$ be a p-group. Then the multiplicity of P as a lower defect group of the block B is the multiplicity of the Scott-module $M_o(P)$ as a direct summand of B under conjugate action.

Burry's work shows that this elementary definition is equivalent to Brauer's original, which is very complicated and consequently was left unattended for quite a while. Note that among the lower defect groups of B, the maximal one (there is only one by the theory) is the defect group of B. Thus there no longer is an excuse for group theorists not to know what a defect group is! For a very short proof of this equivalence, see Landrock (1983), Theorem III, 10.4.

A number of fundamental questions in modular representations can be restated in terms of properties of the module $F[G]$ under conjugate action. As an example, we mention

Theorem. Consider $J_o = F[G]/J(F[G])$ as an $F[G]$-module under conjugate action. Then

a) The number of simple $F[G]$-modules of dimension prime to p is equal to the multiplicity of the trivial $F[G]$-module as a direct summand of J_o.

b) The number of blocks of $F[G]$ of defect 0 is equal to the multiplicity of the projective cover of the trivial $F[G]$-module as a direct summand of J_o.

Proof. a) By Wedderburn's Theorem,

$$J_o = \bigoplus_{i=1}^{\ell} A_i \tag{7}$$

where ℓ is the number of isomorphism classes of simple $F[G]$-modules. Assume for simplicity that F is a splitting field of $F[G]$. Then

$A_i = \mathrm{Mat}_{n_i}(F)$, where n_i is the dimension of the corresponding simple module E_i. Moreover,

$$A_i \cong E_i \otimes E_i^* \tag{8}$$

as an $F[G]$-module under conjugate action, where E_i^* is the dual of E_i. Thus

$$\mathrm{Hom}_{F[G]}(A_i, F) = \mathrm{Hom}_{F[G]}(E_i, E_i) = F \tag{9}$$

i.e. the trivial $F[G]$-module occurs with multiplicity at most 1 as a direct summand of A_i. However, by Corollary II.6.19 of Landrock (1983), F is a direct summand of $E_i \otimes E_i^*$ if and only if n_i is prime to p.

b) Sketch: Using projective maps (maps which factors through a projective module), it appears that the projective cover of the trivial module is a direct summand of $E_i \otimes E_i^*$ if and only if E_i is projective, and the number of projective simple $F[G]$-modules is equal to the number of blocks of defect 0.

This, and a number of other facts suggest that further study of the modular Hecke algebras corresponding to subgroups of the form $C_G(x)$, $x \in G$, could be very useful.

We end our discussion with a very important special case, which provides a beautiful example of what we are looking for.

Let G be a finite group with a split (B,N)-pair of characteristic p. Let $B \geq U \in \mathrm{Syl}_p(G)$. Then $B = N_G(U) = U \cdot H$, where $U \cap H = 1$ and $N = N_G(H)$, while $W = N/H$ is the Weyl group, say of rank n with generators $s_1, \ldots, s_n$, subject to $s_i^2 = 1$ and $s_i s_j \ldots = s_j s_i \ldots$, m_{ij} factors. We are interested in the modular Hecke algebra $\mathcal{H}_F(B) = \mathcal{H}_F$.

Recall that G has exactly one simple module (up to isomorphism) which is projective as well (see Curtis (1970)), the so-called Steinberg module St. As St is a direct summand of $F[G]$ under right translation, there exists a primitive idempotent (actually, many) e_{St} of $F[G]$, such that $St = e_{St}F[G]$. Such an idempotent was given by Steinberg (1957) in case G is a finite group of Lie type. Namely,

$$e_{St} := \frac{1}{|H|}\bar{B}\Big(\sum_{w \in W} \sigma(w) w\Big) \tag{10}$$

where $\sigma(w) = (-1)^{\ell(w)}$, with $\ell(w)$ the length of w (as a minimal word in the generators of W). Set $X_i = U \cap V^{s_i}$, where $V = U^{w_o}$ and w_o is the element of W of maximal length, the root subgroup corresponding to s_i. Let $G_i = \langle X_i, X_i^{s_i} \rangle$. Then G_i is a group of Lie-type of rank 1, and

$$G_i = H_i X_i \cup H_i X_i s_i X_i \tag{11}$$

(see for instance Fong & Seitz (1973), (1G)), where $H_i = G_i \cap H$.

We may now describe H_F. Recall that as the set $B\backslash G/B$ we may choose W by the Bruhat decomposition. Thus $\dim H_F = |W|$, and our basis of H_F consists of $\{A_w | w \in W\}$, where

$$A_w(\bar{B}) = \overline{BwB} = a_w \bar{B} \tag{12}$$

where $a_w = \overline{B\backslash B \cap B^{w^{-1}}}$. At the same time, the Steinberg idempotent e_i of G_i is, by (10),

$$e_i = \frac{1}{|H_i|} \bar{H}_i \bar{X}_i (1 - s_i) \tag{13}$$

(Notice that (11) provides an easy proof of the fact that e_i is an idempotent.)

Next we observe that

$$e_i \bar{B} = -a_{s_i} \bar{B}. \tag{14}$$

Strongly suggested by some recent work of Kuhn (1984), extended by Landrock (1984) and Lyons (1983), we state, in view of our discussion above concerning a natural embedding of $H_F(B)$ into $F[G]$,

<u>Conjecture</u>. Let G be a finite group of Lie type of characteristic p, and use the same notation as above. The subalgebra of $F[G]$ generated by the e_i's, $i = 1,\ldots,n$ (and 1) is isomorphic to $H_F(B)$, through the map

$$A_w \xrightarrow{\Phi} (-1)^{\ell(w)} e_{i_1} \cdots e_{i_{\ell(w)}} \tag{15}$$

where $w = s_{i_1} \cdots s_{i_{\ell(w)}}$. Moreover,

$$\Phi(A_{w_o}) = (-1)^{\ell(w_o)} e_{St} \tag{16}$$

where w_o is the element of W of maximal length, or, in other words, e_{St} is the maximal word in the e_i's.

As pointed out in Kuhn (1984), Iwahori's Theorem on the structure of $H_{\mathbf{Z}}(B)$ yields that it suffices to prove that for all i,j,

$$e_i e_j \ldots = e_j e_i \ldots, \tag{17}$$

m_{ij} factors. Also, $m_{ij} \in \{2,3,4,6,8\}$ for finite groups of Lie type. The Conjecture has been proved by Kuhn (1984) when $m_{ij} \leq 3$, for all i,j; by Landrock (1984) when $m_{ij} \leq 4$; and by Landrock (1984) and Lyons (1983) when $m_{ij} \leq 6$, all i,j.

Apart from being a very important result for finite groups of Lie type, this result has some impressive application in staple homotopy theory as well, as shown in Kuhn (1984).

In relation to this we mention that it is well known that H_F has exactly 2^n simple representations, all 1-dimensional. This, together with Curtis' and Richen's work on modular representations of groups with a split (B,N)-pair, see Curtis (1970), imply that $F[B\backslash G]$ has exactly 2^n indecomposable direct summands, all with simple head and socle. Now, whenever the Conjecture above holds, it should be possible to find 2^n orthogonal idempotents $f_1,\ldots,f_{2^n}$ of $F[G]$ such that

$$F[B\backslash G] = \bigoplus_{i=1}^{2^n} f_i F[B\backslash G] \tag{18}$$

which we hereby suggest as a problem. For a non-trivial example ($A_3 = A_8$!), see Ex. 2.14 of Kuhn (1984).

REFERENCES

Burry, D. (1982). Scott modules and lower defect groups. Comm. Algebra 10, 1855-1872.

Curtis, C.W. (1970). Modular representations of finite groups with split (B,N)-pairs. In Seminar on Algebraic Groups and Related Finite Groups. Lecture Notes in Math. 131, 57-95, Springer, Berlin.

Fong, P. & Seitz, G. (1973). Groups with a (B,N)-pair of rank 2. I, Invent. Math. 21, 1-57.

Kuhn, N.J. (1984). The modular Hecke algebra and Steinberg representation of finite Chevalley groups. To appear in J. Algebra.

Landrock, P. (1984). Appendix to N.J. Kuhn, op. cit. To appear in J. Algebra.

Landrock, P. (1983). Finite Group Algebras and Their Modules. Lon. Math. Soc. Lecture Notes Series, vol. 84, Cambridge University Press, Cambridge.

Lyons, R. (1983). Private communication.

Scott, L.L. (1973). Modular permutation representations. Trans. A.M.S. 175, 101-121.

Steinberg, R. (1957). Prime power representations of finite linear groups II. Canad. J. Math. 9, 347-351.

MODULAR REPRESENTATIONS OF CHEVALLEY GROUPS

Stephen D. Smith
University of Illinois, Chicago, Illinois

Abstract. We describe a number of recent applications of representation-theoretic techniques, in areas related to finite groups. There is a briefer discussion of problems in the general theory, particularly those which might be important for future applications.

INTRODUCTION

The term "modular" in the title indicates there will be no real mention of the considerable activity in characteristic-0 representations of Chevalley groups (as in Deligne-Lusztig [4]), or representations in a prime characteristic differing from that defining the Chevalley group (as in Fong-Srinivasan [5]). Furthermore the more general block theory of Brauer is not of much interest for a Chevalley group in its own characteristic, since (as is well-known) the only non-principal block is that defined by the Steinberg module.

One important reason for considering representations in "natural" characteristic is that they are naturally encountered in the construction of parabolic subgroups - namely the chief factors on the unipotent radical, determined by the action of a Levi complement. Moreover, much recent work on subgroup-geometries is based on the observation that local subgroups in sporadic groups often have the form: (a normal p-subgroup) extended by (a Chevalley group in characteristic p).

A more esthetic reason for this study comes from the observation that rather little representation theory is used in the present classification of finite simple groups - especially when compared with the classification of complex simple Lie algebras.* It seems likely that any success in extending this analogy with representation techniques could both simplify the proof and deepen our understanding of the structure of the groups.

---

*Compare also the remarks of Scott [13] in his Santa Cruz article, in the context of the maximal subgroups/maximal subalgebras problems.

Some specific techniques, intended for applications of this type, have been developed by Ronan & Smith [10]. The present article, however, will aim at a more general survey of some recent uses of representation-theoretic methods.

I SOME APPLICATIONS

A Structure & Classification

i) Cyclic defect theory. Applications in this area don't really fit into the scheme described above: they are mostly not so recent, and involve block theory without real reference to a single "characteristic" prime. Furthermore, the techniques do not seem to give conclusive results except in "low rank" situations.

ii) Recognition problems. It would seem that the greatest relevance of representations is to the identification of groups via specified local configurations, which comes at the conclusion of many classification problems. Some analysis of representations appears in the more general methods of failure of factorization and pushing-up. The reader should refer to the articles (in these Proceedings) of Stellmacher, Stroth, and Timmesfeld for more recent developments in these areas.

A classic example of the use of representation analysis in a recognition problem is provided by the F*-extraspecial work, particularly that of Timmesfeld [14]. Here if $M = C(z)$ for a suitable involution z, with normal extraspecial 2-group $Q = F^*(M)$, then it is crucial to study the action of group $\bar{M} = M/Q$ on its $\mathbb{F}_2$-module $\tilde{Q} = Q/\langle z\rangle$. In the generic cases, M turns out to be a Chevalley group in characteristic 2, and $\tilde{Q}$ a module of fairly simple structure.

A particularly fascinating argument of this type shows why there can be only finitely many non-classical groups of F*-extraspecial type. Timmesfeld shows that outside the configurations which determine the linear and unitary groups, for t a conjugate of z in $M-Q$, the centralizer $C_{\bar{M}}(\bar{t})$ is generically an orthogonal group over $\mathbb{F}_2$, and the commutator $[\tilde{Q},\bar{t}]$ affords a natural orthogonal module. However, away from the orthogonal groups, the study of a conjugate $\bar{t}'$ of $\bar{t}$ shows that $[\tilde{Q},\bar{t}]$ has spin-module sections. Now the dimension of the natural module is <u>linear</u> in the rank of the orthogonal section of $C_{\bar{M}}(\bar{t})$, while the dimension of the spin module is <u>exponential</u> in the rank. So these numbers can be comparable for small rank only. (I have discussed this view of Timmesfeld's argument in [11].)

iii) Maximal subgroups. It is clear that a number of representation-theoretic questions are important for describing maximal subgroups of simple groups (see, for example, Scott [13]). For recent developments, the reader should refer to the articles of Aschbacher and Seitz, in these Proceedings.

B <u>Subgroup-geometries</u>

In a sense, the work of Tits [16] on buildings is in a direction opposite to that being presented here, in that it describes a geometry which need not be visualized in terms of certain subspaces of a projective space. On the other hand, once the building is understood, it can then be "represented" by suitable subspaces in <u>any</u> irreducible representation of the Chevalley group in its own characteristic (not just a "natural" representation). Cf. [10].

i) Local geometries. A similar comment applies to the pioneering work of Buekenhout [3] on diagram geometries for more general simple groups: there is no special emphasis on representations or local subgroups. Such an emphasis appeared in Ronan-Smith [9] and other later work; it is clear that geometries determined by p-local subgroups are conveniently viewed in representations in characteristic p. (At the same time, other recent work shows that many interesting simple-group geometries are not of local-subgroup type.)

ii) Classification of certain group geometries. The recent work of Timmesfeld [15] exhibits some striking uses of representation techniques, particularly in proving non-existence of suitable geometries corresponding to non-spherical diagrams (like the affine extensions $\tilde{A}_n$ of the usual A_n diagrams). In this setting, a geometry is defined by certain "stabilizer" subgroups P_i, and the representation methods can be used in the usual situation where the "kernels" $M_i = O_2(P_i)$ can all be shown to be non-trivial. The initial study focuses on the modules defined by $Z_i \equiv \langle Z(S)^{P_i}\rangle$, where S is a Sylow 2-group; some of these must be failure-of-factorization modules, and often it is possible to show these known modules cannot fit together in a consistent way. We note finally that the case where some of the M_i may be trivial (so that modules are not available) is the one leading to the discovery of the unusual geometries of types $\tilde{A}_2$, $\tilde{A}_3$, $\tilde{A}_4$ [8, 2]. The reader should also refer to the articles of Stellmacher, Stroth, and Timmesfeld (in these Proceedings) to see how the theory of amalgams can be combined with these arguments to extend their range greatly.

C Other areas

It is to be expected that modern group theory will find many of its applications through the representation theory. Perhaps characteristic 0 should be most important, but the modular theory is also finding uses. I mention two areas in particular.

G. Mason [6] and others have achieved a number of new results in areas of finite geometry farther afield than those described above. In particular, there are general results on non-existence of translation planes defined on most irreducible modules for a Chevalley group.

It seems from the work of Ronan-Smith [10] that analysis of representations from the viewpoint of local structure (that is, the parabolics) can lead to some new features of the more general theory, applicable in particular for (split) groups and algebras in characteristic 0. This would be a pleasing development, since much of the existing theory for prime characteristic has been imported (with modifications) from characteristic 0.

II SOME POSSIBLE DIRECTIONS FOR THE THEORY

I will mention a few areas of the theory on which attention is focused by the developing applications mentioned above. I think these questions would be among the most important on intrinsic theoretical grounds alone.

A What are the irreducibles? (in characteristic equal to that of G)

Here we have only a partial description due to Steinberg, with variations given by Curtis and W. Wong. We do not know the modules in any strong sense, or even the degrees of the modules. Apparently the problem is quite difficult - but it is visibly very important.

Techniques from the theory of algebraic groups should still be helpful, but there are problems for the finite groups which seem inaccessible to the method of restricting from rational modules, especially for the crucial small primes. A recursive construction of irreducibles, with the prime characteristic "built in", is described in [10]. (See also Ronan's article in these Proceedings.) However, this is a long way from a natural and illuminating construction, as is desired for real applications.

B <u>What are the projective indecomposables? the Cartan matrix? etc.</u>

In this area, there is also some existing theory, with further recent developments. But the results are still far from comprehensive, especially for applications of the sort described above. Complete results for SL_2 appear in [1], but extensions to larger rank seem so complicated as to be still out of reach.

Probably for most local-subgroup studies, it would be enough to know only the top two layers of the projective indecomposables, as this includes the non-split extensions of most frequent occurrence. The zero-th homology construction of [10] is designed to deal with extensions generated by fixed points of a full unipotent group, and often splitting questions under this hypothesis are easier [12]; but $G_2(q)$ (see (3.3) of [10]) shows that life can still be rather complicated.

C <u>Develop a "theory" of irreducible modules for a sporadic group in a "natural" characteristic</u>

The local geometries studied in [9] and other works indicate a subgroup structure resembling that of parabolics in a Chevalley group. As these latter exhibit distinctive behavior in natural characteristic, it seems desirable to obtain some analogues for suitable sporadic groups. That these analogues must be partial, at best, is shown by a number of counterexamples to obvious conjectures [7]. Nonetheless, there should be a way of describing at least a <u>class</u> of modules for a sporadic group - rather than studying them one at a time, as with present technology. Ronan and I are extending the techniques of [10] to the context of automorphism groups of chamber systems. The reader can refer to Ronan's article (in these Proceedings) for a report on this new work.

More generally, it would be desirable for group theorists to keep an eye open for places where the representation theory can be applied "internally" in a natural and intrinsic way. I believe that many of the experts in representations would be glad to hear of particular problems motivated by the structure theory.

REFERENCES

1. Andersen, Jorgensen, Landrock, P. (1983). The projective indecomposable modules of $SL_2(p^n)$. To appear in Proc. London Math. Soc. (3) 46.

2. Aschbacher, M. & Smith, S.(1983). Tits geometries over GF(2) defined by groups over GF(3). Comm. in Alg. <u>11</u>, 1675-1684.

3. Buekenhout, F. (1979). Diagrams for geometries & groups. J. Comb. Th. A. 27, 121-151.
4. Deligne, P. & Lusztig, G. (1982). Duality for representations of a reductive group over a finite field. J. Alg. 74, 284-291.
5. Fong, P. & Srinivasan, B. (1982). The blocks of finite general linear and unitary groups. Inv. Math. 69, 109-153.
6. Mason, G. (1982). Irreducible translation planes and representations of Chevalley groups in characteristic 2. In Finite Geometries (Proceedings of Ostrom Conference), ed. Kallaher & Long. New York: Marcel Dekker.
7. Mason, G. & Smith, S. (1982). Minimal 2-local geometries for the Held and Rudvalis groups. J. Algebra, 79, 286-306.
8. Ronan, M. (1983). Triangle geometries. (Preprint, U. Ill. Chicago).
9. Ronan, M. & Smith, S. (1980). 2-Local geometries for some sporadic groups. In Proc. Symp. Pure Math. no. 37 (Santa Cruz, 1979, ed. Cooperstein-Mason), pp. 283-289. Providence, A.M.S.
10. Ronan, M. & Smith, S. (1983). Sheaves on buildings and modular representations of Chevalley groups. To appear in J. of Algebra.
11. Smith, S. (1982). Spin modules in characteristic 2. J. Alg. 77, 392-401.
12. ________. (1983). Sheaf homology and complete reducibility. To appear in J. of Algebra.
13. Scott, L.L. In Proc. Symp. Pure Math. no. 37 (Santa Cruz 1979, ed. Cooperstein-Mason). Providence, Amer. Math. Soc.(1981).
14. Timmesfeld, F.G. (1978). Finite simple groups in which the generalized Fitting group of the centralizer of some involution is extra special. Ann. Math. 107, 297-369.
15. ______________. (1983). Tits geometries & parabolic systems for finitely generated groups. Math. Z. 184, 377-396 and 449-487.
16. Tits, J. (1974). Buildings of spherical type and finite BN-pairs. Springer Lecture Notes in Math. no. 386. Berlin: Springer-Verlag.

RIGIDITY AND GALOIS GROUPS

W. Feit
Yale University, New Haven, Connecticut 06520

INTRODUCTION

This chapter contains several results which insure that various classes of finite groups G are Galois groups over $K(T)$, where K is a suitable number field depending on G, and T is an indeterminate. By Hilbert's irreducibility theorem these results imply that G is a Galois group over K. The case that $K = \mathbb{Q}$ is of course of paramount interest.

The relevant ideas from algebraic geometry have apparently been known for some time. See e.g. [1], [2], [4], [5], [6]. However the full force of these ideas was first exploited by Thompson [9] who amongst other things isolated the fundamental concept of rigidity and saw that it could be used in many explicit cases. As a first application he proved the striking result that the Fischer-Griess monster M is a Galois group over $\mathbb{Q}$. (In contrast to this, the smallest sporadic simple group M_{11} is not yet (February 1984) known to be a Galois group over $\mathbb{Q}$.)[1]

A few years ago the only finite simple groups which were known to be Galois groups over $\mathbb{Q}$ were alternating groups, some groups $PSL_2(q)$ [7], [8] and $Sp_6(2)$ which is a homomorphic image of $W(E_7)$. The various results in this chapter show that several new infinite classes of finite groups are Galois groups over $\mathbb{Q}$. However not all of Ribet's results can be handled by these new methods. For instance $(\frac{144169}{29}) = -1$. Thus $PSL_2(29^2)$ is a Galois group over $\mathbb{Q}$, [7] p. 271. None of the results in this chapter imply this fact.

The object of this introduction is to state some of the pertinent definitions and results that are needed for the various papers in this chapter. I am indebted to T. Tamagawa for several illuminating discussions.

[1] (added in proof) H. Matzat has now shown that M_{11} is indeed a Galois group over $\mathbb{Q}$.

Definition. Let $C_1,\ldots,C_k$ be conjugate classes of G and let $x_i \in C_i$. Then

$$K = K_{x_1,\ldots,x_k} = K_{C_1,\ldots,C_k} = Q(\chi_u(x_i)\,|\,i = 1,\ldots,k;\ u = 1,2\ldots)$$

where $\{\chi_u\}$ is the set of all irreducible characters of G. K is the field belonging to $\{C_1,\ldots,C_k\}$ or $\{x_1,\ldots,x_k\}$. If $k = 1$ then x or C is rational if $K_x = K_C = Q$.

Let $C_1,\ldots,C_k$ be conjugate classes of the finite group G. Let

$$\begin{aligned} A = A(C_1,\ldots,C_k) &= A_G(C_1,\ldots,C_k) \\ &= \{(x_1,\ldots,x_k)\,|\,x_i \in C_i,\ \ x_1x_2\ldots x_k = 1\}. \end{aligned}$$

If $y \in G$ and $(x_1,\ldots,x_k) \in A$ then $(x_1^y,\ldots,x_k^y) \in A$. Thus G acts as a permutation group on A.

Definition. $A = A_G(C_1,\ldots,C_k)$ is rigid if

i) $A \neq \emptyset$.

ii) G acts transitively on A.

iii) If $(x_1,\ldots,x_k) \in A$ then $G = \langle x_1,\ldots,x_k\rangle$.

A is rationally rigid if $A = A_G(C_1,\ldots,C_k)$ is rigid and C_i is rational for $i = 1,\ldots,k$. We will also say that G is rigid or rationally rigid if A is.

Lemma 1. Suppose that the center of G is $\{1\}$. Then $A = A_G(C_1,\ldots,C_k)$ is rigid if and only if the following conditions are satisfied.

i) $|G| = |A|$.

ii) If $(x_1,\ldots,x_k) \in A$ then $G = \langle x_1,\ldots,x_k\rangle$.

Proof. Since the center of G has order 1, no element of G fixes any element of A by (ii). Thus each orbit of G on A has cardinality $|G|$. The result follows.

The following well-known formula describes $|A|$ in terms of the irreducible characters of G.

$$(*)\qquad |A_G(C_1,\ldots,C_k)| = \frac{\prod_{i=1}^{k}|C_i|}{|G|}\left(\sum_{\chi}\frac{\chi(x_1)\ldots\chi(x_k)}{\chi(1)^{k-2}}\right).$$

Here $x_i \in C_i$ and χ ranges over all the irreducible characters of G.

In view of (*), condition (i) of Lemma 1 can be read off from the character table of G if that is available. In fact it is sufficient to know only a part of the character table G. To verify condition (ii), it is necessary to survey all possible subgroups of G. Here the classification of finite simple groups is very helpful, though in practice it is usually sufficient to know all simple groups satisfying additional conditions.

The relevance of these concepts is due to the following result. It was proved by Thompson [9] for $k \leq 6$, but this restriction appears to be unnecessary.

Theorem A. Let G be a finite group with center $\{1\}$. Let $A = A_G(C_1,\ldots,C_k)$, let $K = K_{C_1,\ldots,C_k}$ and let T be an indeterminate over K. Assume that A is rigid. Then there exists a finite Galois extension L of K(T) such that the following hold:

i) G is the Galois group of L over K(T).

ii) K is algebraically closed in L.

iii) There exist places $P_1,\ldots,P_k$ in K(T) such that every place distinct from all P_i is unramified in L.

iv) The index of ramification e_i of P_i equals the order of $x_i \in C_i$. The residue class degree f_i of P_i equals 1 for all i. (Thus there are exactly $[L:K(T)]/e_i$ places in L over P_i.)

The Hilbert Irreducibility Theorem now implies

Corollary. With assumptions and notations as in Theorem A, G is a Galois group over K.

Let G, H be finite groups such that $A_G(C_1,\ldots,C_k)$ and $A_H(C_1',\ldots,C_m')$ are rigid. Then it is easily seen that

$$A_{G\times H}(C_i\times C_j' \mid 1 \leq i \leq k,\ 1 \leq j \leq m)$$

is rigid. Hence one can get rigidity with arbitrarily large k. However

$k = 3$ in all known cases of rigidity of $A_G(C_1,\ldots,C_k)$, where $G \subseteq \mathrm{Aut}(G_0)$ for a simple group G_0 and $C_i \neq \{1\}$ for all i.

Let F_0 be a function field of genus g_0 over the complex numbers C and let F be a finite extension of F_0 with $[F:F_0] = n$. Let g be the genus of F. The following formula of Hurwitz relates g and g_0. See e.g. Lang [3] p. 26.

$$2g - 2 = n(2g_0 - 2) + \Sigma(e_i - 1)$$

where P_i ranges over all ramified places in F and e_i is the corresponding index of ramification. If $g_0 = 0$, F/F_0 is a Galois extension and $k = 3$ this implies that

$$2g = 2 - 2n + n \sum_{i=1}^{3} (1 - \frac{1}{e_i}).$$

In particular this implies the following result.

If $n = 2m$, $e_1 = e_2 = 2$, $e_3 = m$ then $g = 0$.

By using results from algebraic geometry, see e.g. Matzat [4], this formula implies the following

Theorem B. Suppose that the center of G is $\{1\}$ and $A_G(C_1,C_2,C_3)$ is rigid. Let $H \lhd G$ so that either $|G:H| \leq 2$ or G/H is a dihedral group. Let $K \subseteq L$ be the fields described in Theorem A and let $K(T) \subseteq M \subseteq L$ where M corresponds to H. Then $M \approx K(T)$ and H is a Galois group over M.

By combining Theorem A and Theorem B we get

Theorem C. Let G be a finite group with center $\{1\}$. Let $A = A_G(C_1,C_2,C_3)$, let $K = K_{C_1,C_2,C_3}$ and let T be an indeterminate over K. Assume that A is rigid. Let $H \lhd G$ so that either $|G:H| \leq 2$ or G/H is dihedral. Then H is a Galois group over $K(T)$.

Once again Hilbert's Irreducibility Theorem can be applied to get

Corollary. Let H, K be as in Theorem C. Then H is a Galois group over K.

As an example let $G = \Sigma_n$, let x_1 be an n-cycle, x_2 an (n-1)-cycle and x_3 a transposition. If $x_i \in C_i$ for $i = 1,2,3$ then it is easily seen that $A_G(C_1,C_2,C_3)$ is rationally rigid. Thus by the Corollary to Theorem C the alternating group A_n is a Galois group over $\mathbb{Q}$. This argument seems to have been noticed by several people independently. The result is of course due to Hilbert.

In his paper in this chapter Fried generalizes the concept of rigidity, however his generalization coincides with rigidity in case k = 3 and another technical condition is satisfied. Most of the other papers in this chapter have Theorem C and its Corollary as their starting point.

REFERENCES

1. Belyi, G.V. (1980). On Galois extensions of a maximal cyclotomic field. Math. USSR Izvestia A.M.S. Translation 14, 247-256.
2. Fried, M. (1977). Fields of definition of function fields and Hurwitz families - Groups as Galois groups. Comm. Alg. 5, 17-82.
3. Lang, S. (1982). Introduction to algebraic and abelian functions. 2nd ed. Springer Verlag Graduate Texts in Math. 89.
4. Matzat, B.H. (1979). Konstruktion von Zahlkörpern mit der Galoisgruppe M_{11} über $\mathbb{Q}(\sqrt{-11})$. Manuscripta, Math. 27, 103-111.
5. __________. (1980). Zur Konstruktion von Zahl-und Funktionenkörpern mit vorgegebener Galoisgruppe. Thesis Karlsruhe.
6. __________. (1983). Konstruktion von Zahlkörpern mit der Galoisgruppe M_{12} über $\mathbb{Q}(\sqrt{-5})$. Archiv der Math 40, 245-254.
7. Ribet, K.A. (1975). On ℓ-adic representations attached to modular forms. Invent. Math. 28, 245-275.
8. Shih, K. (1974). On the construction of Galois extensions of function fields and number fields. Math. Ann. 207, 99-120.
9. Thompson, J.G. Some finite groups which appear as Gal L/K, where $K \subseteq Q(\mu_n)$. To appear in J. of Algebra.

ON REDUCTION OF THE INVERSE GALOIS GROUP PROBLEM TO SIMPLE GROUPS

M. Fried
Mathematics Department, University of California,
Irvine, CA 92664

INTRODUCTION AND NOTATION

John Thompson has initiated a program for realizing each finite group as a Galois group over $\mathbb{Q}$. His contributions have two parts. Firstly: the production of rationally rigid generators (§1) for some simple groups. A group with rationally rigid generators can be realized as a Galois group over $\mathbb{Q}$ ([T,1] and a special case of [Fr,1]). But rational rigidity requires a group to be generated by two elements in a special way. We generalize rational rigidity to super rational (and unirational) connectivity (§3 and §4) in order to realize more groups as Galois groups over $\mathbb{Q}$ (again, a special case of [Fr,1]). The essential additional hypothesis asks that certain projective variety covers of $\mathbb{P}^r$, $r \geq 4$, be $\mathbb{Q}$-rational. Along with a numerical test for $\mathbb{Q}$-rationality, §4 connects to the abelian case of the Noether problem.

Secondly: Thompson's program attempts to reduce the inverse problem of Galois theory to the case of simple groups. The essential step considers exact sequences of finite groups, $1 \to M \to G \xrightarrow{\phi} H \to 1$ satisfying these conditions:

1) M is a minimal normal subgroup of G; and

2) $H \simeq G(F/\mathbb{Q})$ for some Galois extension $F/\mathbb{Q}$.

From (1), $M \simeq M_1^k$ with M_1 simple, for some k.

THE FUNDAMENTAL QUESTION: What conditions on M_1 and the sequence allow us to assert that $G \simeq G(L/\mathbb{Q})$ with L containing F so that restriction to F gives $\phi: G \to H$?

All results to date require two conditions:

3) the sequence $1 \to M \to G \xrightarrow{\phi} H \to 1$ splits; and

4) $M_1 \simeq G(L_1/\mathbb{Q}(x))$ with L_1 a regular Galois extension of $\mathbb{Q}(x)$.

These follow from rational rigidity (or its generalization in §3). If (3) holds, then G is a Galois group over $\mathbb{Q}$ in the desired way if either M is abelian (Theorem 2.5) or, (4) holds and $\text{Aut}(M_1) = \text{Inn}(M_1)$ (Theorem 2.2). The latter, in particular, includes the case [T,2] that M_1 is $G_2(p)$, a Chevalley group of type G_2 over the prime finite field $\mathbb{F}(p)$ with $p \geq 5$ (Corollary 2.3).

Added in proof. B. Matzat informed the author that Theorem 2.5 already appears in [N].

1 RATIONAL RIGIDITY FOR TWO GENERATOR GROUPS

Let G be a group and $C_1,\ldots,C_r$ nontrivial conjugacy classes of G. Denote the group generated by the entries of $(\sigma_1,\ldots,\sigma_r) = \underset{\sim}{\sigma} \in G^r$ by $G(\underset{\sim}{\sigma})$. The Nielsen class of $(\underset{\sim}{C},G)$ is the collection (assumed nonempty) $\text{Ni}(\underset{\sim}{C}) =$

$$\{\underset{\sim}{\tau} \in G^r \mid G(\underset{\sim}{\tau}) = G,\ \tau_1\cdots\tau_r = \text{Id. and } \tau_{(i)\alpha} \in C_i,\ i = 1,\ldots,r \text{ for some } \alpha \in S_r\}.$$

From this point on assume that G is a subgroup of S_n - not necessarily the regular representation. When an element $\underset{\sim}{\sigma} \in S_n^r$ represents the Nielsen class, denote it by $\text{Ni}(\underset{\sim}{\sigma})$. The normalizer, $N_{S_n}(\underset{\sim}{C})$ of the Nielsen class is $\{\tau \in S_n \mid \text{conjugation by } \tau \text{ permutes } C_1,\ldots,C_r\}$. When appropriate abbreviate this to $N_{S_n}(\underset{\sim}{\sigma})$. It acts on the elements of $\text{Ni}(\underset{\sim}{C})$ by conjugation: $\gamma \in N_{S_n}(\underset{\sim}{C})$ maps $\underset{\sim}{\tau} \in \text{Ni}(C)$ to the result, $\gamma^{-1}\underset{\sim}{\tau}\cdot\gamma$, of conjugating each entry of $\underset{\sim}{\tau}$ by γ. Denote the quotient of this action by $\text{Ni}(\underset{\sim}{C})^{ab}$, the absolute Nielsen classes.

The Hurwitz monodromy group, $H(r)$, of degree r [Bi] acts on $\text{Ni}(\underset{\sim}{C})^{ab}$. It suffices to know that $H(r)$ has $r-1$ generators, $Q_1,\ldots,Q_{r-1}$, and the ith of these acts on a representative $\underset{\sim}{\tau} \in \text{Ni}(\underset{\sim}{C})$ by this formula [BFr]:

$$(\underset{\sim}{\tau})Q_i = (\tau_1,\ldots,\tau_{i-1},\tau_i\cdot\tau_{i+1}\cdot\tau_i^{-1},\tau_i,\tau_{i+2},\ldots,\tau_r),\quad i = 1,\ldots,r-1. \tag{1.1}$$

Call $\text{Ni}(\underset{\sim}{C})$ connected if $H(r)$ has only one orbit on $\text{Ni}(\underset{\sim}{C})^{ab}$. This is a weakening of the rigid condition of [T,1]. If, however, $r = 3$, C_1, C_2 and C_3 are distinct conjugacy classes and G is embedded in S_n in the regular representation, then rigidity and connectedness are the same. In the language of [BFr], connectedness is equivalent to the <u>connectedness of the moduli space of covers of</u> $\mathbb{P}^1$ of Nielsen class $\mathrm{Ni}(\underset{\sim}{C})$.

With $N = |G|$, let ζ_N be a primitive Nth root of 1. Identify the group of the Galois extension $\mathbb{Q}(\zeta_N)/\mathbb{Q}$ with $(\mathbb{Z}/(N))^*$, the invertible integers modulo N. For $\alpha \in (\mathbb{Z}/(N))^*$ and C a conjugacy class of G denote the conjugacy class of αth powers of C by C^α. Similarly, denote $(C_1^\alpha,\ldots,C_r^\alpha)$ by $\underset{\sim}{C}^\alpha$. Call a Nielsen class <u>rational</u> if $\mathrm{Ni}(\underset{\sim}{C}^\alpha) = \mathrm{Ni}(\underset{\sim}{C})$ for all $\alpha \in (\mathbb{Z}/(N))^*$, and call it <u>rationally connected</u> if it is also connected.

For the remainder of this section $r = 3$ (for $r > 3$ see §3).

<u>Theorem 1.1</u>. Assume the centralizer of G, $\mathrm{Cen}_{S_n}(G)$, in S_n is trivial (e.g., G is a primitive group). Let x be an indeterminate. If $\mathrm{Ni}(\underset{\sim}{C})$ is rationally connected, then there exists a regular extension $L/\mathbb{Q}(x)$ of degree n for which the Galois closure $\hat{L}$ satisfies

$$\text{a)}\quad G \subseteq G(\hat{L}/\mathbb{Q}(x)) \subseteq N_{S_n}(\underset{\sim}{C}), \quad \text{and} \tag{1.2}$$
$$\text{b)}\quad G(\hat{L}\cdot\mathbb{C}/\mathbb{C}(x)) = G.$$

If, in addition, $N_{S_n}(\underset{\sim}{C}) = G$, then $\hat{L}$ is a regular Galois extension of $\mathbb{Q}(x)$ with group G.

<u>Proof</u>. The majority of [Fr,1] handles the case $r > 3$. The existence of $L/\mathbb{Q}(x)$, a regular extension with $G(\hat{L}\cdot\mathbb{C}/\mathbb{C}(x)) = G$ follows just from "the branch cycle argument" [Fr,1; p. 62-64]. In addition, the major accomplishment in this is treatment of the case where the entries of $\underset{\sim}{C}$ are nontrivially permuted by $(\mathbb{Z}/(N))^*$, $i = 1,2,3$. Expression (1.2)a) is an entirely elementary observation on inertia groups [Fr,1; p. 33, 1st statement of Proposition 2].

<u>Remark 1.2</u>. If $N_{S_n}(\underset{\sim}{C}) = G$ we can in Theorem 1.1 replace the triviality of $\mathrm{Cen}_{S_n}(G)$ by the assumption that G has no center. Then, the argument of Theorem 1.1 gives a Galois extension $F/\mathbb{Q}$ and a regular Galois extension $\hat{L}/F(x)$ with these properties: $L/\mathbb{Q}(x)$ is Galois; and conjugation of $G(\hat{L}/\mathbb{Q}(x))$ on $G(\hat{L}/F(x)) = G$ maps the former group into

$N_{S_n}(\underset{\sim}{C}) = G$. Thus, $G(\hat{L}/\mathbb{Q}(x)) \cong G\times G(F/\mathbb{Q})$. The fixed field, $\hat{L}_1$, of $G(F/\mathbb{Q})$ in $G(\hat{L}/\mathbb{Q}(x))$ is therefore a regular Galois extension of $\mathbb{Q}(x)$ with group G. These hypotheses are only slightly weaker than those on G and $\underset{\sim}{C}$ that [T,1] calls <u>rational rigidity</u>.

2 SPLIT SEQUENCES

Let $1 \to M \to G \xrightarrow{\phi} H \to 1$ be an exact sequence of finite groups where $M = M_1^k$, a product of k copies of M_1. Let S_k act as permutations of the coordinates of M_1^k and denote the semidirect product of M and S_k (with this action on M) by $M \times^s S_k$.

With $C = \mathrm{Cen}_G(M)$ there is natural exact sequence:

$$1 \to C \to G \xrightarrow{\psi} G/C \to 1. \tag{2.1}$$

Finally, with $U_1 = \{(\sigma_1,\mathrm{Id.},\dots,\mathrm{Id.}) \in M \mid \sigma \in M_1\}$ let $N_G(U_1) = N$ be the normalizer in G of U_1.

<u>Lemma 2.1</u>. Suppose that M_1 has no center. Identify M and $\mathrm{Inn}(M)$. Then the image, M^ψ, of M under ψ is isomorphic to M, and G^ψ acts faithfully on M^ψ by conjugation. Thus G^ψ embeds in $\mathrm{Aut}(M)$. Denote the homomorphism induced on G by $\alpha: G \to \mathrm{Aut}(M)$.

Suppose that $G^\alpha \subseteq M \times^s S_k$ (e.g., if $\mathrm{Inn}(M_1) = \mathrm{Aut}(M_1)$). Then G contains a subgroup H_1 that maps isomorphically to H by ϕ (i.e., ϕ splits) and conjugation by H_1 on M maps H_1 into $\mathrm{Id.} \times S_k \subseteq M \times^s S_k$.

<u>Proof</u>. Since G^α contains $M = \mathrm{Inn}(M)$, the second paragraph assumption implies that $G^\alpha = M \times^s H'$. Take $H_1 = \psi^{-1}(H')$. The rest is obvious.

With no loss, for what follows, assume that H_1 in Lemma 2.1 has transitive image in S_k. In particular, $(H_1 : H_1 \cap N) = k$.

<u>Theorem 2.2</u>. Assume for the sequence $1 \to M \to G \xrightarrow{\phi} H \to 1$ the following: M_1 has no center and $G^\alpha \subseteq M \times^s S_k$ (the hypotheses of Lemma 2.1); $H = G(F/\mathbb{Q})$ (condition (3)); and $M_1 \simeq G(L_1/\mathbb{Q}(x))$ with L_1 a regular extension of $\mathbb{Q}(x)$ (condition (4)). Then there exist indeterminates $x_o,\dots,x_{k-1}$ and a Galois extension $\Omega/\mathbb{Q}(x_o,\dots,x_{k-1})$ whose Galois group is isomorphic to G. In addition, the algebraic closure of $\mathbb{Q}$ in Ω

is F. In particular $G \simeq G(L/\mathbb{Q})$ with $F \subseteq L$ so that restriction to F gives $\phi: G \to H$.

Proof. Use the notation of Lemma 2.1, and identify H_1 with H. Denote the fixed field of $H_1 \cap N$ in F by F_1. Let λ be a primitive generator of $F_1/\mathbb{Q}$ so that $F_1 = \mathbb{Q}(\lambda)$. And let $\lambda_1,\ldots,\lambda_k$ be the conjugates of λ over $\mathbb{Q}$. The action of H_1 on $\lambda_1,\ldots,\lambda_k$ naturally identifies it with its image under α (in Lemma 2.1) in S_k. We are now prepared to form the field Ω.

Let t be a primitive generator for $L_1/\mathbb{Q}(x)$ and let $g(x,y) \in \mathbb{Q}[x,y]$ be the irreducible polynomial for t over $\mathbb{Q}(x)$. Regularity of $L_1/\mathbb{Q}(x)$ means that $g(x,y)$ is absolutely irreducible.

Introduce new variables $z_i = x_0 + \lambda_i \cdot x_1 + \ldots + \lambda_i^{k-1} \cdot x_{k-1}$, $i = 1,\ldots,k$. These are algebraically independent over $\mathbb{Q}$ and $F(x_0,\ldots,x_k) = F(z_1,\ldots,z_k)$. Let y_i be a zero of $g(z_i,y)$, $i = 1,\ldots,k$. Denote $F(z_1,\ldots,z_k)$ (resp., $\mathbb{Q}(x_0,\ldots,x_k)$, $F(z_1,\ldots,z_k,y_1,\ldots,y_k)$, etc.) by $F(\underset{\sim}{z})$ (resp., $\mathbb{Q}(\underset{\sim}{x})$, $F(\underset{\sim}{z},\underset{\sim}{y})$, etc.). An easy induction shows that $F(\underset{\sim}{z},\underset{\sim}{y})/F(\underset{\sim}{z})$ is a Galois extension with group equal to M_1^k. Let H_1 act as automorphisms of $F(\underset{\sim}{z},\underset{\sim}{y})$ through the following formula. If $\sigma \in H_1$ has image $\alpha(\sigma) \in S_k$ (Lemma 2.1), then σ maps z_i to $z_{(i)\alpha(\sigma)}$ and y_i to $y_{(i)\alpha(\sigma)}$, $i = 1,\ldots,k$. Also, σ acts on F through the identification of H_1 with H. As remarked above, this implies that H_1 leaves $x_0,\ldots,x_{k-1}$ fixed and thus identifies the group of the Galois extension $F(\underset{\sim}{z},\underset{\sim}{y})/\mathbb{Q}(\underset{\sim}{x})$ with the semidirect product of M_1^k and H_1. That is, with $F(\underset{\sim}{z},\underset{\sim}{y}) = \Omega$ we have produced the desired Galois extension of $\mathbb{Q}(\underset{\sim}{x})$ with group G. Apply Hilbert's irreducibility theorem to get the last line of the theorem.

Corollary 2.3 [T,2]. Consider an exact sequence $1 \to M \to G \to H \to 1$ of finite groups where $M = G_2(p)^k$ with $G_2(p)$ the Chevalley group of type G_2 over $\mathbb{F}(p)$, $p \geq 5$. If H is a Galois group over $\mathbb{Q}$, then so is G.

Proof. Thompson in [T,3] and, later, Feit and Fong in [FF] give rationally rigid generators $\sigma_1,\sigma_2,\sigma_3 \in G_2(p)$ (Remark 1.2) of $G_2(p)$. From §1, $G_2(p)$ is isomorphic to a regular Galois extension of $\mathbb{Q}(x)$. Also, it is known that $\mathrm{Inn}(G_2(p)) = \mathrm{Aut}(G_2(p))$. Theorem 2.2 now applies.

Remark 2.4. (i) In Lemma 2.1 (and Theorem 2.2) we can drop the assumption that M_1 has no center if we know, a priori, that $G \subseteq M \times^s S_k$ with S_k acting as permutations of the M_1 factors of $M = M_1^k$. (ii) Furthermore, we may weaken (4) in Theorem 2.2 to the condition that $M_1 = G(L_1/F_1(x))$ with L_1 a regular extension of $F_1(x)$, and F_1 the fixed field of $H_1 \cap N$ that appears in the proof of Theorem 2.2. Then $g(x,y) \in F_1(x,y)$ and for each conjugate λ_i of λ there is a new polynomial $g_i(x,y)$. The proof is the same, then, if we let y_i be a zero of $g_i(z_i,y)$, $i = 1,\ldots,k$.

Now we do the case when the kernel M is abelian.

Theorem 2.5. If $1 \to M \to G \xrightarrow{\phi} H \to 1$ is split exact. M is abelian and $H \cong G(F/\mathbb{Q})$, then $G = G(L/\mathbb{Q})$ with $F \subseteq L$.

Proof (influenced by Uchida). With no loss assume that the abelian group M is isomorphic to $(\mathbb{Z}/(q))^k$ with $q = \ell^n$, ℓ a prime; and that M is indecomposable for the action of H. Let ζ_q be a primitive qth root of 1. We divide the proof into 3 parts. Identify G with $M \times^s H$.

Part 1. Cyclotomic extension. Let $H_1 = G(F(\zeta_q)/\mathbb{Q}) \subseteq G(F/\mathbb{Q}) \times G(\mathbb{Q}(\zeta_q)/\mathbb{Q})$. Write elements of H_1 as $(\tau,\alpha) = \tilde{\tau}$ with $\tau \in G(F/\mathbb{Q})$, $\alpha \in G(\mathbb{Q}(\zeta_q)) \simeq (\mathbb{Z}/(q))^*$ and $\tau|_{F \cap \mathbb{Q}(\zeta_q)} = \alpha|_{F \cap \mathbb{Q}(\zeta_q)}$.

Denote the group ring of H_1 with coefficients in $\mathbb{Z}/(q)$ by $R = \mathbb{Z}/(q)[H_1]$. Let M' be R as an abelian group under addition. But let R act on M' through this formula: with

$$m = \sum_{\tilde{\sigma} \in H_1} v_{\tilde{\sigma}} \cdot \tilde{\sigma} \in M', \; v_{\tilde{\sigma}} \in \mathbb{Z}/(q) \text{ and } \tilde{\tau} = (\tau,\alpha) \in H_1$$
$$m^{\tilde{\tau}} = \sum v_{\tilde{\sigma}} \cdot \alpha \cdot \tilde{\sigma} \cdot \tilde{\tau}. \tag{2.2}$$

Compute that $m^{\tilde{\tau}_1 \cdot \tilde{\tau}_2} = (m^{\tilde{\tau}_1})^{\tilde{\tau}_2}$. With this action, M' is a 1-dimensional R module, therefore isomorphic to R.

Part 2. An extending sequence. Since H_1 projects naturally onto $G(F/\mathbb{Q}) = H$ we regard M as an R module through the action of H on M. Let $m_1,\ldots,m_t$ be generators of M (as an R module). For each $i = 1,\ldots,t$ let $R = M_i'$ be a copy of M' and consider the map $\phi_i : M_i' \to M$

by $r \to m_i^r$. With $M_1 = \bigoplus_{i=1}^{t} M_i'$, extend these R module homomorphisms to an R module homomorphism $\phi: M_1 \to M$ by $\phi(r_1,\dots,r_t) = \sum_{i=1}^{t} m_i^{r_i}$. Now take G_1 to be $M_1 \times^s H_1$. Clearly, if G_1 is a Galois group over $\mathbb{Q}$, then the quotient G_1 of G is also.

Part 3. G_1 is a Galois group over $\mathbb{Q}$. From Galois theory we are easily reduced to the case where the action of H_1 on M_1 is indecomposible: where $M_1 = M'$ (i.e., $t = 1$). Since $M' = R = (\mathbb{Z}/(q))^k$ with $k = |H_1|$, H_1 acts by permutations of the copies of $\mathbb{Z}/(q)$ and the hypothesis of Remark 2.4 (i) holds. In addition, the fixed field F_1 of $H_1 \cap N$ contains $\mathbb{Q}(\zeta_q)$. Thus, $L_1 = F_1(x^{1/q})$ is a regular Galois extension of $F_1(x)$ with group $\mathbb{Z}/(q)$, and the hypothesis of Remark 2.4 (ii) holds. This concludes the proof.

3 SUPER RATIONAL CONNECTIVITY

Let K be a field. For X a nonempty Zariski K-open subset of a projective K-variety Y, denote by $K(X)$ the field of K-rational functions on X. Then $K(X) = K(Y)$. Call X a K-rational variety if $K(X)$ is a purely transcendental extension of K.

Let G be a subgroup of S_n and $C_1,\dots,C_r$ nontrivial conjugacy classes of G. Use the notation of §1. Assume that the (nonempty) absolute Nielsen class $\mathrm{Ni}(\underset{\sim}{C})$ is connected. Attached to $\mathrm{Ni}(\underset{\sim}{C})$ is an algebraic variety, $\mathcal{H}(\mathrm{Ni}(\underset{\sim}{C})^{ab})$, and a natural covering map of an open subset of projective r-space, $\mathbb{P}^r$, by $\mathcal{H}(\mathrm{Ni}(\underset{\sim}{C})^{ab})$. We describe this.

Projective 1-space, $\mathbb{P}^1$, is just $\mathbb{C} \,\dot\cup\, \{\infty\}$. Regard $\mathbb{P}^r$ as $\mathbb{C}^{r+1} - \{\underset{\sim}{0}\}$ modulo the relation that equivalences $\underset{\sim}{x}_1$ and $\underset{\sim}{x}_2 \in \mathbb{C}^{r+1}$ if there exists $\lambda \in \mathbb{C}$ with $\lambda \cdot \underset{\sim}{x}_1 = \underset{\sim}{x}_2$. Then there is a natural map $\Theta_r: (\mathbb{P}^1)^r \to \mathbb{P}^r$, called the Noether cover, that presents $\mathbb{P}^r$ as the quotient of $(\mathbb{P}^1)^r$ by the action of S_r as permutations of the copies of $\mathbb{P}^1$. Indeed, for $\underset{\sim}{z} = (z_1,\dots,z_r) \in (\mathbb{P}^1)^r$ form the polynomial $\prod_{i=1}^{r} (x - z_i) = \sum_{j=0}^{r} x_i \cdot x^i$ in x with the proviso that for $z_i = \infty$ we write 1 instead of $x - z_i$. Then $\Theta_r(\underset{\sim}{z})$ is the element of $\mathbb{P}^r$ represented by $(x_0,\dots,x_r)$. Let Δ_r be the subset of $(\mathbb{P}^1)^r$ consisting of elements with 2 or more equal entries. Then the discriminant locus, D_r, of Θ_r is the image of Δ_r under Θ_r. The Hurwitz monodromy group $H(r)$ (§1) is the fundamental group of $\mathbb{P}^r - D_r$. Thus the action of $H(r)$ on $\mathrm{Ni}(\underset{\sim}{C})^{ab}$ (expression (1.1)) - a transitive permutation representation of the fundamental group of $\mathbb{P}^r - D_r$ - gives rise to an equivalence class of (topological) covers

$$\Theta(\mathcal{H}): \mathcal{H}(\mathrm{Ni}(\underset{\sim}{C})^{ab}) \to \mathbb{P}^r - D_r. \tag{3.1}$$

Call $\mathcal{H}(\mathrm{Ni}(\underset{\sim}{C})^{ab})$ the Hurwitz space of covers of Nielsen class $\mathrm{Ni}(\underset{\sim}{C})$ [BFr].

Now we state the analogue of Theorem 1.1 for $r > 3$. It is a special case of [Fr,1; Theorem 5.1]. The complex manifold $\mathcal{H}(\mathrm{Ni}(\underset{\sim}{C})^{ab})$ is a Zariski open subset of some projective variety.

<u>Theorem 3.1</u>. Assume the centralizer of G in S_n is trivial. Assume also that $\mathrm{Ni}(\underset{\sim}{C})$ is rationally connected. Then $\mathcal{H}(\mathrm{Ni}(\underset{\sim}{C})^{ab})$ and $\Theta(\mathcal{H})$ are defined over $\mathbb{Q}$. Finally, assume that $\mathcal{H}(\mathrm{Ni}(\underset{\sim}{C})^{ab})$ is a $\mathbb{Q}$-rational variety. Then, there exists a regular extension $L/\mathbb{Q}(x)$ of degree n for which the Galois closure $\hat{L}$ satisfies the conclusion of Theorem 1.1. In particular, if $N_{S_n}(\underset{\sim}{C}) = G$ then $G(\hat{L}/\mathbb{Q}(x)) = G$.

<u>Definition 3.2</u>. Call the rationally connected Nielsen class $\mathrm{Ni}(\underset{\sim}{C})$ <u>super rationally connected</u> if $\mathcal{H}(\mathrm{Ni}(\underset{\sim}{C})^{ab})$ is a $\mathbb{Q}$-rational variety. Theorem 3.1 says that if, in addition, $N_{S_n}(\underset{\sim}{C}) = G$, then G is realized as a Galois group of a regular extension of $\mathbb{Q}(x)$. With this last hypothesis we may, as in Remark 1.2, replace the triviality of $\mathrm{Cen}_{S_n}(G)$ by the assumption that G has no center.

Theorem 3.1 effectively generalizes Theorem 1.1 only if there are reasonable tests for $\mathbb{Q}$-rationality (or unirationality - see §4) of $\mathcal{H}(\mathrm{Ni}(\underset{\sim}{C})^{ab})$. First, however, we give examples of the varieties $\mathcal{H}(\mathrm{Ni}(\underset{\sim}{C})^{ab})$ for different pairs $(\underset{\sim}{C}, G)$. Let H be any subgroup of S_r. Then, up to equivalence as a cover of $\mathbb{P}^r - D_r$, there is a unique complex manifold $W(H)$ that fits in a sequence of covers

$$(\mathbb{P}^1)^r \to \Delta_r \to W(H) \to \mathbb{P}^r - D \tag{3.2}$$

where the first part of the sequence is a Galois cover with group H.

Ex. 3.3. Groups H with $W(H) = \mathcal{H}(\mathrm{Ni}(\underset{\sim}{C})^{ab})$ for some $(\underset{\sim}{C}, G)$. Let $A = G(\underset{\sim}{\sigma})$ be an abelian group where $\underset{\sim}{\sigma} = (\sigma_1, \ldots, \sigma_r)$ and $\sigma_1 \cdots \sigma_r = \mathrm{Id}$. Suppose that H is the exact set of permutations of the entries of $\underset{\sim}{\sigma}$ that induce automorphisms of A. With $n = |A|$, embed A in S_n by its right regular representation. From (1.1), Q_i acts on an element $\underset{\sim}{\tau} \in \mathrm{Ni}(\underset{\sim}{\sigma})$ by switching the i and i+1th entries, $i = 1, \ldots, r$. That is, this gives a homomorphism $H(r) \to S_r$ that maps Q_i to $(i\ i{+}1)$, $i = 1, \ldots, r$. But $\mathrm{Ni}(\underset{\sim}{\sigma})^{ab}$ is the quotient of $\mathrm{Ni}(\underset{\sim}{\sigma})$ by the action of H, so fundamental group theory identifies $W(H)$ with $\mathcal{H}(\mathrm{Ni}(\underset{\sim}{\sigma})^{ab})$.

The condition on H that allows us to find $\underset{\sim}{\sigma}$ as above is this. As H is a subgroup of S_r, it naturally acts on $(\mathbb{Z}/(p))^r = V$, p any prime. Suppose that V has a subspace V_1 for which H is the maximal subgroup of S_r for which V_1 is an invariant subspace. Let V_2 be the subspace of V generated by V_1 and $(1, \ldots, 1)$ and let σ_i be the image in V/V_2 of e_i, the vector with 1 in the ith position and 0's elsewhere, $i = 1, \ldots, r$. Then $\mathcal{H}(\mathrm{Ni}(\underset{\sim}{\sigma})^{ab}) = W(H)$.

4 TESTS FOR SUPER UNIRATIONAL CONNECTIVITY

Assume that $\mathrm{Ni}(\underset{\sim}{C})$ is a rationally connected Nielsen class. How do we test for super rational connectivity (or some suitable weakening of this)? Two observations tie us to the literature. In the case that $\underset{\sim}{\sigma} \in S_n^r$ has all its entries 2-cycles, let $g = r/2 - 2\cdot(n-1)$. If $n \geq 2\cdot g-1$, then there is a surjective map from $\mathcal{H}(\mathrm{Ni}(\underset{\sim}{\sigma})^{ab})$ to the moduli space $\mathcal{M}_g$ of curves of genus g (e.g., [Fr,1; Ex. 1] or [Fu]). If $\mathcal{H}(\mathrm{Ni}(\underset{\sim}{\sigma})^{ab})$ were $\mathbb{Q}$-rational, then $\mathcal{M}_g$ would be, at least, $\mathbb{C}$-unirational. But, for g odd and suitably large, it is not [HM]: rationality of $\mathcal{H}(\mathrm{Ni}(\underset{\sim}{\sigma})^{ab})$ is a special phenomenon.

We use the thinking of Ex. 3.3 for our next point about a Hurwitz class $\mathrm{Ni}(\underset{\sim}{\sigma})$. Let $H_{\underset{\sim}{\sigma}}$ be the subgroup of S_r consisting of elements β for which there exists $\tau \in N_{S_n}(\underset{\sim}{\sigma})$ with $\tau^{-1}\cdot\sigma_i\cdot\tau$ conjugate in $G(\underset{\sim}{\sigma})$ to $\sigma_{(i)\beta}$, $i = 1, \ldots, r$. Then, with $H_{\underset{\sim}{\sigma}}$ replacing H in Ex. 3.3, there is a natural commutative diagram of covers

$$\begin{array}{ccc} \mathcal{H}(\mathrm{Ni}(\underset{\sim}{\sigma})^{ab}) & \longrightarrow & W(H_{\underset{\sim}{\sigma}}) \\ & \searrow^{\Theta(\mathcal{H})} & \downarrow \\ & & \mathbf{P}^r - D_r \end{array} \qquad (4.1)$$

Of course, if $W(H)$ were $\mathbb{Q}$-rational for every H, then every group would already be a Galois group over $\mathbb{Q}$. It is sometimes easier to deal with a connected component H' of the pullback of $H(\mathrm{Ni}(\underset{\sim}{\sigma})^{ab})$ to $(\mathbb{P}^1)^r$. Then $\mathbb{Q}$-rationality of H' implies $\mathbb{Q}$-unirationality (i.e., $\mathbb{Q}(H(\mathrm{Ni}(\underset{\sim}{\sigma})^{ab}))$ is contained in a purely transcendental field over $\mathbb{Q}$) of $H(\mathrm{Ni}(\underset{\sim}{\sigma})^{ab}))$. Is this sufficient to get the conclusion of Theorem 3.1? Our first result: The answer is not quite, but almost.

Let $H'_{\underset{\sim}{\sigma}}$ be the subgroup of S_r generated by $H_{\underset{\sim}{\sigma}}$ and those elements β for which there exists $\alpha \in (\mathbb{Z}/(N))^*$ (notation of Theorem 1.1) with σ_i^{α} conjugate in $G(\underset{\sim}{\sigma})$ to $\sigma_{(i)\beta}$, $i = 1,\dots,r$.

<u>Theorem 4.1</u>. The conclusion of Theorem 3.1 holds if the hypothesis that $H(\mathrm{Ni}(\sigma)^{ab})$ is $\mathbb{Q}$-rational is weakened to this: $\mathbb{Q}(H(\mathrm{Ni}(\underset{\sim}{\sigma})^{ab}))$ is contained in a field L purely transcendental over $\mathbb{Q}$ for which - disjointness hypothesis -

$$L\cdot\mathbb{Q}(W(H_{\underset{\sim}{\sigma}})) \cap \mathbb{Q}(W(H'_{\underset{\sim}{\sigma}})) = \mathbb{Q}(W(H_{\underset{\sim}{\sigma}})). \tag{4.2}$$

In particular, (4.2) holds if $H_{\underset{\sim}{\sigma}} = H'_{\underset{\sim}{\sigma}}$.

<u>Proof</u>. Without (4.2), the pullback of the family of covers that $H(\mathrm{Ni}(\underset{\sim}{\sigma})^{ab})$ parametrizes, in the proof of [Fr,1; Theorem 5.1], to the variety with function field L would break up into components that are not defined over $\mathbb{Q}$. Without this circumstance, however, the proof is the same as that of Theorem 3.1.

Call $\mathrm{Ni}(\underset{\sim}{\sigma})$ <u>super unirationally connected</u> if in addition to rational connectivity, condition (4.2) holds. Practical tests for the unirationality of $H(\mathrm{Ni}(\underset{\sim}{\sigma})^{ab})$ therefore demand definitive tests for the rationality of $W(H)$. Let $U = H'_{\underset{\sim}{\sigma}}/H_{\underset{\sim}{\sigma}}$. In the special case that $H'_{\underset{\sim}{\sigma}}$ contains U' which maps isomorphically to U under the natural map $H'_{\underset{\sim}{\sigma}} \to U$, we are led to consider the rationality of $W(U')$. Since U' is abelian, this connects us with the work of [L], [Sa] and [Sw], among others. We conclude with a numerical test for super unirationality.

For $\sigma \in S_n$, $\mathrm{ind}(\sigma) = n - d(\sigma)$ where $d(\sigma)$ is the number of disjoint cycles in σ. Consider these elements of the Hurwitz monodromy group: $A_{12} = Q_1^{-2}$, $A_{13} = Q_1 \cdot Q_2^{-2} \cdot Q_1^{-1}, \dots, A_{1r} = Q_1 \cdots Q_{r-2} \cdot Q_{r-1}^{-2} \cdot Q_{r-2}^{-1} \cdots Q_1^{-1}$. These elements generate a subgroup of $H(r)$, denoted here as $P(r)$.

Choose any orbit of $P(r)$ on $Ni(\underset{\sim}{g})^{ab}$ and thus, if n' is the cardinality of this orbit, obtain a transitive permutation representation $P(r) \to S_{n'}$. Denote by a_{1j} the image of A_{1j}, $j = 2,\dots,r$.

<u>Theorem 4.2</u>. Assume that $Ni(\underset{\sim}{g})$ is rationally connected. Use the notation above to assume the following (explicitly testable) conditions:

$$\text{a) } H'_{\underset{\sim}{g}} = H_{\underset{\sim}{g}}; \tag{4.3}$$

$$\text{b) } n' - 1 = \Big(\sum_{j=2}^{r} \mathrm{ind}(a_{1j})\Big)/2; \text{ and}$$

c) in a listing of the lengths of all disjoint cycles in $a_{12},\dots,a_{1r}$, some integer appears an odd number of times.

Then $Ni(\underset{\sim}{g})$ is super unirationally connected. Note that (4.3) c) holds if n' and $r-1$ are both odd.

<u>Proof</u>. From Theorem 4.1, thanks to condition (4.3)a), we have only to demonstrate unirationality of $\mathcal{H}(Ni(\underset{\sim}{g})^{ab})$. But this follows immediately if we demonstrate rationality of a connected component $\mathcal{H}'$ of the pullback of $\mathcal{H}(Ni(\underset{\sim}{g})^{ab})$ over $(\mathbb{P}^1)^r - \Delta_r$ via (3.2). We tacitly use the proof of Theorem 4.1 to assume that $\mathcal{H}'$ and its map to $(\mathbb{P}^1)^r - \Delta_r$ are defined over $\mathbb{Q}$.

Let $(z_1,\dots,z_{r-1}) = \underset{\sim}{z}_{r-1} \in (\mathbb{P}^1)^{r-1} - \Delta_{r-1}$. Consider the set $L(\underset{\sim}{z}_{r-1}) = \{(z_1,\dots,z_{r-1},z) \in (\mathbb{P}^1)^r - \Delta_r \mid z \in \mathbb{P}^1\}$. Denote the set of points of $\mathcal{H}'$ lying over $L(\underset{\sim}{z}_{r-1})$ by $\mathcal{H}'(\underset{\sim}{z}_{r-1})$. This is an open subset of an algebraic curve that is defined over the field $\mathbb{Q}(\underset{\sim}{z}_{r-1})$: $\mathbb{Q}$ with the coordinates of $\underset{\sim}{z}_{r-1}$ adjoined. Condition 4.3)b) is a restatement, for the cover $\mathcal{H}'(\underset{\sim}{z}_{r-1}) \to L(\underset{\sim}{z}_{r-1})$, of this consequence of the Riemann-Hurwitz formula [BFr; p. 94 - Figure 1.9]: the function field of $\mathcal{H}'(\underset{\sim}{z}_{r-1})$ over $\mathbb{Q}(\underset{\sim}{z}_{r-1})$ is of genus 0. If this function field is pure transcendental, over $\mathbb{Q}(\underset{\sim}{z}_{r-1})$, then the $\mathbb{Q}$-rationality of $\mathcal{H}'$ follows easily (e.g., [R; p. 43]).

Let m be the integer given by (4.3)c). Define the divisor D to be the formal sum of places on the function field of $\mathcal{H}'(\underset{\sim}{z}_{r-1})$ that correspond to disjoint cycles of $a_{12},\dots,a_{1r}$ of length m. Then D is of odd degree. In addition, we show easily that D is $\mathbb{Q}(\underset{\sim}{z}_{r-1})$-rational. Indeed, let $\overline{\mathbb{Q}(\underset{\sim}{z}_{r-1})}$ be an algebraic closure of $\mathbb{Q}(\underset{\sim}{z}_{r-1})$. Then $\sigma \in G(\overline{\mathbb{Q}(\underset{\sim}{z}_{r-1})}/\mathbb{Q}(\underset{\sim}{z}_{r-1}))$ naturally acts on the support of D to give a divisor D whose support consists, as does D, of all the places over branch points with ramification index m in the extension $\mathbb{Q}(\underset{\sim}{z}_{r-1})(H'(\underset{\sim}{z}_{r-1}))/\mathbb{Q}(\underset{\sim}{z}_{r-1})(L(\underset{\sim}{z}_{r-1}))$. Therefore $D^{\sigma} = D$ and D is defined over $\mathbb{Q}(\underset{\sim}{z}_{r-1})$. From such an odd degree divisor, a famous argument of Hilbert-Hurwitz yields the $\mathbb{Q}(\underset{\sim}{z}_{r-1})$-rationality of $H'(\underset{\sim}{z}_{r-1})$. The proof is complete.

A special case of Theorem 4.2 in [Fr,2; last section] shows that the space of Davenport polynomials of degree 13 - a source of many arithmetic anomalies (as Feit puts it) - is a rational variety. A later paper will give more examples using an improvement of Theorem 4.2 that is based on the following ideas. Condition (4.3) presents the $\mathbb{Q}$-unirationality of $H(\mathrm{Ni}(\underset{\sim}{\sigma})^{ab})$ as a two part conclusion: first it is $\mathbb{C}$-rational (from (4.3)b)); then we can descend to $\mathbb{Q}$ (from (4.3)c)). In theory we can compute the $\mathbb{Q}$-cohomology of a compactification of $H(\mathrm{Ni}(\underset{\sim}{\sigma})^{ab})$ from ideas of [Mi]. If $H(\mathrm{Ni}(\underset{\sim}{\sigma})^{ab})$ is $\mathbb{C}$-rational, then its first cohomology group is zero. Assuming this latter we can, again in theory, compute the Picard group of this compactification of $H(\mathrm{Ni}(\underset{\sim}{\sigma})^{ab})$. And with this, and the explicit Galois action of $G(\bar{\mathbb{Q}}/\mathbb{Q})$ on the Picard group (which can be derived from [Fr,1; §5]) the universal torseurs of Colliot-Thélène and Sansuc ([C-TS] or [Sw]) provide a more general test for $\mathbb{Q}$-rationality.

REFERENCES

[BFr] Biggers, R. & Fried, M. (1982). Moduli spaces of covers and the Hurwitz monodromy group. J. fur die reine und angew., 335, 87-121.

[Bi] Birman, J. (1975). Braids, Links and Mapping Class Groups, based on lecture notes of James Cannon. Princeton U. Press.

[C-TS] Colliot-Thélène, J.-L. & Sansuc, J.-J. La descente sur les varietes rationnelles, Algebraic Geometry Angers 1979, ed. by A. Beauville, Sijthoff and Noordhoff 1980, Alphen aon den Rijn, the Netherlands and Rockville, Md.

[FF] Feit, W. and Fong, P. Rational rigidity of $G_2(p)$ for any prime $p > 5$. These Proceedings.

[Fr,1] Fried, M. (1977). Fields of definition of function fields and Hurwitz families, and groups as Galois groups. Comm. in Alg. 5 (1), 17-82.

[Fr,2] Fried, M. Applications of the classification of simple groups to monodromy. Preprint.

[Fu] Fulton, W. (1969). Hurwitz schemes and irreducibility of moduli of algebraic curves. Annals of Math. 90, 542-575.

[HM] Harris, J. & Mumford, D. (1982). On the Kodaira dimension of the moduli spaces of curves. Invent. Math. 67, 23-86.

[L] Lenstra Jr., H.W. (1974). Rational functions invariant under a finite abelian group. Invent. Math. 25, 299-325.

[Mi] Milgram, R.J. (1969). The homology of Symmetric products. TAMS 138, 251-265.

[R] Roth, L. (1955). Algebraic threefolds. Berlin, Springer-Verlag.

[Sa] Saltman, D. (1982). Generic Galois extensions and problems in field theory. Adv. in Math. 43, 250-283.

[Sw] Swan, R. (1981). Noether's problem in Galois theory, a preprint expansion of "Galois Theory". In Emmy Noether: A tribute to her life and work, ed. J.W. Brewer & M.K. Smith. New York: Marcel Dekker.

[T,1] Thompson, J. Some finite groups which appear as Gal(L/K), where $K \subseteq \mathbb{Q}(\mu_n)$. To appear in J. Alg.

[T,2] Thompson, J. Some finite groups which appears as Galois groups over $\mathbb{Q}$. To appear in J. Alg.

[T,3] Thompson, J. Some finite groups of type G_2 which appear as Galois groups over Q. Preprint.

Added in proof. [N] Neukirch, J. (1973). Über das Einbettungsproblem der algebraischen Zahlentheorie. Inv. Math. 21, 59-116.

RIGIDITY OF $SL_n(q)$ AND CERTAIN SUBGROUPS FOR $(n,q-1) = 1$ AND $n > 2$

W. Feit[1]
Yale University, New Haven, CT 06520

P. Fong[2]
University of Illinois at Chicago, Chicago, IL

B. Srinivasan[3]
University of Illinois at Chicago, Chicago, IL

1 INTRODUCTION

The results in this paper were proved independently by the first author and the other two authors. The work grew out of attempts to answer a question of John G. Thompson.

Suppose that q is a prime power, $n > 2$ is an integer and $(n,q-1) = 1$. Thus $SL_n(q) \approx PSL_n(q)$ and $GL_n(q) = SL_n(q)\times Z$, where Z is the center of $GL_n(q)$.

Let C_1 be the conjugate class of $SL_n(q)$ which contains a transvection.

Let C_2 be the conjugate class of $SL_n(q)$ which contains a regular unipotent element.

Let z be an element in $SL_n(q)$ such that $\mathbb{C}_{SL_n(q)}(z)$ is cyclic of order $(q^n-1)/(q-1)$. Let C_3 be the conjugate class of $SL_n(q)$ which contains z.

For instance z can be chosen to be an element of order $(q^n-1)/(q-1)$. As another example, let p be a prime with $p|(q^n-1)$, $p \nmid (q^i-1)$ for $0 < i < n$ and let z be any p-singular element. (If $n > 2$ then such a prime p always exists except when $q = 2$ and $n = 6$.)

The main object of this note is to prove the following result.

__Theorem 1__. Suppose that $(n,q-1) = 1$. Then there are exactly $|SL_n(q)|$ ordered triples (w_1,w_2,w_3) with $w_i \in C_i$ and $w_1w_2w_3 = 1$.

Before proving Theorem 1 we deduce the following consequences.

__Theorem 2__. Suppose that $(n,q-1) = 1$. Let $w_i \in C_i$ for $i = 1,2,3$ with $w_1w_2w_3 = 1$ and let $H = \langle w_1,w_2,w_3\rangle$. Assume that $H = \mathbb{N}(H)$, the normalizer

[1]Supported by NSF grant MCS-8201333.

[2]Supported by NSF grant MCS-8101689.

[3]Supported by NSF grant MCS-9117533.

of H in $SL_n(q)$. Let C_i' be the conjugate class of H with $w_i \in C_i'$. Then $A_H(C_1',C_2',C_3')$ is rigid.

<u>Proof</u>. Let $S = SL_n(q)$. As $\mathbb{C}_S(w_3)$ contains no unipotent elements Theorem 1 implies that S acts transitively on $A_S(C_1,C_2,C_3)$. Thus if (w_1,w_2,w_3) and (w_1',w_2',w_3') are in $A_H(C_1',C_2',C_3')$, there exists $v \in S$ with $w_i' = w_i^v$ for $i = 1,2,3$. Hence $H^v = \langle w_1^v, w_2^v, w_3^v\rangle \subseteq H$ and so $v \in N(H) = H$ as required.

<u>Corollary 3</u>. Suppose that $(n,q-1) = 1$. Let z be an element of order $(q^n-1)/(q-1)$. Then $A_{SL_n(q)}(C_1,C_2,C_3)$ is rigid.

<u>Proof</u>. Let H be as in Theorem 2. Let Z be the center of $GL_n(q)$. Then $Z\times H$ contains an element of order q^n-1. Thus by the Theorem in [3] $GL_{n/s}(q^s) \lhd Z\times H$ for some s. Since $Z\times H$ contains a regular unipotent element it follows that $s = 1$. Hence $H = SL_n(q)$ and the result follows from Theorem 2.

<u>Corollary 4</u>. Suppose that $(n,q-1) = 1$. Let ε be a primitive $(q^n-1)/(q-1)$-root of 1. Let $\eta = \sum_{i=0}^{n-1} \varepsilon^{q^i}$. Then $SL_n(q)$ is a Galois group over $\mathbb{Q}(\eta)$.

<u>Proof</u>. This follows from Corollary 3 and the Corollary to Theorem A in the introduction.

2 PROOF OF THEOREM 1

We will freely use the results and notation of [1], especially Lemma 4.4, Theorem 14 and p. 445.

Let $G = GL_n(q)$ for some $n > 1$. Let $\Phi_n(t)$ be the nth cyclotomic polynomial. We will consider irreducible characters θ of G which satisfy

$$\theta(1) \not\equiv 0 \pmod{q}. \tag{I}$$

$$\theta(1) \not\equiv 0 \pmod{\Phi_n(q)}. \tag{II}$$

Let $\psi_n = \psi_n(q) = \prod_1^n (q^i-1)$.

Let $\phi_n = \phi_n(q) = (-1)^n\psi_n = \prod_1^n (1-q^i)$.

For λ, ρ partitions of n let $Q_\rho^\lambda = Q_\rho^\lambda(q)$ denote the Green function. If $\rho = \{1^{r_1}, 2^{r_2}, \ldots\}$ let $e_\rho(t) = \prod_i (1-t^i)^{r_i}$.

See [1] and [5] Lemma 3 for the following results:

$$Q_\rho^{\{1^n\}} = \frac{\phi_n}{e_\rho(q)} \tag{1}$$

$$Q_\rho^{\{2,1^{n-2}\}} = \frac{\phi_{n-2}}{e_\rho(q)} \{(r_1-1)q^n - r_1q^{n-1} + 1\}. \tag{2}$$

In particular

$$Q^{\{2,1^{n-2}\}} = \frac{\phi_{n-2}}{e_\rho(q)} (1-q^n) \quad \text{if} \quad r_1 = 0. \tag{3}$$

Lemma 1. Suppose that θ satisfies (I) and (II). Then there exist positive integers s, d with $sd = n$ such that $\theta = \theta_d$, where

$$\theta_d(1) = \frac{\psi_n(q)}{\psi_s(q^d)} . \tag{4}$$

Proof. If (I) is satisfied, there exist natural numbers s_i, d_i for $1 \le i \le k$ with $\Sigma s_i d_i = n$ such that $\theta = (\ldots g^{\nu_i} \ldots)$, where ν_i is the partition $\{s_i\}$ and

$$\theta(1) = \psi_n(q)\{\prod_i \psi_{s_i}(q^{d_i})\}^{-1}.$$

Thus if θ also satisfies (II), then $k = 1$, $\theta = g^{\{s\}}$ with $sd = 1$ and (4) holds.

Lemma 2. Suppose that θ satisfies (I) and (II). Let x be a unipotent element of type $\{1^{n-2},2\}$. Then one of the following occurs

a) $d = 1$, $\theta(1) = \theta(x) = 1$.

b) $d > 1$, $\theta(1) = (1-q^{n-1})\theta(x)$.

Proof. Let ρ, λ be partitions of n. Let B^ρ denote the basic character of type ρ and let u be a unipotent element of type λ in G. Then $B^\rho(u) = Q_\rho^\lambda(q)$, see [4], p. 153. Thus in particular

$B^\rho(x) = Q^{\{1^{n-2},2\}}(q)$. Hence if every part of ρ is bigger than 1, it follows from (1) and (3) that $B^\rho(1) = (1-q^{n-1})B^\rho(x)$.

If $d = 1$, the result is clear. Suppose that $d > 1$. Then $\theta = \theta_d$ is a complex linear combination of basic characters B^ρ, where each part of ρ is a multiple of d, and hence is bigger than 1. Thus the result follows from the previous paragraph.

It is easily seen that the following hold.

$$|C_1| = (q^n-1)(q^{n-1}-1) \tag{5}$$

$$|C_2| = \frac{|G|}{q^{n-1}(q-1)} \tag{6}$$

$$|C_3| = \frac{|G|}{(q^n-1)(q-1)} \; . \tag{7}$$

<u>Lemma 3</u>. Let $y \in C_2$, $z \in C_3$. Let θ be an irreducible character of G with $\theta(y)\theta(z) \neq 0$. Then θ satisfies (I) and (II).

<u>Proof</u>. If $\theta(z) \neq 0$, then (II) is satisfied. If $\theta(y) \neq 0$, then (I) is satisfied by [2].

<u>Lemma 4</u>. Let $x \in C_1$, $y \in C_2$, $z \in C_3$. Then

$$\frac{|C_1||C_2||C_3|}{|G|} \sum_\theta \frac{\theta(x)\theta(y)\theta(z)}{\theta(1)} = \frac{|G|}{q-1} \, ,$$

where θ ranges over all irreducible characters of G.

<u>Proof</u>. By Lemma 3

$$\frac{\theta(x)\theta(y)\theta(z)}{\theta(1)} = \begin{cases} 1 & \text{if } \theta(1) = 1 \\ \dfrac{\theta(y)\theta(z)}{1-q^{n-1}} & \text{if } \theta \text{ satisfies (I) and (II) and } \theta(1) \neq 1 \\ 0 & \text{otherwise} \end{cases}$$

Therefore

$$\Sigma \frac{\theta(x)\theta(y)\theta(z)}{\theta(1)} = (q-1) - \frac{1}{q^{n-1}-1} \sum_{\theta(1)\neq 1} \theta(y)\theta(z)$$

$$= (q-1) + \frac{(q-1)}{q^{n-1}-1} - \frac{1}{q^{n-1}-1} \sum_{\theta} \theta(y)\theta(z)$$

$$= (q-1) + \frac{(q-1)}{q^{n-1}-1} = \frac{(q-1)q^{n-1}}{q^{n-1}-1} .$$

By (5), (6), and (7) this implies the result.

Lemma 5. Suppose that $(n,q-1) = 1$. Then $G = G_0 \times Z$ where $G_0 = SL_n(q)$, Z is the center of G. Furthermore C_i is a conjugate class of G_0 for $i = 1,2,3$.

Proof. Clear.

Proof of Theorem 1. By Lemma 5 $A_G(C_1,C_2,C_3) = A_{G_0}(C_1,C_2,C_3)$, where $G_0 = SL_n(q)$. The result follows from Lemma 4.

REFERENCES

1. Green, J.A. (1955). The characters of the finite general linear groups. Trans. A.M.S. 80, 402-447.
2. Green, J.A., Lehrer, G.I. & Lusztig, G. (1976). On the degrees of certain group characters. Quart. J. Mat., Oxford (II) 27, 1-4.
3. Kantor, W.M. (1980). Linear groups containing a Singer cycle. J. Algebra 62, 232-234.
4. Macdonald, I.G. (1979). Symmetric functions and Hall polynomials. Clarendon Press, Oxford.
5. Morris, A.O. (1963). The characters of the group GL(n,q). Math. Z. 81, 112-123.

PSL_3 AND GALOIS GROUPS OVER Q

J.G. Thompson
Cambridge University

It seems to me to be worthwhile to study rational rigidity in a systematic way. This is a massive task. In this note, I nibble at an edge.

<u>Theorem</u>. If p is a prime and

$$p \equiv 1 \pmod 4,$$

then there is a normal extension $E(p)$ of Q such that

$$\mathrm{Gal}\ E(p)/Q \cong PSL(3,p).$$

Let $x \longmapsto x^*$ be the transpose-inverse automorphism of $GL(n,q)$, and let $GL(n,q)^+$ be the splitting extension of $GL(n,q)$ by a group $\langle\rho\rangle$ of order 2 such that

$$\rho^{-1}x\rho = x^* \quad (x \in GL(n,q)).$$

I will use both matrices and a root system, and work in the group

$$G = G(p) = SL(3,p)\langle\rho\rangle,$$

where p is a prime $\equiv 1 \pmod 4$.

Let $\Pi = \{a,b\}$ be a root system of type A_2, let $\Sigma^+ = \{a,b,a+b\}$, $\Sigma = \Sigma^+ \cup -\Sigma^+$. Set

$$x_a(t) = \begin{pmatrix} 1 & t & 0 \\ 0 & 1 & 0 \\ 0 & 0 & 1 \end{pmatrix}, \quad x_b(t) = \begin{pmatrix} 1 & 0 & 0 \\ 0 & 1 & t \\ 0 & 0 & 1 \end{pmatrix}, \quad x_{a+b}(t) = \begin{pmatrix} 1 & 0 & t \\ 0 & 1 & 0 \\ 0 & 0 & 1 \end{pmatrix},$$

so that

$$[x_a(t), x_b(u)] = x_{a+b}(tu) \quad (t,u \in F_p).$$

Let $x_{-r}(t)$ be the transpose of $x_r(t)$, and set

$$\omega_a = \begin{pmatrix} 0 & 1 & 0 \\ -1 & 0 & 0 \\ 0 & 0 & 1 \end{pmatrix}, \qquad \omega_b = \begin{pmatrix} 1 & 0 & 0 \\ 0 & 0 & 1 \\ 0 & -1 & 0 \end{pmatrix},$$

$$h_a(t) = \begin{pmatrix} t & 0 & 0 \\ 0 & t^{-1} & 0 \\ 0 & 0 & 1 \end{pmatrix}, \quad h_b(t) = \begin{pmatrix} 1 & 0 & 0 \\ 0 & t & 0 \\ 0 & 0 & t^{-1} \end{pmatrix},$$

$h_{a+b}(t) = h_a(t)h_b(t)$, $h_{-r}(t) = h_r(t^{-1})$ $(t \in F_p^{\times}, r \in \Sigma^+)$. Set $w_r(s) = s - (r,s)\cdot r$, and check that

$$\omega_r^{-1} h_s(t)\omega_r = h_{w_r(s)}(t) \quad (r \in \Pi,\ s \in \Sigma,\ t \in F_p^{\times}).$$

We check that

$$\omega_r^{-1} x_s(t)\omega_r = x_{w_r(s)}(\eta_{r,s} t), \quad (r,s \in \Sigma^+)$$

where the $\eta_{r,s}$ are as follows:

r \ s	a	b	a+b
a	-1	-1	1
b	1	-1	-1

Also, we check that

$$h_r(t)x_s(u)h_r(t)^{-1} = x_s(t^{(r,s)}u) \quad (r,s \in \Sigma^+).$$

If $x \in SL(3,p)$, let $Tr(x)$ be the trace of x. If

$$x = \begin{pmatrix} a & b & c \\ a' & b' & c' \\ a'' & b'' & c'' \end{pmatrix} \in SL(3,p),$$

then from Cramer's rule, we get

$$\mathrm{Tr}(x^{-1}) = \begin{vmatrix} a & b \\ a' & b' \end{vmatrix} + \begin{vmatrix} a & c \\ a'' & c'' \end{vmatrix} + \begin{vmatrix} b' & c' \\ b'' & c'' \end{vmatrix} . \qquad (1)$$

Let

$$C_3' = \{x \in SL(3,p) \,|\, \mathrm{Tr}(x) = \mathrm{Tr}(x^{-1}) = 0\}.$$

Suppose $x \in C_3'$ and $f = X^3 + \alpha X^2 + \beta X + \gamma$ is the characteristic polynomial of x. Since $\det x = 1$, we have $\gamma = -1$; since $\mathrm{Tr}(x) = 0$, we have $\alpha = 0$. Since the characteristic polynomial of x^{-1} is

$$\gamma^{-1}(\gamma X^3 + \beta X^2 + \alpha X + 1),$$

and since $\mathrm{Tr}(x^{-1}) = 0$, we have $\beta = 0$, whence

$$x^3 = 1.$$

Since $p \equiv 1 \pmod 4$, we have $p \neq 3$, and it follows that C_3' is a conjugacy class of $SL(3,p)$ and of G.

Let

$$\pi = \omega_a\omega_b = \begin{pmatrix} 0 & 0 & 1 \\ -1 & 0 & 0 \\ 0 & -1 & 0 \end{pmatrix}.$$

Then $\pi\rho = \rho\pi$ and $\pi \in C_3'$. Set

$$x_3 = \pi\rho,$$

$$C_3(p) = C_3 = \text{the conjugacy class of } G \text{ which contains } x_3.$$

The elements of C_3 have order 6 and it is straightforward to check that x_3 and x_3^{-1} are G-conjugate, so C_3 is a rational class. Moreover, C_3 is the set of all elements of G which have order 6 and are not contained in $SL(3,p)$, so if $z \in G$, then $z \in C_3$ if and only if the following conditions hold:

α) $z \notin SL(3,p)$.

β) $\mathrm{Tr}(z^2) = \mathrm{Tr}(z^{-2}) = 0$.

Let $x_1 = \omega_a\rho$, and note that $\omega_a\rho = \rho\omega_a$, so that x_1 has order 4. Also, $h_{a+b}(-1)$ inverts x_1. Let

$$C_1(p) = C_1 = \text{the conjugacy class of } G \text{ containing } x_1,$$

$$C_2(p) = C_2 = \text{the conjugacy class of } G \text{ containing } x_a(1).$$

Thus, C_i is rational for $i = 1,2,3$.

The main problem is to show that

$$A_G(C_1,C_2,C_3) \text{ is rigid.} \tag{2}$$

Suppose (2) holds. In this case, I proceed to show that the theorem holds, too. There are 2 cases.

Case 1. $p \equiv -1 \pmod 3$.

In this case, $PSL(3,p) = SL(3,p)$ and the theorem follows from the main result of [1], together with "descent" to the subgroup of G of index 2.

Case 2. $p \equiv 1 \pmod 3$.

In this case, let z be a generator for $Z(SL(3,p))$, and let $\bar{X} = X\langle z\rangle/\langle z\rangle$ for each subset X of G. If $(\xi,\eta,\zeta) \in A_G(C_1,C_2,C_3)$, then

$$(\bar{\xi},\bar{\eta},\bar{\zeta}) \in A_{\bar{G}}(\bar{C}_1,\bar{C}_2,\bar{C}_3),$$

so we have a projection map

$$\text{pr}: A_G(C_1,C_2,C_3) \to A_{\bar{G}}(\bar{C}_1,\bar{C}_2,\bar{C}_3).$$

Since x_1 and x_3 invert z, and z has odd order, it also follows that

$$zC_1 = C_1, \quad zC_3 = C_3.$$

On the other hand, $zC_2 \cup z^{-1}C_2$ consists of elements of order $3p$, and so $zC_2 \cap C_2 = \emptyset$, $z^{-1}C_2 \cap C_2 = \emptyset$.

Suppose $(\bar{u},\bar{v},\bar{w}) \in A_{\bar{G}}(\bar{C}_1,\bar{C}_2,\bar{C}_3)$. Then there are precisely 27 triples (ξ,η,ζ) of elements of G such that

$$\bar{\xi} = \bar{u}, \quad \bar{\eta} = \bar{v}, \quad \bar{\zeta} = \bar{w}.$$

Of these 27 triples precisely 9 satisfy the conditions

$$\xi \in C_1, \quad \eta \in C_2, \quad \zeta \in C_3,$$

and for such a triple, there is a unique element z^i of $\langle z \rangle$ such that

$$\xi\eta\zeta = z^i.$$

Then $(z^{-i}\xi,\eta,\zeta) \in A_G(C_1,C_2,C_3)$, so

$$\mathrm{card}(\mathrm{pr}^{-1}(\bar{u},\bar{v},\bar{w})) = 3, \tag{*}$$

whence

$$\mathrm{card}\ A_{\bar{G}}(\bar{C}_1,\bar{C}_2,\bar{C}_3) = \frac{1}{3}\,|G| = |\bar{G}|.$$

Since $\langle u,v,w \rangle = G$ for all $(u,v,w) \in A_G(C_1,C_2,C_3)$, (*) implies that

$$A_{\bar{G}}(\bar{C}_1,\bar{C}_2,\bar{C}_3) \text{ is rigid.} \qquad (2)'$$

Since C_i is rational, so is $\bar{C}_i$, and so the theorem of this paper is a consequence of (2).

Since $x_a(1)$ and $x_{a+b}(1)$ are conjugate in $SL(3,p)$, it follows that every element of C_2 is of the shape

$$g^{-1}x_{a+b}(1)g, \quad (g \in G).$$

Let

$$C = C_G(x_{a+b}(1)), \quad D = C_G(x_1).$$

The G-conjugacy class of $x_1 \cdot g^{-1}x_{a+b}(1)g$ depends only on the double coset CgD. So the procedure is to obtain a good set of elements of G which contains a representative from each (C,D)-double coset.

Let $G_o = SL(3,p)$, $C_o = C \cap G_o$, $D_o = D \cap G_o$. Since $C_o \subset C$, it follows that every (C,D)-double coset in G contains an element of G_o, so the first problem is to study the double cosets $C_o g_o D_o$, $g_o \in G_o$. Note that $C_o g_o D_o \subseteq C g_o D$.

Visibly,

$$C_o = \langle x_a(1),\ x_b(1),\ h_a(t)h_b(t^{-1})\ |t \in F_p^\times\rangle, \tag{3}$$

and

$$D_o \supseteq \langle \omega_a,\ x_a(1),\ h_a(t) | t \in F_p^\times\rangle. \tag{4}$$

Set

$$U = \langle x_a(1),\ x_b(1)\rangle,$$

$$H = \langle h_a(t),\ h_b(t) | t \in F_p^\times\rangle,$$

$$\Omega = \{1,\ \omega_a,\ \omega_b,\ \omega_a\omega_b,\ \omega_b\omega_a,\ \omega_a\omega_b\omega_a\}.$$

Each element g of G_o is of the shape

$$g = u(g)h(g)\omega(g)u'(g),\quad u(g),u'(g) \in U,$$

$$h(g) \in H,\ \omega(g) \in \Omega.$$

We say that g is <u>normalized</u> if and only if

i) $u(g) = 1$.

ii) $u'(g) = x_b(t_1)x_{a+b}(t_2)$ $(t_1,t_2 \in F_p)$.

iii) $\omega(g) \in \{1,\ \omega_b,\ \omega_a\omega_b\}$.

iv) $t_1 \neq 0 \Longrightarrow \omega(g)x_b(t_1)\omega(g)^{-1} \notin U$;

$t_2 \neq 0 \Longrightarrow \omega(g)x_{a+b}(t_2)\omega(g)^{-1} \notin U$.

v) $h(g) = 1$ or $\omega(g) = \omega_b$.

From (3) and (4), we see that if $g_o \in G_o$, then $C_o g_o D_o$ contains at least one element g which satisfies (i) through (iv). We check that if $\omega \in \{1,\ \omega_a\omega_b\}$, then

$$H = \langle h_a(t)h_b(t^{-1}),\ \omega h_a(u)\omega^{-1} | t,\ u \in F_p^\times\rangle,$$

and so each double coset $C_o g_o D_o$ contains a normalized element. Let

$$N_\omega = \{g \in G_o | g \text{ is normalized and } \omega(g) = \omega\}.$$

Pick $g \in N_\omega$ and set

$$z = \rho\omega_a \cdot g^{-1} x_{a+b}(1) g.$$

Case 1. $\omega(g) = 1$.

Since $N_1 = \{1\}$, we have

$$z = \rho\omega_a x_{a+b}(1),$$
$$z^2 = \omega_a x_{-a-b}(-1)\omega_a x_{a+b}(1),$$

and $Tr(z^2) = -1$, so $z \notin C_3$.

Case 2. $\omega(g) = \omega_b$.

Here $g = h(g)\omega_b x_b(t)$ for some $h(g) \in H$, $t \in F_p$. Define $u \in F_p^\times$ by

$$h(g)^{-1} x_{a+b}(1) h(g) = x_{a+b}(u).$$

Then $$g^{-1} x_{a+b}(1) g = x_b(-t) x_a(-u) x_b(t),$$

$$z = \rho\omega_a x_b(-t) x_a(-u) x_b(t),$$

$$z^2 = \omega_a x_{-b}(t) x_{-a}(u) x_{-b}(-t) \omega_a x_b(-t) x_a(-u) x_b(t)$$

$$= \begin{pmatrix} -1 & 2u & tu \\ 0 & -1 & 0 \\ 0 & tu & 1 \end{pmatrix},$$

so

$$Tr(z^2) = -1.$$

Thus, in this case, also, $z \notin C_3$.

Case 3. $\omega(g) = \omega_a \omega_b$.

In this case, we have more work. We have $x_b(t) \in C_o$, $x_a(t) \in D_o$ and $x_b(t)\omega_a\omega_b = \omega_a\omega_b x_a(-t)$, whence

$$g' = x_b(t)gx_a(t) \in C_o g D_o,$$

and

$$g' = \omega_a\omega_b x_b(t_1)x_{a+b}(t_2 - tt_1),$$

so that g' is normalized. If $t_1 \neq 0$, we take $t = t_1^{-1}t_2$, and get $g' = \omega_a\omega_b x_b(t_1)$. Thus, if we set

$$N'_{\omega_a\omega_b} = \{\omega_a\omega_b x_{a+b}(t)\} \cup \{\omega_a\omega_b x_b(t)\} \quad (t \in F_p),$$

then

$$C_o N_{\omega_a\omega_b} D_o = C_o N'_{\omega_a\omega_b} D_o.$$

Case 3a. $g = \omega_a\omega_b x_{a+b}(t)$.

We have

$$\begin{aligned} z &= \rho\omega_a g^{-1} x_{a+b}(1)g \\ &= \rho\omega_a x_{a+b}(-t)\omega_b^{-1}\omega_a^{-1} x_{a+b}(1)\omega_a\omega_b x_{a+b}(t), \\ &= \rho\omega_a x_{a+b}(-t)x_{-b}(-1)x_{a+b}(t), \\ z^2 &= \omega_a x_{-a-b}(t)x_b(1)x_{-a-b}(-t)\omega_a x_{a+b}(-t)x_{-b}(-1)x_{a+b}(t) \\ &= \begin{pmatrix} -1 & -1-2t & 1 \\ 0 & -1 & 0 \\ 0 & -1 & 1 \end{pmatrix}, \\ \mathrm{Tr}(z^2) &= -1, \end{aligned}$$

so $z \notin C_3$.

Case 3b. $g = \omega_a\omega_b x_b(t)$.

We may assume that $t \neq 0$, for otherwise we are in Case 3a. We check that

$$h_a(-t)h_b(t^{-1})\rho\omega_a\omega_b\omega_a^{-1} \in C,$$

$$\rho h_a(t^2) \in D,$$

and in addition, we compute that

$$g' = h_a(-t)h_b(-t^{-1})\rho\omega_a\omega_b\omega_a^{-1}\cdot\omega_a\omega_b x_b(t)\rho\cdot h_a(t^2) \in U\omega_a\omega_b x_b(-t).$$

Thus, if we set $g(t) = \omega_a\omega_b x_b(t)$, then for $t \in F_p$, we get

$$Cg(t)D = Cg(-t)D. \tag{5}$$

We have

$$z = \rho\omega_a\cdot x_b(-t)\omega_b^{-1}\omega_a^{-1}x_{a+b}(1)\omega_a\omega_b x_b(t),$$

$$z^2 = \omega_a x_{-b}(t)\omega_b^{-1}\omega_a^{-1}x_{-a-b}(-1)\omega_a\omega_b x_{-b}(-t)\omega_a\cdot x_b(-t)\cdot$$
$$\omega_b^{-1}\omega_a^{-1}x_{a+b}(1)\omega_a\omega_b x_b(t)$$

$$= \begin{pmatrix} t-1 & -1 & -t+1 \\ 0 & -t-1 & -t^2 \\ t^2 & -t-1 & -t^2+1 \end{pmatrix}.$$

Hence

$$Tr(z^2) = -1-t^2,$$

and from (1),

$$Tr(z^{-2}) = \begin{vmatrix} t-1 & -1 \\ 0 & -t-1 \end{vmatrix} + \begin{vmatrix} t-1 & -t+1 \\ t^2 & -t^2+1 \end{vmatrix} + \begin{vmatrix} -t-1 & -t^2 \\ -t-1 & -t^2+1 \end{vmatrix}$$
$$= -1-t^2,$$

and so, with $g = g(t)$,

$$(x_1, g^{-1}x_{a+b}(1)g, z^{-1}) \in A_G(C_1,C_2,C_3) \Longleftrightarrow t^2 + 1 = 0.$$

By hypothesis, $p \equiv 1 \pmod 4$, so there is $i \in F_p$ with $i^2 + 1 = 0$, and from (5), we get that

$$\text{card}\{(u,v,w) \in A_G(C_1,C_2,C_3),\ u = x_1\}$$

is the number of cosets of C in $Cg(i)D$. In particular, G permutes transitively $A_G(C_1,C_2,C_3)$. Thus, to complete the proof of (2), we set

$$H = \langle x_1,\ g(i)^{-1}x_{a+b}(1)g(i)\rangle,$$

and proceed to show that $H = G$.

Set

$$y = g(i)^{-1}x_{a+b}(1)g(i),\quad H_o = H \cap G_o.$$

We check that $g(i)x_1 = \omega_a\omega_b\omega_a\rho x_{-a-b}(i)$. Set

$$K = \langle y,\ x_1^{-1}yx_1\rangle \quad (\subsetneq H_o).$$

We compute that

$$g(i)x_{-a-b}(-i) \in UH\omega_a\omega_b\omega_aU,$$

and since

$$K = \langle x_{a+b}(1)^{g(i)x_{-a-b}(-i)},\ x_{a+b}(-1)\rangle^{x_{-a-b}(i)},$$

we get

$$K \cong SL(2,p).$$

Suppose by way of contradiction that $H_o \neq G_o$. Since $|H:H_o| = 2$, it follows that the involution j of K is in the center of H. Since $x_1^2 = h_a(-1)$ is visibly not in the center of H, we have

$$j = \begin{pmatrix} J & 0 \\ 0 & -1 \end{pmatrix},$$

where $J = \begin{pmatrix} \alpha & \beta \\ \gamma & \delta \end{pmatrix}$ has determinant -1 and is of order 2. But then

$$x_1^{-1} j x_1 = j = \begin{pmatrix} -a & -b & 0 \\ -c & -d & 0 \\ 0 & 0 & -1 \end{pmatrix},$$ and so $J = 0$, which is false, as $\det J = -1$.

So (2) holds.

Remark. Setting

$$\Gamma = SL(3, \mathbf{Z}[i]),$$

it seems of interest to set $\Gamma^+ = \Gamma\langle\sigma\rangle$, where σ is of order 2 and induces the transpose-inverse automorphism of Γ. Then let Δ be the subgroup of Γ^+ generated by $\omega_a\sigma$ and $g(i)^{-1}x_{a+b}(1)g(i)$, where $g(i) = \omega_a\omega_b x_b(i)$, and where ω_a, ω_b, $x_b(i)$, $x_{a+b}(1)$ are the corresponding elements of Γ. Then Δ has the property that if $(p) = \mathfrak{g}\mathfrak{g}'$ is a prime ideal of $\mathbf{Z}$ which splits into two distinct primes in $\mathbf{Z}[i]$, then the map $\mathbf{Z}[i] \to \mathbf{Z}[i]/\mathfrak{g} \cong F_p$ induces a surjection

$$\Delta \to SL(3,p)\langle\sigma\rangle.$$

REFERENCES

1. Thompson, J.G. Some Finite Groups which appear as Gal L/K, where $K \subseteq Q(\mu_n)$. To appear in J. Alg.

RATIONAL RIGIDITY OF $G_2(5)$

J.G. Thompson
Cambridge University

In the following discussion, p denotes a prime ≥ 5.

Let $\{a,b\}$ be a set of fundamental roots for G_2 with a short, b long. Let

$C_1(p)$ = the set of involutions of $G_2(p)$,

$C_2(p)$ = the conjugacy class of $G_2(p)$ which contains $x_a(1)$,

$C_3(p)$ = the conjugacy class of $G_2(p)$ which contains $x_a(1)x_b(1)$.

It is well known that $C_1(p)$ is a conjugacy class of $G_2(p)$. Furthermore, for each $i = 1,2,3$, $C_i(p)$ is rational, that is, $x \in C_i(p)$ and m prime to $|G_2(p)|$ imply that $x^m \in C_i(p)$. Let

$$A(p) = A_{G_2(p)}(C_1(p),C_2(p),C_3(p))$$

$$= \{(x_1,x_2,x_3) | x_1x_2x_3 = 1,\ x_i \in C_i(p)\}.$$

<u>Conjecture</u>. $A(p)$ is rigid.

<u>Result</u>. $A(5)$ is rigid.

I shan't give a complete proof, but will indicate the basic trick. Pick $(x_1,x_2,x_3) \in A(p)$ (assuming for the time being that $A(p) \neq \emptyset$). Let $S = \langle x_1,x_2,x_3\rangle$. Suppose $S \subset G_2(p)$. Let M be a maximal subgroup of $G_2(p)$ which contains S.

<u>Lemma 1</u>. M is not a p-local subgroup.

<u>Proof</u>. Suppose false. By a basic result of Chevalley group theory, M is

a parabolic subgroup of $G_2(p)$. Let M_o be the subgroup of M generated by its p-elements. Since x_2, x_3 are p-elements, and since $x_1^{-1} = x_2 \cdot x_3$, it follows that $S \subseteq M_o$. Since we are in $G_2(p)$, we have

$$\bar{M}_o \underset{def}{=} M_o/O_p(M_o) \cong SL_2(p).$$

In particular, every involution of M_o is contained in $Z^*(M_o)$, by inspection. Let $\bar{x}_i = x_i \cdot O_p(M_o)$. Then $\bar{x}_1$ is an involution, and $\bar{x}_2$, $\bar{x}_3$ are elements of $\bar{M}_o$ of order 1 or p, whose product is $\bar{x}_1$. By inspection, $SL(2,p)$ has no such triples of elements. The lemma is proved.

When $p = 5$, J. McKay used the character table of $G_2(5)$ and discovered that

$$\text{card } A(5) = |G_2(5)|.$$

To exploit this result, I record another lemma about $G_2(p)$.

<u>Lemma 2</u>. If L_1, L_2 are non Abelian finite simple groups, then $G_2(p)$ does not contain any subgroup isomorphic to $L_1 \times L_2$.

<u>Proof</u>. If false, then since each L_i contains a 4-group, it would follow that $G_2(p)$ contains an elementary Abelian subgroup of order 2^4. This is not the case.

Returning to the case $p = 5$, we have

$$|G_2(5)| = 2^6 \cdot 3^3 \cdot 5^6 \cdot 7 \cdot 31.$$

Also, $x_a(1) \cdot x_b(1)$ has order 25. It follows easily from the structure of q-local subgroups of $G_2(5)$ that $O_q(S) = 1$ for all primes q (including, of course, $q = 5$, by Lemma 1, together with the fact that every p-local subgroup of $G_2(p)$ is contained in a parabolic subgroup). By Lemma 2, $F^*(S) = L$ is a non Abelian simple group, and so $S \cong S^*$, with $\text{Inn}(L) \subseteq S^* \subseteq \text{Aut}(L)$. Since $\text{Aut}(SL(3,5))$ has no elements of order 25, we have $L \ncong SL(3,5)$. Inspection of simple groups whose order divides $|G_2(5)|$ forces $S = G_2(5)$, and we have rigidity.

RATIONAL RIGIDITY OF $G_2(p)$ FOR ANY PRIME $p > 5$

W. Feit[1]
Yale University, New Haven, CT
P. Fong[2]
University of Illinois at Chicago, Chicago, IL

The result in this paper was proved independently by the two authors and also by J.G. Thompson, who first conjectured it. Thompson's proof is somewhat longer since, in contrast to the proof given here, it uses neither the classification of finite simple groups nor the character table of $G_2(p)$.

Let p be a prime, $p > 5$. Let $G = G_2(p)$.

Let C_1 be the conjugate class of G which contains an involution.

Let C_2 be the conjugate class of G which contains $x_a(1)$.

Let C_3 be the conjugate class of G which contains $x_a(1)x_b(1)$.

<u>Theorem</u>. $A_G(C_1,C_2,C_3)$ is rationally rigid. Hence by Theorem C and its Corollary in the Introduction G is a Galois group over Q.

<u>Proof</u>. By character theory (See Appendix).

$$|A_G(C_1,C_2,C_3)| = |G| = p^6(p^6-1)(p^2-1).$$

Let $x \in C_1$, $y \in C_2$, $z \in C_3$ with $xyz = 1$. Let $H = \langle x,y,z\rangle$. We will show in a series of steps that $H = G$.

The first 5 statements are well known.

1) $|\langle y\rangle| = |\langle z\rangle| = p$. $|C_G(z)| = p^2$.

2) $\langle y\rangle$, $\langle z\rangle$ are not conjugate in G. In particular H has a noncyclic S_p-group.

3) H has a faithful 7-dimensional representation over $\mathbb{F}_p$.

4) If $r \neq p$ is an odd prime, then the r-rank of H is at most 2.

[1]Supported by NSF grant MCS-8201333.
[2]Supported by NSF grant MCS-8101689.

5) The 2-rank of H is at most 3. Thus H does not contain a direct product of 2 simple groups.

6) H is not p-local.

<u>Proof</u>. Suppose not. H is contained in a maximal parabolic subgroup P. Let P_0 be the subgroup of P generated by all the p-elements. Hence $H \subseteq P_0$. Let $\bar{P}_0 = P_0/\mathbf{O}_p(P_0)$ and let $\bar{u}$ denote the image of an element u. Then $\bar{x}$ has order 2 and so is in the center of $\bar{P}_0 \simeq SL_2(p)$. Thus $\bar{x}\bar{y}$ has even order and so $\bar{x}\bar{y}\bar{z} \neq 1$.

7) H is not local.

<u>Proof</u>. Suppose not. Let $\langle 1\rangle \neq R \triangleleft H$ with R an elementary abelian r-group for some prime r. By (6) $r \neq p$. If $|R| = 8$ or r^2 for $r \neq 2$ then $p \nmid |\mathbf{N}_G(R)|$. By (1) z acts without fixed points on $R-\{1\}$. Thus $r \neq 2$ as $p > 3$. Hence $|R| = r$ and $p|(r-1)$. Therefore $r \nmid (p^2-1)$ and so an S_r-group of G is cyclic. Hence $p \nmid |\mathbf{N}_G(R)|$.

8) H is simple.

<u>Proof</u>. Let $\langle 1\rangle \neq H_0$ be a minimal normal subgroup of H. By (5) and (7) H_0 is simple. As $H_0 \times \mathbf{C}_H(H_0) \triangleleft H$, (7) implies that $\mathbf{C}_H(H_0) = \langle 1\rangle$ and $H \subseteq \mathrm{Aut}(H_0)$. If $H \neq H_0$, then $p\,|\,|\mathrm{Aut}(H_0):H_0|$.

If H_0 is sporadic or alternating, then $|\mathrm{Aut}(H_0):H_0| \leq 4$. If H_0 is of Lie type, then it is defined over a field F_{r^i} for some prime r and some $i \leq 3$ as the r-rank of G is at most 3. Hence $p \nmid |\mathrm{Aut}(H_0):H_0|$. Thus $H = H_0$.

9) H is not sporadic.

<u>Proof</u>. Suppose not. As $p > 5$ (3) implies that H is not a Mathieu group. Thus (5) implies that $H \simeq J_1$ or Mc. As $p > 5$ this contradicts (2).

10) H is not an alternating group.

<u>Proof</u>. As $p \geq 7$ this follows from (2) and (3).

11) H is not of Lie type in characteristic $r \neq p$.

Proof. Suppose not. By (2) $H \not\approx PSL_2(r^i)$ for any i. Hence in particular an S_r-group of H is not abelian. Thus $r = 2$ or 3.

Suppose that $r = 3$. An S_3-group of G has an abelian subgroup of index 3 and is of rank 2. Thus H is defined over $\mathbb{F}_3$. The only possibilities are $PSL_3(3)$, $PSU_3(3)$ or $PSp_4(3)$. These all contradict (2).

Suppose that $r = 2$. The 2 rank of H is at most 3. Thus G is one of the following: $PSL_3(2^i)$; $PSU_3(2^i)$ with $i = 1,2$; or $Sz(8)$. These all contradict (2).

12) $H = G$.

Proof. Suppose not. By (9), (10) and (11) H is of Lie type in characteristic p. By (2) $H \neq PSL_2(p^i)$ for odd i. As $|H| \mid p^6(p^6-1)(p^2-1)$ it follows that $H \neq PSL_2(p^{2i})$ for all i and so $H \simeq PSL_3(p)$ or $PSU_3(p)$. Thus $H = H_1/\mathbf{Z}(H_1)$ with $H_1 \simeq SL_3(p)$ or $SU_3(p)$.

Let V be the underlying vector space over $\mathbb{F}_p$. H_1 has exactly 2 conjugate classes of cyclic unipotent subgroups. Thus one of these is generated by an element u of Jordan type $(2,1)$. Let x_1 be an involution in H_1. Then u maps onto y or z and x_1 maps onto x. Thus x_1u has order p or order $3p$. The group $\langle u,x_1\rangle$ has an invariant 1-dimensional subspace W with $x_1 = -1$ on W and $u = 1$ on W. Thus ux_1 has even order and so does the image of ux_1 in H, contrary to assumption.

APPENDIX

Let $\varepsilon = \pm 1$, $\delta = \pm 1$ where

$$q \equiv \delta \pmod 4), \quad q \equiv \varepsilon \pmod 3).$$

The following table consists of all the irreducible characters of $G = G_2(q)$ which are simultaneously nonzero on k_2, u_2, u_6. This is copied directly from [1]; the only additional information consists of the number of characters of each type.

If we set $k_2 = x_1$, $u_2 = x_a(1)$, $u_6 = x_a(1)\, x_b(1)$, then the table yields that $|A_G(C_1,C_2,C_3)| = G$ by direct computation. This completes the proof of the Theorem.

REFERENCES

1. Chang, B. & Ree, R. (1974). The characters of $G_2(q)$. Symposia Mathematica XIII, 395-413. Academic Press, London.

TABLE

	1	u_2	u_6	k_2	# such characters
$\chi_1(\pi_1)$	$\frac{(q^2-1)(q^6-1)}{(q+1)^2}$	$(q+1)(2q+1)$	1	$3(q+1)^2$	$\frac{1}{48}(q^2-14q+47+6\delta+8\varepsilon)$
				$-(q+1)^2$	$\frac{1}{48}(3q^2-18q+21-6\delta)$
$\chi_2(\pi_2)$	$\frac{(q^2-1)(q^6-1)}{(q+1)^2}$	$(q-1)(2q-1)$	1	$3(q-1)^2$	$\frac{1}{48}(q^2-10q+23-6\delta-8\varepsilon)$
				$-(q-1)^2$	$\frac{1}{48}(3q^2-6q-3+6\delta)$
$\chi_a(\pi_a)$	q^6-1	$-q-1$	-1	q^2-1	$\frac{1}{8}(q^2-4q+3)$
				$-(q^2-1)$	$\frac{1}{8}(q^2-1)$
$\chi_b(\pi_b)$	q^6-1	$q-1$	-1	q^2-1	$\frac{1}{8}(q^2-4q+3)$
				$-(q^2-1)$	$\frac{1}{8}(q^2-1)$
$\chi'_{1a}(\pi_1,\pi_a)$	$\frac{q^6-1}{q-1}$	$(q+1)^2$	1	$(q+2)(q+1)$	$\frac{1}{4}(q-6-2\varepsilon-\delta)$
				$-q(q+1)$	$\frac{1}{4}(q-2+\delta)$
$\chi'_{1b}(\pi_1,\pi_b)$	$\frac{q^6-1}{q-1}$	q^2+q+1	1	$(q+2)(q+1)$	$\frac{1}{4}(q-4-\delta)$
				$-q(q+1)$	$\frac{1}{4}(q-2+\delta)$
$\chi'_{2a}(\pi_2,\pi_a)$	$\frac{q^6-1}{q+1}$	$-q^2+q-1$	-1	$-(q-2)(q-1)$	$\frac{1}{4}(q-2+\delta)$
				$q(q-1)$	$\frac{1}{4}(q-\delta)$
$\chi'_{2b}(\pi_2,\pi_b)$	$\frac{q^6-1}{q+1}$	$-(q-1)^2$	-1	$-(q-2)(q-1)$	$\frac{1}{4}(q-4+2\varepsilon+\delta)$
				$q(q-1)$	$\frac{1}{4}(q-\delta)$
χ_{22}	$\frac{q^6-1}{q^2-1}$	q^2+1	1	$2\delta q+1$	1
χ_{32}	$q^3+\varepsilon$	$q+\varepsilon$	ε	$q+\varepsilon$	1
χ_{11}	1	1	1	1	1

PRIMITIVE ROOTS AND RIGIDITY

J.G. Thompson
Cambridge University

1 THE BASIC CONFIGURATION

Suppose p, ℓ are distinct primes,

$$p \text{ is a primitive root } \pmod{\ell}, \tag{1.1}$$

and

$$\ell \neq 2. \tag{1.2}$$

Let

$$q_o = p^{\frac{\ell-1}{2}}, \quad q = q_o^2, \tag{1.3}$$

$$F = GF(q), \tag{1.4}$$

and let τ be the automorphism of F whose fixed field is $GF(q_o)$, so that

$$\tau^2 = 1, \quad x^\tau = x^{q_o} \quad (x \in F). \tag{1.5}$$

Let

$$B = \{\zeta \in F \mid \zeta^\ell = 1, \zeta \neq 1\}. \tag{1.6}$$

Thus

$$B \text{ is a normal basis for } F \text{ over } F_p. \tag{1.7}$$

Viewing F as an $(\ell-1)$-dimensional vector space $/F_p$, set $G = GL(F)$, so that

$$G \cong GL(\ell-1,p), \quad |G| = p^{(\ell-1)(\ell-2)/2} \prod_{i=1}^{\ell-1} (p^i-1). \tag{1.8}$$

There is an injection $h: F^\times \to G$ given by $\xi \longmapsto h(\xi)$, where $h(\xi)$ is the F_p-linear transformation of F such that

$$x \circ h(\xi) = x\xi.$$

This is the familiar regular representation of $F^\times$, and $h(F^\times)$ is a non-split torus in G. Let

$$T: F \to F_p, \; T(\xi) = tr_{F_q/F_p}(\xi), \; F_o = \ker T. \tag{1.9}$$

Then F_o is a hyperplane of F, and every hyperplane of F is of the form $F_o \circ h(\xi)$ for some $\xi \in F^\times$. Furthermore, if $\lambda \in F_o \backslash \{0\}$, then we let $v(F_o,\lambda)$ be the map of F into itself given by

$$x \circ v(F_o,\lambda) = x + T(x)\cdot\lambda. \tag{1.10}$$

One checks easily that $v(F_o,\lambda)$ is a transvection and that the set of fixed points of $v(F_o,\lambda)$ on F is F_o. Furthermore, if $\xi \in F^\times$, one checks that

$$h(\xi)^{-1}v(F_o,\lambda)h(\xi) = v_\xi(F_o,\lambda)$$

satisfies

$$x \circ v_\xi(F_o,\lambda) = x + T(\xi^{-1}x)\cdot\lambda\xi. \tag{1.11}$$

There are precisely $(p^{\ell-2}-1)(p^{\ell-1}-1)/(p-1)$ transvections in G, and so every transvection is of the shape $v_\xi(F_o,\lambda)$. Note that $v_\xi(F_o,\lambda) = v(F_o,\lambda)$ if and only if $\xi \in F_p^\times$.

Let

$$C_2 = \{v_\xi(F_o,\lambda) \mid \xi \in F^\times, \; \lambda \in F_o\backslash\{0\}\}, \tag{1.12}$$

so that C_2 is a conjugacy class in G. If m is an integer prime to p, then $v(F_o,\lambda)^m$ is a transvection, and so C_2 is rational, that is,

$$C_2 = \{x^m | x \in C_2\} \tag{1.13}$$

for all integers relatively prime to p.

If S is any subset of G, set

$$Q_G([S]) = Q(\{\chi(s) | s \in S, \chi \in \text{Char } G\}),$$

where χ ranges over Char G, the set of characters of G. Thus, (1.13) may be written as

$$Q = Q_G([C_2]). \tag{1.14}$$

By (1.1) and (1.8), the Sylow ℓ-subgroups of G are cyclic, and $h(F^\times)$ contains a Sylow ℓ-subgroup of G. Thus, every element of G of order ℓ is conjugate in G to an element of h(B). Furthermore, $\mathrm{Gal}(F_q/F_p)$ is cyclic of order $\ell-1$, and permutes B transitively. Since the map: $F \to F$ given by $x \longmapsto x^p$ is F_p-linear, (1.1) implies that

$$C_1 = \{g \in G | g^\ell = 1, g \neq 1\} \tag{1.15}$$

is a single conjugacy class in G.

Next, let

$$C_3 = \{u \in G | u \text{ is unipotent and } u \text{ acts indecomposably on } F\}. \tag{1.16}$$

Then C_3 is the conjugacy class of regular unipotent elements, or in more old-fashioned language, the set of unipotent elements with just one Jordan block.

Set

$$A = \{(u,v,w) | u,v,w \in G,\ uvw = 1,\ o(u) = \ell,\ v \text{ is a transvection, } w \text{ is unipotent}\}. \tag{1.17}$$

Suppose $(u,v,w) \in A$. Let F(v), F(w) be the set of fixed points of v, w, respectively, on F. Then $F(v) \cap F(w)$ is fixed by $vw = u^{-1}$. Since $o(u^{-1}) = \ell$, and since u^{-1} is conjugate to $h(\zeta)$ in G, we see that 0 is the set of fixed points of u^{-1} on F, and so

$$F(v) \cap F(w) = 0.$$

Since $F(v)$ is a hyperplane, v being a transvection, we have

$$\dim_{F_p} F(w) \leq 1.$$

By hypothesis, w is unipotent, and so $F(w)$ has dimension ≥ 1, whence $\dim_{F_p} F(w) = 1$. Equivalently, $w \in C_3$. Hence

$$A = A_G(C_1,C_2,C_3) = \{(x_1,x_2,x_3) | x_i \in C_i,\ x_1x_2x_3 = 1\}. \tag{1.18}$$

Furthermore, as already discussed,

$$Q = Q([C_i]),\quad i = 1,2,3. \tag{1.19}$$

Pick $\zeta \in B$, once and for all. Let

$$A_\zeta = \{(x_1,x_2,x_3) \in A | x_1 = h(\zeta)\}. \tag{1.20}$$

Since card $C_1 = |G:C_G(h(\zeta))| = |G|/(p^{\ell-1}-1) = |G|/|F^\times|$, we have

$$\text{card } A = \frac{|G|}{|F^\times|} \cdot \text{card } A_\zeta. \tag{1.21}$$

Suppose $(h(\zeta),x_2,x_3) \in A_\zeta$. Thus, as already remarked, there are $\xi \in F^\times$, $\lambda \in F_o \setminus \{0\}$, such that

$$x_2 = v_\xi(F_o,\lambda) = h(\xi)^{-1}v(F_o,\lambda)h(\xi).$$

Since $h(\zeta)$ and $h(\xi)$ commute, and since F has $|F^\times|/|F_p^\times|$ hyperplanes, it follows that

$$\text{card } A = \frac{|G|}{|F_p^\times|} \text{card } \Lambda, \tag{1.22}$$

where

$$\Lambda = \{\lambda \in F_o \setminus \{0\} | h(\zeta)\cdot v(F_o,\lambda) \in C_3\}. \tag{1.23}$$

I propose to show that

$$\Lambda = \{\frac{(1-\zeta)^{\ell}}{\ell}\} \text{ has cardinal } 1. \tag{1.24}$$

In any case, if $\lambda \in \Lambda$, we have

$$\lambda \neq 0, \ T(\lambda) = 0. \tag{1.25}$$

We grapple with Λ by picking $\lambda \in \Lambda$, and constructing a sequence $(f_1, f_2, \ldots, f_{\ell-1})$ of elements of F such that

$$\begin{aligned} &\text{i)} \quad f_{\ell-1} \neq 0. \\ &\text{ii)} \quad f_{\ell-1} \circ h(\zeta)v(F_o,\lambda) = f_{\ell-1}. \\ &\text{iii)} \quad \text{for } i = 1,2,\ldots,\ell-2, \\ &\qquad f_i \circ h(\zeta)v(F_o,\lambda) = f_i + f_{i+1}. \end{aligned} \tag{1.26}$$

Condition (i), and an easy induction argument, show that every such sequence $(f_1, \ldots, f_{\ell-1})$ is a sequence of elements of F which are linearly independent over F_p, and so form a basis. Conditions (ii), (iii) then ensure that $h(\zeta)v(F_o,\lambda) \in C_3$, whence $w = \{h(\zeta)v(F_o,\lambda)\}^{-1} \in C_3$. The constraints put upon λ so that $h(\zeta)v(F_o,\lambda) \in C_3$ are thus so tight that there is a unique survivor, given in (1.24).

Conversely, of course, if $h(\zeta)v(F_o,\lambda) \in C_3$, then a sequence satisfying (1.26) necessarily exists, because of the Jordan normal form of a Jordan block. I construct the sequence $(f_1, \ldots, f_{\ell-1})$ in the reverse order, that is, I construct successively, $f_{\ell-1}, f_{\ell-2}, \ldots, f_1$, so that (1.26) holds.

The first step, then, is to search for an element $f'_{\ell-1} \neq 0$ such that

$$f'_{\ell-1}h(\zeta)v(F_o,\lambda) = f'_{\ell-1}. \tag{1.27}$$

From the definition of $h(\zeta)$ and $v(F_o,\lambda)$, the left hand side is

$$(f'_{\ell-1}\zeta)v(F_o,\lambda) = f'_{\ell-1}\zeta + T(f'_{\ell-1}\zeta)\cdot\lambda. \tag{1.28}$$

Since $\zeta \neq 1$, a necessary condition that (1.27) holds is that $T(f'_{\ell-1}\zeta) = c_1 \in F_p^{\times}$. Set

$$f_{\ell-1} = c_1^{-1} f'_{\ell-1}.$$

Multiply (1.27) by c_1^{-1}, and multiply (1.28) by c_1^{-1}, too, so that the element $f_{\ell-1}$ is constrained by

$$f_{\ell-1} h(\zeta) v(F_o, \lambda) = f_{\ell-1}, \tag{1.29}$$

$$f_{\ell-1} h(\zeta) v(F_o, \lambda) = f_{\ell-1}\zeta + \lambda, \tag{1.30}$$

$$T(f_{\ell-1}\zeta) = 1. \tag{1.31}$$

From (1.29), (1.30), we get

$$f_{\ell-1} = \frac{\lambda}{1-\zeta}, \tag{1.32}$$

and from (1.31), we get

$$T\left(\frac{\lambda\zeta}{1-\zeta}\right) = 1. \tag{1.33}$$

Now (1.25), (1.33) have given us 2 constraints on λ. However, if both of these conditions are satisfied, then (1.32) does indeed define a fixed point of $h(\zeta)v(F_o, \lambda)$.

We now begin a search for $f_{\ell-2}$. Suppose we have an element $f'_{\ell-2}$ such that

$$f'_{\ell-2} h(\zeta) v(F_o, \lambda) = f'_{\ell-2} + f_{\ell-1}. \tag{1.34}$$

Let

$$c_2 = T(f'_{\ell-2}\zeta), \quad (c_2 \in F_p), \tag{1.35}$$

and set

$$f_{\ell-2} = f'_{\ell-2} - c_2 f_{\ell-1}. \tag{1.36}$$

Since $f_{\ell-1}$ is fixed by $h(\zeta)v(F_o, \lambda)$, it follows from (1.34) that

$$\begin{aligned} f_{\ell-2} h(\zeta) v(F_o, \lambda) &= f'_{\ell-2} + f_{\ell-1} - c_2 f_{\ell-1} \\ &= f_{\ell-2} + f_{\ell-1}. \end{aligned} \tag{1.37}$$

Moreover, by (1.31), (1.35), (1.36), we have

$$\begin{aligned} T(f_{\ell-2}\zeta) &= T(f'_{\ell-2}\zeta) - c_2T(f_{\ell-1}\zeta) \\ &= c_2 - c_2 = 0. \end{aligned} \tag{1.38}$$

Thus, in our search for $f_{\ell-2}$, we are entitled to assume that

$$f_{\ell-2}h(\zeta)v(F_o,\lambda) = f_{\ell-2} + f_{\ell-1}, \tag{1.39}$$

$$T(f_{\ell-2}\zeta) = 0. \tag{1.40}$$

From (1.39), (1.40) and from the definitions of $h(\zeta)$ and $v(F_o,\lambda)$, we have

$$\begin{aligned} f_{\ell-2}h(\zeta)v(F_o,\lambda) &= (f_{\ell-2}\zeta)v(F_o,\lambda) \\ &= f_{\ell-2}\zeta + T(f_{\ell-2}\zeta)\lambda \\ &= f_{\ell-2}\zeta = f_{\ell-2} + f_{\ell-1}. \end{aligned} \tag{1.41}$$

Now (1.41), (1.32) give

$$f_{\ell-2} = f_{\ell-1}/(\zeta-1) = -\frac{\lambda}{(1-\zeta)^2}, \tag{1.42}$$

and (1.40) puts yet another constraint on λ, namely,

$$T(-\frac{\lambda\zeta}{(1-\zeta)^2}) = 0. \tag{1.43}$$

If $\ell = 3$, the following argument may be omitted. Suppose $\ell \geq 5$. This sets up the correct induction situation. Suppose $\lambda \in F_o\backslash\{0\}$, and $3 \leq \nu \leq \ell-1$ satisfy:

$$f_{\ell-\mu} = \frac{(-1)^{\mu-1}\lambda}{(1-\zeta)^{\mu}}, \ \mu = 1,2,\ldots,\nu-1, \tag{1.44}$$

$$T(\lambda) = 0, \ T(\frac{\lambda\zeta}{1-\zeta}) = 1, \ T(\frac{\lambda\zeta}{(1-\zeta)^{\mu}}) = 0, \ \mu = 2,\ldots,\nu-1. \tag{1.45}$$

$$f_{\ell-1}h(\zeta)v(F_o,\lambda) = f_{\ell-1}, \tag{1.46}$$

$$f_{\ell-\mu}h(\zeta)v(F_o,\lambda) = f_{\ell-\mu} + f_{\ell-\mu+1}, \ \mu = 2,\ldots,\nu-1. \tag{1.47}$$

We then begin our search for $f_{\ell-\nu}$. Suppose

$$f'_{\ell-\nu}h(\zeta)v(F_o,\lambda) = f'_{\ell-\nu} + f_{\ell-\nu+1}, \tag{1.48}$$

and

$$c_\nu = T(f'_{\ell-\nu}\zeta) \qquad (c_\nu \in F_p). \tag{1.49}$$

Set

$$f_{\ell-\nu} = f'_{\ell-\nu} - c_\nu f_{\ell-1}.$$

Since $f_{\ell-1}$ is fixed by $h(\zeta)v(F_o,\lambda)$, (1.31), (1.48), (1.49) yield

$$f_{\ell-\nu}h(\zeta)v(F_o,\lambda) = f_{\ell-\nu} + f_{\ell-\nu+1}, \tag{1.50}$$

$$T(f_{\ell-\nu}\zeta) = 0. \tag{1.51}$$

Now the definition of $h(\zeta)$, $v(F_o,\lambda)$, together with (1.50), (1.51), give

$$f_{\ell-\nu}\zeta = f_{\ell-\nu} + f_{\ell-\nu+1}, \tag{1.52}$$

whence

$$\begin{aligned} f_{\ell-\nu} &= \frac{f_{\ell-\nu+1}}{(\zeta-1)} = \frac{f_{\ell-(\nu-1)}}{(\zeta-1)} \\ &= \frac{(-1)^{\nu-2}\lambda}{(\zeta-1)(1-\zeta)^{\nu-1}} = \frac{(-1)^{\nu-1}\lambda}{(1-\zeta)^\nu}, \end{aligned} \tag{1.53}$$

with the restraint on λ that

$$T\left(\frac{(-1)^{\nu-1}\lambda\zeta}{(1-\zeta)^\nu}\right) = 0. \tag{1.54}$$

Since $(-1)^{\nu-1} \in F_p$, (1.54) is equivalent to

$$T\left(\frac{\lambda\zeta}{(1-\zeta)^\nu}\right) = 0. \tag{1.55}$$

This is the induction step, and so

$$\Lambda = \{\lambda \in F^{\times} | T(\lambda) = 0,$$
$$T(\frac{\lambda\zeta}{1-\zeta}) = 1, \quad T(\frac{\lambda\zeta}{(1-\zeta)^{\nu}}) = 0, \quad \nu = 2,\ldots,\ell-1\}. \tag{1.56}$$

In order to obtain (1.24), I first show that

$$\{(1-\zeta)^{\mu} | \mu = 1,2,\ldots,\ell-1\} \text{ is a basis for } F/F_p. \tag{1.57}$$

To see this, suppose

$$\sum_{\mu=1}^{\ell-1} c_{\mu}(1-\zeta)^{\mu} = 0 \qquad (c_{\mu} \in F_p).$$

Since $1-\zeta \neq 0$, we get

$$\sum_{\mu=1}^{\ell-1} c_{\mu}(1-\zeta)^{\mu-1} = 0.$$

Since the minimal polynomial for $1-\zeta$ over F_p has degree $\ell-1$, by (1.1), we get $c_1,c_2,\ldots,c_{\ell-1} = 0$, and (1.57) holds. From (1.57), we get

$$\{(1-\zeta)^{-\mu} | \mu = 1,2,\ldots,\ell-1\} \tag{1.58}$$

is a basis for F/F_p. For if $\sum_{\mu=1}^{\ell-1} d_{\mu}(1-\zeta)^{-\mu} = 0$ $(d_{\mu} \in F_p)$, then, upon multiplication by $(1-\zeta)^{\ell}$, we get an F_p-linear relation on $\{(1-\zeta)^{\mu} | \mu = 1,2,\ldots,\ell-1\}$, against (1.57). Since $\lambda\zeta \neq 0$, (1.58) gives

$$\{\frac{\lambda\zeta}{(1-\zeta)^{\nu}} | \nu = 1,2,\ldots,\ell-1\} \text{ is a basis for } F, \text{ for each } \lambda \in F^{\times}. \tag{1.59}$$

Since T is non-singular, (1.56), (1.59), imply that

$$\text{card } \Lambda = 0 \text{ or } 1. \tag{1.60}$$

So it only remains to check that

$$T(\frac{(1-\zeta)^{\ell}}{\ell}) = 0, \quad T(\frac{(1-\zeta)^{\ell-1}\zeta}{\ell}) = 1,$$
$$T(\frac{(1-\zeta)^{\ell}\zeta}{\ell(1-\zeta)^{\nu}}) = 0, \ \nu = 2,\ldots,\ell-1.$$

For each element $\zeta_o \in B$, we have

$$T(\zeta_o) = \sum_{\zeta' \in B} \zeta' = -1,$$

since B is a normal basis for F over F_p. Moreover,

$$\zeta^\tau = \zeta^{-1} \quad (\text{as } \zeta^\tau = \zeta^{p^{\frac{\ell-1}{2}}}),$$

and so

$$\{\frac{(1-\zeta)^\ell}{\ell}\}^\tau = \frac{(1-\zeta^{-1})^\ell}{\ell} = \frac{(\zeta-1)^\ell}{\ell} = -\frac{(1-\zeta)^\ell}{\ell},$$

so

$$T_{F/F_{q_o}}(\frac{(1-\zeta)^\ell}{\ell}) = 0. \tag{1.61}$$

Since T factors through F_{q_o} via

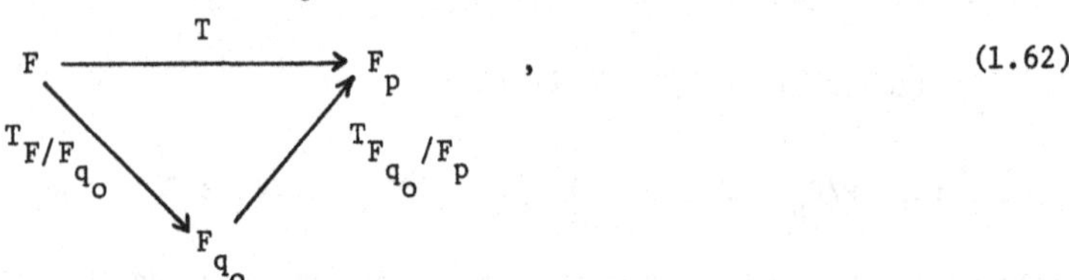

, (1.62)

it follows that

$$T(\frac{(1-\zeta)^\ell}{\ell}) = 0.$$

Suppose $2 \leq \nu \leq \ell-1$. Then $1 \leq \ell-\nu \leq \ell-2$. Set $\mu = \ell-\nu$. Then

$$T(\frac{(1-\zeta)^\ell\zeta}{\ell(1-\zeta)^\nu}) = \frac{1}{\ell}\, T((1-\zeta)^\mu\zeta)$$

$$= \frac{1}{\ell}\, T(\sum_{n=0}^{\mu} \binom{\mu}{n}(-1)^n\zeta^{n+1})$$

$$= \frac{1}{\ell}\, (\sum_{n=0}^{\mu} \binom{\mu}{n}(-1)^n(-1))$$

$$= \frac{1}{\ell}\,(-1)\sum_{n=0}^{\mu} \binom{\mu}{n}(-1)^n = \frac{(-1)}{\ell}\,(1-1)^\mu = 0.$$

So it all comes down to checking that

$$T((1-\zeta)^{\ell-1}\zeta) = \ell.$$

We have

$$(1-\zeta)^{\ell-1}\zeta = \sum_{m=0}^{\ell-1} \binom{\ell-1}{m}(-1)^m\zeta^{m+1},$$

so

$$T((1-\zeta)^{\ell-1}\zeta) = \sum_{m=0}^{\ell-2} \binom{\ell-1}{m}(-1)^m(-1) + \ell-1,$$

as $T(1) = \ell-1$. Since

$$\sum_{m=0}^{\ell-2} \binom{\ell-1}{m}(-1)^m + (-1)^{\ell-1} = 0,$$

we have

$$\sum_{m=0}^{\ell-2} \binom{\ell-1}{m}(-1)^m = (-1)^{\ell},$$

whence

$$\begin{aligned} T((1-\zeta)^{\ell-1}\zeta) &= \sum_{m=0}^{\ell-2} \binom{\ell-1}{m}(-1)^m(-1) + \ell-1 \\ &= (-1)^{\ell+1} + \ell-1 = \ell, \end{aligned}$$

and (1.24) is proved. Now (1.24), (1.22), give

$$\text{card } A(C_1,C_2,C_3) = \frac{|G|}{|F_p^\times|}. \qquad (1.63)$$

The next step is to show that

$$\text{if} \quad (u,v,w) \in A, \qquad \text{then} \qquad (1.64)$$

$$C_G(u) \cap C_G(v) \cap C_G(w) = Z(G) \quad (= h(F_p^\times)).$$

To see this, we may assume that $u = h(\zeta)$. Since $h(\zeta)$ acts irreducibly on F, we get

$$C_G(h(\zeta)) = h(F^\times).$$

Suppose $\xi \in F^\times$ and $h(\xi)$ centralizes $v(F_o,\lambda)$. Then $h(\xi)$ acts on $F(1 - v(F_o,\lambda)) = F_p \cdot \lambda$. Since $h(F^\times)$ acts transitively and regularly on $F^\times$, we get $\xi \in F_p^\times$. Conversely, of course, $h(F_p^\times)$, being the center of G, centralizes u, v, and w.

From (1.63), (1.64), we get

$$G \text{ permutes } A(C_1,C_2,C_3) \text{ transitively by conjugation.} \tag{1.65}$$

For each $(u,v,w) \in A$, let $L(u,v) = \langle u,v \rangle$. By (1.65), the groups $L(u,v)$ form a single conjugacy class in G.

The remainder of the paper is devoted to the complete determination of the group

$$S = L(h(\zeta), v(F_o, \frac{(1-\zeta)^\ell}{\ell})).$$

Here, of course, the relevant element w of C_3 is given by $w = \{h(\zeta)v(F_o, \frac{(1-\zeta)^\ell}{\ell})\}^{-1}$.

Define an inner product $(\ ,\ )$ on F, with values in F_p, via

$$(\alpha,\beta) = T(\alpha^\tau \lambda^{-1}\beta), \qquad \lambda = \frac{(1-\zeta)^\ell}{\ell}. \tag{1.66}$$

Since $\alpha^{1+\tau} \in F_{q_o}$ for all $\alpha \in F$, (1.61) implies that

$$(\alpha,\alpha) = 0 \quad \text{for all } \alpha \in F. \tag{1.67}$$

Since $(\lambda^{-1}) = -\lambda^{-1}$, we also get, by direct calculation, that

$$(\alpha,\beta) = -(\beta,\alpha), \text{ for all } \alpha,\beta \in F. \tag{1.68}$$

If $\alpha \in F^\times$, we take $\beta = \lambda\alpha^{-\tau}\zeta$, and get

$$(\alpha,\lambda\alpha^{-\tau}\zeta) = T(\zeta) = -1,$$

and so

$$(\ ,\ ) \text{ is non-singular.}$$

For the next few pages, I assume that

$$p \neq 2. \tag{1.69}$$

As we have already shown,

$$C_G(S) = h(F_p^\times),$$

and so

the F_pS-module F is absolutely irreducible. (1.70)

Thus (S,F) is a quadratic pair, since $v(F_o,\lambda)$ acts quadratically on F. By a theorem of Baer which has also become folklore, there is a conjugate v' of $v(F_o,\lambda)$ in S such that $\langle v(F_o,\lambda),v'\rangle$ is not a p-group. By inspection,

$$\langle v(F_o,\lambda),v'\rangle \cong SL(2,p),$$

and moreover,

$$F = V \oplus W,$$

$\dim V = 2$, $\langle v(F_o,\lambda),v'\rangle$ is the identity on W, and induces $SL(2,p)$ on V. Let j be the unique involution of $\langle v(F_o,\lambda),v'\rangle$. Thus, $C_S(j)$ acts on V, and on W. Hence

$$\langle v(F_o,\lambda),v'\rangle \lhd C_S(j),$$

and so j is a classical involution in the sense of Aschbacher. From [1], it is straightforward to deduce that

S is the group of all elements of G which preserve the form $(\; , \;)$, so (1.71)

$$S \cong Sp(\ell-1,p), \qquad S/\langle -I\rangle \cong C_{\frac{\ell-1}{2}}(p). \tag{1.72}$$

It is well known that if

$$p^* = (-1)^{\frac{p-1}{2}} p, \tag{1.73}$$

then the following hold:

$$S \cap C_2 = C_2' \cup C_2'', \; S \cap C_3 = C_3' \cup C_3'',$$

where C_2', C_2'' and C_3', C_3'' are conjugacy classes of S, and that

$$Q_S([C_2']) = Q_S([C_3']) = Q(\sqrt{p^*}). \tag{1.74}$$

Let $A_S = \{(u',v',w') \in A \mid u',v',w' \in S\}$, and let

$$N_o = h(F_p^\times)S, \; N = N_G(S). \tag{1.75}$$

We have

$$N_o \cap S = \langle -1 \rangle, \quad |N_o:S| = \frac{p-1}{2}, \tag{1.76}$$

$$|N:N_o| = 2, \quad |N:S| = p-1. \tag{1.77}$$

Now (1.77), (1.65) imply that

$$S \text{ has precisely 2 orbits on } A_S. \tag{1.78}$$

Let

$$S \cap C_1 = C_1' \cup \ldots,$$

where the notation is chosen so that

$$A_S(C_1',C_2',C_3') \neq \emptyset.$$

Let m be an integer $\equiv 1 \pmod{\ell}$, such that m is a non square (modulo p). Then

$$C_2' \circ \bar{m} = C_2'', \; C_3' \circ \bar{m} = C_3'', \; C_1' \circ \bar{m} = C_1',$$

and

$$A(C_1', C_2'', C_3'') \neq \emptyset.$$

Since S has just 2 orbits on A_S, we conclude that

$$S \cap C_1 \text{ is a class of } S\text{, that is, all elements of } S \text{ of order } \ell \text{ are conjugate, and also that} \tag{1.79}$$

$$A(C_1', C_2'', C_3') = A(C_1', C_2', C_3'') = \emptyset. \tag{1.80}$$

Thus, the notion of rigidity gives us a definite linkage

$$C_2' \longleftrightarrow C_3'$$
$$C_2'' \longleftrightarrow C_3'',$$

where the "link" is defined as follows:

$$\text{the conjugacy classes } X, Y \text{ of } S \text{ are linked} \Longleftrightarrow \exists x \in X, y \in Y \text{ such that } o(xy) = \ell. \tag{1.81}$$

The precise significance of this relation is not yet apparent, but it can scarcely be fortuitous.

Now we set

$$G_o = S/\langle -I\rangle \cong C_{\frac{\ell-1}{2}}(p).$$

Let D_i be the image of C_i' in G_o. If $x \in C_i'$, then $o(x)$ is odd, being either ℓ, which by hypothesis is odd, or a power of p, which, in the case at hand, is also odd. Hence $x\langle -I\rangle \cap C_i' = \{x\}$, and

$$\text{card } D_i = \text{card } C_i', \; i = 1,2,3.$$

Furthermore,

$$A_{G_o}(D_1, D_2, D_3) \text{ is rigid.}$$

Let

$$D = \ell.c.m.(\ell, p^h),$$

where $p^h = o(w)$. Let

$$U = U_D.$$

Then

$$V = \{\bar{n} \in U | D_i \circ \bar{n} = D_i,\ i = 1,2,3\}$$

is easy to identify:

$$V = \{\bar{n} \in U | n \text{ is a square mod } p\}.$$

In particular, $|U:V| = 2$.
Since $|N:N_o| = 2$, it also follows that if

$$W = \{\bar{n} \in U | \exists \alpha \in \mathrm{Aut}(G_o) \ni D_i \circ \bar{n} = D_i \circ \alpha\},$$

then

$$W = U.$$

It follows that

$$F^*(G_o, A_{G_o}(D_1, D_2, D_3)) = A = \mathrm{Aut}\ G_o,$$

and that $A/\mathrm{Inn}\ G_o$ is of order 2. By the main theorem of [3], it follows that there is a normal extension E of Q such that

$$E \supsetneq Q(\sqrt{p^*}),$$

$$\mathrm{Gal}\ E/Q \cong \mathrm{Aut}\ C_{\frac{\ell-1}{2}}(p),$$

and the fixed field of $\mathrm{Inn}\ C_{\frac{\ell-1}{2}}(p)$ is $Q(\sqrt{p^*})$.

2 THE CASE $p = 2$, $\ell \geq 11$

Here more work is necessary, but in the end we get to Q. We already know that S is contained in the symplectic group given by the form $(\alpha,\beta) = T(\alpha^\tau\lambda^{-1}\beta)$. And the possibility arises that S is a proper subgroup. Indeed it is, and I propose to determine a quadratic form

$$f: F \to F_2 \tag{2.1}$$

such that

i) f admits S.

ii) If $\alpha,\beta \in F$, then (2.2)

$$f(\alpha+\beta) + f(\alpha) + f(\beta) = (\alpha,\beta) = T(\alpha^\tau\lambda^{-1}\beta).$$

Once this is done, I propose to show that S is the largest subgroup of G which fixes f, and furthermore, that A_S is permuted transitively by S.

Set

$$\mu = \zeta + \zeta^3 + \zeta^5 + \dots + \zeta^{\ell-2}. \tag{2.3}$$

We have

$$\mu + \mu^\tau = 1,\ \lambda + \lambda^\tau = 0,\ \lambda^{-1} + (\lambda^{-1})^\tau = 0, \tag{2.4}$$

since we are now in characteristic 2.

Set

$$\varepsilon = T_{F_{q_o}/F_2}(\lambda^{-1}). \tag{2.5}$$

I argue that

$$\varepsilon = T(\lambda^{-1}\mu). \tag{2.6}$$

Namely, by what I have just shown

$$\begin{aligned} T_{F_q/F_{q_o}}(\lambda^{-1}\mu) &= \lambda^{-1}\mu + \lambda^{-\tau}\mu^\tau \\ &= \lambda^{-1}\mu + \lambda^{-1}(1+\mu) = \lambda^{-1}, \end{aligned} \tag{2.7}$$

and (2.6) follows from (1.62), (2.5), (2.7).

I do not know the value of ε. Since λ^{-1} is an explicit element of F_{q_o}, ε is determined by ℓ (given that 2 is a primitive root). When $\ell = 11$, direct calculation shows that $\varepsilon = 0$ (since $\lambda^{-1} = \zeta + \zeta^5 + \zeta^6 + \zeta^{10}$ in this case). In any case I define

$$f(\zeta') = \varepsilon \quad \forall\ \zeta' \in B. \tag{2.8}$$

Since B is a basis for F over F_2, it follows that to each element ξ of $F^\times$, there is a non-empty subset $B(\xi)$ of B such that

$$\xi = \sum_{\zeta' \in B(\xi)} \zeta'.$$

Now (2.8), (2.2) force

$$f(\xi) = \varepsilon \cdot \text{card } B(\xi) + \sum_{\substack{\zeta', \zeta'' \in B(\xi) \\ \zeta' < \zeta''}} T(\zeta'^{-1}\lambda^{-1}\zeta'').$$

Here the order relation $<$ is arbitrary, but fixed. Since

$$B(1) = B, \text{ and } \text{card } B = 0 \text{ in } F_2,$$

we get

$$f(1) = \sum_{\substack{\zeta', \zeta'' \in B \\ \zeta' < \zeta''}} T(\zeta'^{-1}\lambda^{-1}\zeta'').$$

If we take the ordering $\zeta < \zeta^2 < \zeta^3 < \ldots < \zeta^{\ell-1}$, we get

$$f(1) = T(\lambda^{-1}\mu) = T_{F_{q_o}/F_2}(\lambda^{-1}) = \varepsilon,$$

according to (2.5). This tells us that our given quadratic form admits $h(\zeta)$.

In order to check that f admits $v(T_o, \lambda)$, I need to check that

$$f(\lambda) = 1. \tag{2.9}$$

We have

$$\lambda = (1+\zeta)^{\ell} = \sum_{\mu=1}^{\ell-1} \binom{\ell}{\mu}\zeta^{\mu},$$

as the terms corresponding to $\mu = 0, \ell$ contribute $1+1 = 0$. Since $\binom{\ell}{\mu} = \binom{\ell}{\ell-\mu}$, we have card $B(\lambda) = 0$ in F_2, and so

$$f(\lambda) = \sum_{\substack{\mu,\nu=1\\ \nu<\mu}}^{\ell-1} T(\binom{\ell}{\mu}\binom{\ell}{\nu}\lambda^{-1}\zeta^{\mu-\nu}).$$

To compute this, we first compute $T_{F_q/F_{q_o}}$. We have

$$\sum_{\substack{\mu,\nu=1\\ \nu<\mu}}^{\ell-1} T_{F_q/F_{q_o}}(\binom{\ell}{\mu}\binom{\ell}{\nu}\lambda^{-1}\zeta^{\mu-\nu}) =$$

$$\sum_{\substack{\mu,\nu=1\\ \nu<\mu}}^{\ell-1} \binom{\ell}{\mu}\binom{\ell}{\nu}\lambda^{-1}(\zeta^{\mu-\nu} + \zeta^{\nu-\mu}) \quad \text{as} \quad \lambda^{-1} \in F_{q_o}.$$

We can rewrite the right hand side as

$$\sum_{\substack{\mu,\nu=1\\ \mu\neq\nu}}^{\ell-1} \binom{\ell}{\mu}\binom{\ell}{\nu}\lambda^{-1}\zeta^{\mu-\nu}. \tag{2.10}$$

Since

$$\sum_{\mu=1}^{\ell-1} \binom{\ell}{\mu}^2\lambda^{-1}\zeta^{\mu-\mu} = \lambda^{-1}\{\sum_{\mu=1}^{\ell-1}\binom{\ell}{\mu}\} = \lambda^{-1}(2^{\ell}-2) = 0,$$

(2.10) is just

$$\sum_{\mu,\nu=1}^{\ell-1} \binom{\ell}{\mu}\binom{\ell}{\nu}\lambda^{-1}\zeta^{\mu-\nu} = \lambda^{-1}\{\sum_{\mu=1}^{\ell-1}\binom{\ell}{\mu}\zeta^{\mu}\}\cdot\{\sum_{\nu=1}^{\ell-1}\binom{\ell}{\nu}\zeta^{-\nu}\} = \lambda,$$

and so

$$f(\lambda) = T_{F_{q_o}/F_2}(\lambda).$$

We have, yet again,

$$\lambda = \sum_{\mu=1}^{\ell-1}\binom{\ell}{\mu}\zeta^{\mu} = \sum_{\mu=1}^{\frac{\ell-1}{2}}\binom{\ell}{\mu}(\zeta^{\mu}+\zeta^{-\mu}),$$

and

$$T_{F_{q_o}/F_2}(\zeta^{\mu} + \zeta^{-\mu}) = T(\zeta^{\mu}) = 1,$$

whence

$$f(\lambda) = \sum_{\mu=1}^{\frac{\ell-1}{2}} \binom{\ell}{\mu}\cdot 1 = 2^{\ell-1}-1 = 1.$$

This is (2.9).

We can now show that f admits $v(F_o,\lambda)$. To do this, it is necessary and sufficient to check that

$$f(\zeta^{\mu}) = f(\zeta^{\mu} \circ v(F_o,\lambda)), \quad \mu = 1,2,\ldots,\ell-1.$$

The left hand side is ε. The right hand side is

$$f(\zeta^{\mu} + \lambda) = f(\zeta^{\mu}) + f(\lambda) + (\zeta^{\mu},\lambda)$$
$$= \varepsilon + 1 + T(\zeta^{-\mu}\lambda^{-1}\lambda) = \varepsilon + 1 + T(\zeta^{-\mu}) = \varepsilon.$$

So S is a subgroup of the group S^* of all elements of G ($\cong GL(\ell-1,2)$) which preserve f.

Since S contains a transvection, and since S^* contains a simple subgroup of index 2 with no transvections, it follows that

$S^* \cap S$ is of index 2 in S.

I argue that

$S^* \cap S$ is a non Abelian simple group. (2.11)

Let M be a minimal normal subgroup of S. Suppose we can show that $h(\zeta) \in M$. With this assumption, we can conclude that one of the following holds:

a) M is an elementary Abelian ℓ-group.

b) M is the direct product of non Abelian pairwise isomorphic simple groups, each of which has order divisible by ℓ.

Since the Sylow ℓ-subgroups of G are cyclic, case (a) implies that M is a cyclic ℓ-group, and so $\langle h(\zeta)\rangle \triangleleft S$. Since $\ell > 3$, this is false.

In case (b), M is forced to be simple since, once again, the

Sylow ℓ-subgroups of G are cyclic. Thus, (2.11) is equivalent to showing that every minimal normal subgroup of S contains $h(\zeta)$. Suppose false for the minimal normal subgroup M of S, so that $\ell \nmid |M|$.

Case 1. M is an elementary Abelian r-group for some prime r.

Case 2. M is the direct product of pairwise isomorphic non Abelian simple groups.

In Case 1, we certainly have $r \neq 2$, as S acts irreducibly on F. By hypothesis, $r \neq \ell$. Let $M_o = [M,h(\zeta)]$. Suppose $M_o \neq 1$. Then $M_o\langle h(\zeta)\rangle$ acts faithfully on F, and $\dim_{F_2} F < \ell$. This is impossible, since $M_o\langle h(\zeta)\rangle$ is a Frobenius group whose Frobenius complements have order ℓ.

So $h(\zeta)$ centralizes M, whence $M \subseteq h(F^\times)$, a cyclic group. Since $S = \langle h(\zeta), v(F_o,\lambda)\rangle$, and S is centerless, it follows that $v(F_o,\lambda)$ does not centralize M. Since $v(F_o,\lambda)$ is a transvection, it follows that $r = 3$, and so $|M| = 3$, as M is cyclic. But then $F = [F,M]$ is a 4-group, and $\ell = 3$, an excluded case.

As for Case 2, since $C_G(h(\zeta)) = h(F^\times)$, a cyclic group, it follows that $h(\zeta)$ normalizes every simple normal subgroup of M. Obviously, $v(F_o,\lambda)$ does, too, and so M is a non Abelian simple group. Since $(|M|,\ell) = 1$, $h(\zeta)$ normalizes some Sylow 2-subgroup U of M. This is not possible, since $[F,U]$ is a non zero proper subspace of F which admits $h(\zeta)$, while $h(\zeta)$ acts irreducibly on F. So (2.11) holds. Since $S = \langle h(\zeta), v(F_o,\lambda)\rangle$, and $h(\zeta) \in M$, we have $|S:M| = 2$. Hence

$$S = \langle S \cap C_2\rangle \text{ is a 3-transposition group.} \tag{2.12}$$

We are thus in a position to use the deep results of Fischer [2]. Set $d = v(F_o,\zeta)$.

Suppose X is an elementary Abelian subgroup of S of order 3^2, and X admits d. Since d is a transvection, $[F,d] = \langle\lambda\rangle$ has order 2, and so d does not invert X.

Suppose X is a non Abelian subgroup of S of order 3^3, which admits d. Since d is a transvection, d is forced to centralize X'. Thus, if d acts non trivially on X, then d inverts X/X'. Let Y be a subgroup of X of order 3 which is inverted by d. Then since $X \subseteq G$, it follows that $[F,Y]$ has order at least 2^4, whence $[F,Y,d]$ has order $\geq 2^2$. This is false, and so d centralizes every non Abelian subgroup of S of order 3^3 which it normalizes.

By the main Theorem of [2], one of the following holds for some integer n. (We use $\ell \geq 11$ again at this point to eliminate the symmetric group S_6.)

i) $S \cong S_n$ and d is a transposition.

ii) $S \cong Sp(2n,2)$, d a (symplectic) transvection.

iii) $S \cong O^{\mu}(2n,2)$ $\mu \in \{1,-1\}$, and d is a transvection in (ii) which leaves the corresponding quadratic form invariant.

I propose to show that we are in case (iii), that $\mu = -1$, and that $n = \frac{\ell-1}{2}$.

First suppose that $S \cong S_n$. Then $n \geq \ell$, since $h(\zeta) \in S$. Since $C_S(h(\zeta))$ is of odd order, we have $n \leq \ell+1$.

i)$_a$ $\ell = n$.

It is well known that in this case, the module F is the "even" submodule of the standard n-dimensional permutation module, taken (mod 2). Since S contains an element of order 2^h with just 1 Jordan block, a moment's thought convinces that $\ell = 2^h+1$, and that w corresponds to an element of S_ℓ with just 2 cycles, of lengths 1 and 2^h. Hence $2^{2h} \equiv 1 \pmod{\ell}$, whence $2h \equiv 0 \pmod{\ell-1}$, since 2 is a primitive root (mod ℓ). Thus, $h = \frac{\ell-1}{2} \cdot m$ for some natural number m. Since $\ell = 2^{\frac{\ell-1}{2}m} + 1$, we get $m = 1$, $\ell = 3$ or 5, excluded cases.

i)$_b$ $n = \ell+1$.

In this case, let V be the standard F_2-permutation module for S_n. Since $\dim_{F_2} F = n-2$, it follows that $F \cong V^{even}/\langle \sum_{\omega\in\Omega} \omega \rangle$. Since w exists, we get $n-2 = 2^h = O(w)$, and again $\ell-1 = 2^h$, giving $\ell = 3$ or 5. This kills S_n.

ii) Suppose $S \cong Sp(2n,2)$. We have

$$|Sp(2n,2)| = 2^{n^2} \prod_{i=1}^{n} (2^{2i}-1).$$

Since $\ell \,|\, |S|$, and $S \cong Sp(2n,2)$, we get $n \geq \frac{\ell-1}{2}$. In particular, $|S| \geq |Sp(\ell-1,2)|$. This is false, since the quadratic form f exists. This kills the symplectic groups.

iii) Since $\ell \,|\, |S|$, and $S \cong O^{\mu}(2n,2)$, while 2 is a primitive root (mod ℓ), and since the form f is trivially of minus type (as $|O^{+}(\ell-1,2)|$ is not divisible by ℓ), we have

$$O^{\mu}(2n,2) \subseteq O^{-}(\ell-1,2),$$

and one final application of the fact that 2 is a primitive root (mod ℓ) forces

$$n = \frac{\ell-1}{2}, \quad \mu = -1,$$

$$M \cong {}^{2}D_{\frac{\ell-1}{2}}(2),$$

$$|S:M| = 2.$$

Furthermore, it now follows that S is its own normalizer in G, and that

(*) For each $i = 1,2,3$, $C_i \cap S = C_i'$ is a conjugacy class of S with

$$Q([C_i']) = Q.$$

Also, of course, $A_S(C_1',C_2',C_3')$ is rigid, and so there is a normal extension E of Q with

$$\mathrm{Gal}(E/Q) \cong S. \tag{2.13}$$

Note, however, that S is <u>not</u> of the shape $F^*(M,A(C_1',C_2',C_3'))$ (which is a meaningless abuse in any case), since C_1', C_2', C_3' are classes of S, not of M, and two of the classes (C_2',C_3') are even disjoint from M. Nevertheless, (2.13) holds, and so does

$$\mathrm{Gal}(E_o/Q) \cong M \cong {}^{2}D_{\frac{\ell-1}{2}}(2), \tag{2.14}$$

for some normal extension E_o of Q.

One way to see (2.14) is to use the group Γ of [3], and to take the homomorphism

$$\phi : \Gamma \to S$$

$$u_o \circ \phi = h(\zeta),\ u_1 \circ \phi = v(F_o,\lambda),\ u_2 \circ \phi = w,$$

$$u_i \circ \phi = 1, \text{ otherwise.}$$

Then the complete inverse image under ϕ of M is a subgroup Γ_0 of Γ of index 2 and of genus 0 (K_{Γ_0} has a generator U with $U^2 = (T - \xi_1)/(T - \xi_2)$), such that $\Gamma_0/\ker \phi$ is stable under Aut $\mathbb{C}$, from which (2.14) follows directly. I acknowledge indebtedness to J.-P. Serre for the remark that the passage from S to M is straightforward. I find it well nigh impossible to believe that the appearance of the element $(1-\zeta)^{\ell}/\ell$ (in $F_{p^{\ell-1}}$) is without any bearing on $Q(\mu_\ell)$.

To recapitulate, we see that if ℓ is an odd prime and p is a prime which is a primitive root (mod ℓ), there are isomorphisms

$$\text{Gal } E/Q \cong F^*$$

in the following cases:

1) $p \neq 2$, $\ell \geq 5$, and $F^* = \text{Aut } C_{\frac{\ell-1}{2}}(p)$, where the fixed field of $C_{\frac{\ell-1}{2}}(p)$ is $Q(\sqrt{p^*})$, $p^* = (-1)^{\frac{p-1}{2}} p$,

2) $p = 2$ and $F^* = O^-_{\ell-1}(2)$, ($\ell \geq 11$)

3) $p = 2$ and $F^* = {}^2D_{\frac{\ell-1}{2}}(2)$, ($\ell \geq 11$).

REFERENCES

1. Aschbacher, M. (1977). A characterization of Chevalley groups over fields of odd order. Annals of Mathematics, (2) 106, no. 2, 353-398; II, Annals of Mathematics, (2) 106, no. 3, 399-468.
2. Fischer, B. Finite Groups Generated by 3-transpositions. Lecture Notes University of Warwick.
3. Thompson, J.G. Some Finite Groups which appear as Gal L/K, where $K \subseteq Q(\mu_n)$. To appear in J. Alg.

RIGIDITY OF $\mathrm{Aut}(PSL_2(p^2))$, $p \equiv \pm 2 \pmod 5$, $p \neq 2$

W. Feit
Yale University, New Haven, Connecticut 06520

Let p be an odd prime, $p \equiv \pm 2 \pmod 5$. Let $q = p^2$.

Let $A = \mathrm{Aut}(PSL_2(q))$. Let $I = \mathrm{Inn}(PSL_2(q)) \simeq PSL_2(q)$. $I \triangleleft A$.

Let $p \equiv \varepsilon \pmod 4$, $\varepsilon = \pm 1$.

Let γ be an element of order $2(p-\varepsilon)$ in $F_q^\times$. Thus γ is not a square in $F_q^\times$.

Let σ be the automorphism of $PSL_2(q)$ defined by conjugation by $\begin{pmatrix} 0 & -1 \\ \gamma & 0 \end{pmatrix}$. Let $A_0 = \langle I, \sigma \rangle \simeq PGL_2(q)$. $A_0 \triangleleft A$.

Let τ be the automorphism of $PSL_2(q)$ defined by the field automorphism $u \to u^p$. Thus

$$A = I \cup I\sigma \cup I\tau \cup I\sigma\tau$$

As $p^2+1 \equiv 0 \pmod 5$ there exists an element of order 5 in $PSL_2(q)$ fixed by σ. Let π be the corresponding element in I. Then $\sigma\pi = \pi\sigma$ is of order 10 in A.

We next define an element ν of order 4 in $I\sigma\tau$. The definition depends on ε.

Suppose that $\varepsilon = 1$. Let $\nu = \sigma\tau$. Then

$$\nu^2 \text{ corresponds to } \begin{pmatrix} -\gamma^p & 0 \\ 0 & -\gamma \end{pmatrix}. \tag{I}$$

Hence ν^4 corresponds to $\begin{pmatrix} \gamma^{2p} & 0 \\ 0 & \gamma^2 \end{pmatrix}$. As $\gamma^p \neq \gamma$ but $\gamma^{2p} = \gamma^2$ this implies that ν has order 4.

Suppose that $\varepsilon = -1$. Let μ in A_0 correspond to $\begin{pmatrix} 1 & 0 \\ 0 & \gamma \end{pmatrix}$. Let $\nu = \tau\mu$. Then $\mu \notin I$ and so $\nu \in I\sigma\tau$. Also

$$\nu^2 \text{ corresponds to } \begin{pmatrix} 1 & 0 \\ 0 & \gamma^{p+1} \end{pmatrix} = \begin{pmatrix} 1 & 0 \\ 0 & -1 \end{pmatrix}. \tag{II}$$

Thus ν has order 4.

Define the following conjugate classes of A:

$\tau \in C_2$

$\nu \in C_4$

$\sigma\pi \in C_{10}$.

<u>Lemma 1</u>. For $j = 2, 4, 10$, C_j is a rational conjugate class of A containing elements of order j.

<u>Proof</u>. By definition C_j contains elements of order j. Thus C_2 is rational. It is easily seen that π, and hence $\sigma\pi$, is rational in A.

If $\varepsilon = 1$ then $\sigma^{-1}\nu\sigma = \nu^{-1}$ and so ν is rational.

If $\varepsilon = -1$ choose $\alpha \in F_q$ with $\alpha^{p-1} = -1$ then $\begin{pmatrix} 1 & 0 \\ 0 & \alpha \end{pmatrix}$ conjugates ν into its inverse.

<u>Theorem</u>. A is rigid with respect to C_2, C_4, C_{10}.

<u>Corollary</u>. A, and hence $I \approx PSL_2(q)$ is a Galois group over $\mathbb{Q}$.

The fact that the classes C_2, C_4, C_{10} are candidates for rigidity was first suggested to the author by J.G. Thompson.

The Theorem will be proved in a series of Lemmas.

<u>Lemma 2</u>. Let $x_j \in C_j$ for $j = 2, 4, 10$ with $x_2\, x_4\, x_{10} = 1$. Then $A = \langle x_2, x_4, x_{10}\rangle$.

<u>Proof</u>. Let $H = \langle x_2, x_4, x_{10}\rangle$. As S_5 contains no element of order 10, H is not isomorphic to a subgroup of S_5.

There exists $y \in A$ of order $q+1$ with $N = \mathbf{N}_A(\langle y\rangle) = \langle y,z\rangle$, where $z^4 = 1$ and $z^{-1}yz = y^p$. Furthermore any maximal subgroup of A of order divisible by 5 which is not isomorphic to S_5 is conjugate to N. If $K \subseteq N$ then $K \subseteq \langle y,z^2\rangle$, which is dihedral. Thus $\langle x^4\rangle \triangleleft \langle y,z^2\rangle$. This is impossible and the result is proved.

<u>Lemma 3</u>. Let $x_j \in C_j$ for $j = 2, 4, 10$. Then

i) $|\mathbf{C}_A(x_2)| = 2p(p^2-1)$

ii) $|\mathbf{C}_A(x_{10})| = p^2+1$

iii) $|\mathbf{C}_A(x_4)| = 4(p+\varepsilon)$

<u>Proof.</u> i) $\mathbf{C}_{A_0}(\tau) \simeq PGL_2(p)$.

ii) $\mathbf{C}_A(x_{10}) = \mathbf{C}_{A_0}(x_{10})$ and the result follows.

iii) $|\mathbf{C}_A(\nu): \mathbf{C}_{A_0}(\nu)| = 2$ and $\mathbf{C}_{A_0}(\nu) \subseteq \mathbf{C}_{A_0}(\nu^2)$. Let $\begin{pmatrix} a & b \\ c & d \end{pmatrix}$ in $GL_2(q)$ correspond to an element of $\mathbf{C}_{A_0}(\nu)$. Then either $a = d = 0$ or $b = c = 0$ by (I) (II). It may be assumed that $a = 1$ or $b = 1$ in the respective cases.

Suppose that $\varepsilon = 1$. Then for some $k \in F_q^\times$

$$\begin{pmatrix} -\gamma d, & c \\ \gamma^2 b, & -\gamma a \end{pmatrix} = \begin{pmatrix} 0, & -1 \\ \gamma, & 0 \end{pmatrix}\begin{pmatrix} a & b \\ c & d \end{pmatrix}\begin{pmatrix} 0 & -1 \\ \gamma & 0 \end{pmatrix} = k\begin{pmatrix} a^p & b^p \\ c^p & d^p \end{pmatrix}.$$

Hence

$$-\gamma d = ka^p, \quad -\gamma a = kd^p$$
$$\gamma^2 b = kc^p, \quad c = kb^p.$$

If $b = c = 0$ and $a = 1$, this implies that $-\gamma d = k$ and $-\gamma = kd^p$. Thus $d^{p+1} = 1$ and there are $p+1$ choices for d.

If $a = d = 0$ and $b = 1$ then $\gamma^2 = kc^p$ and $c = k$. Thus $c^{p+1} = \gamma^2$ and there are $p+1$ choices for c.

Hence $|\mathbf{C}_{A_0}(\nu)| = 2(p+1)$ as required.

Suppose that $\varepsilon = -1$. Then for some $k \in F_q^\times$

$$\begin{pmatrix} a, & \gamma^{-1} b \\ \gamma c, & d \end{pmatrix} = \begin{pmatrix} 1 & 0 \\ 0 & \gamma \end{pmatrix}\begin{pmatrix} a & b \\ c & d \end{pmatrix}\begin{pmatrix} 1 & 0 \\ 0 & \gamma^{-1} \end{pmatrix} = k\begin{pmatrix} a^p & b^p \\ c^p & d^p \end{pmatrix}.$$

Hence

$$a = ka^p, \quad d = kd^p$$
$$\gamma^{-1} b = kb^p, \quad \gamma c^p.$$

If $b = c = 0$ and $a = 1$ then $k = 1$ and $d^{p-1} = 1$. Thus there are $p-1$ choices for d.

If $a = d = 0$ and $b = 1$, $\gamma^{-1} = k$ and $\gamma c = kc^p$. Thus $c^{p-1} = \gamma k^{-1} = \gamma^2$ and there are $p-1$ choices for c.

Hence $|\mathbf{C}_{A_0}(\nu)| = 2(p-1)$ as required.

<u>Lemma 4</u>. Let χ be an irreducible character of A with χ_I irreducible. Then $\chi(1) = 1$, q or $q+1$.

<u>Proof</u>. The irreducible characters of $I \approx \mathrm{PSL}_2(q)$ have degrees 1, q, $q+1$, $q-1$, $\frac{q+1}{2}$. Let θ be an irreducible character of I which has A as an inertia group. Then $\theta(1) \neq \frac{q+1}{2}$, as the two characters of degree $\frac{q+1}{2}$ fuse in A_0. If the result is false then $\theta(1) = q-1$. Let y be an element of order $\frac{q+1}{2}$ in I. Then $\theta_{\langle y\rangle} = 2\rho - (\lambda+\lambda^{-1})$, where ρ is afforded by the regular representation of $\langle y\rangle$ and λ is a linear character of y with $\lambda^2 \neq 1$. Thus $\lambda(y) = e^{2\pi ik/m}$ with $(k,m) = 1$, $m|\frac{q+1}{2}$, $m \nmid 2$. As $\theta = \theta^\tau$, $\lambda + \lambda^{-1} = \lambda^p + \lambda^{-p}$. Thus $\lambda^p - \lambda = \lambda^{-p} - \lambda^{-1}$ is real and so $\sin\frac{2\pi j}{m} = \sin\frac{2\pi pj}{m}$ for all j. Hence $m|p-1$. As $m|p^2+1$ this implies that $m|2$ which is not the case.

Let Γ be an extension to A of the Steinberg character of I. Let ϕ_i, $1 \leq i \leq 4$ be the linear characters of A. Let $x_j \in C_j$ for $j = 2, 4, 10$.

<u>Lemma 5</u>. i) $\phi_i(x_2)\,\phi_i(x_4)\,\phi_i(x_{10}) = 1$ for $1 \leq i \leq 4$.

ii) If χ is an irreducible character of A with $\chi(x_2)\,\chi(x_4)\,\chi(x_{10}) \neq 0$ then $\chi = \phi_i$ or $\Gamma\phi_i$ for some $i = 1,\ldots,4$.

<u>Proof</u>. i) Clear.

ii) If the result is false then Lemma 4 implies that $\chi(1) = q+1$. However in that case χ is of defect 0 for 5 and so $\chi(x_{10}) = 0$ contrary to assumption.

<u>Lemma 6</u>. $\dfrac{|A_A(C_2, C_4, C_{10})|}{|A|} = \dfrac{1}{p(p+\varepsilon)}\,[p^2 + \Gamma(x_2)\,\Gamma(x_4)\,\Gamma(x_{10})]$

<u>Proof</u>. This follows directly from Lemmas 3 and 5.

For a p'-element $x \in A$ let $p^{c(x)}$ be the power of p which divides $|\mathbf{C}_A(x)|$.

<u>Lemma 7</u>. Let x be a p'-element in A. Then $\mathbf{C}_A(x)$ contains exactly $p^{2c(x)}$ p-elements.

Proof. $\mathbb{C}_A(x)$ is either reductive or reductive times $\langle\tau\rangle$ and the result is known.

Lemma 8. Let x be a p'-element in A then $\Gamma(x) = p^{c(x)}n(x)$ for some nonzero rational integer $n(x)$.

Proof. Γ is not algebraically conjugate to $\Gamma\phi_i \neq \Gamma$. Thus Γ is rational valued by Lemma 5.

$$\frac{|A|}{|\mathbb{C}_A(x)|}\ \frac{\Gamma(x)}{\Gamma(1)} = \frac{|A|}{|\mathbb{C}_A(x)|}\ \frac{\Gamma(x)}{p^2}$$

is an integer. Hence $p^{c(x)} | \Gamma(x)$. It remains to show that $\Gamma(x) \neq 0$.

If $x \in I$ then it is known that $\Gamma(x) = \pm p^{c(x)}$. Suppose that $x \notin I$. Then $x^2 \in I$ and so x has even order. Thus $x = uv = vu$, where u has even order and v has odd order. Hence $v \in I$ and so $\Gamma(v) = \pm p^k$ for some natural number k. Thus

$$\Gamma(x) \equiv \Gamma(v) = \pm p^k \equiv 1 \pmod 2.$$

Hence $\Gamma(x) \neq 0$.

Lemma 9. Let x be a p'-element in A. Then $\Gamma(x) = \pm p^{c(x)}$.

Proof. For any p'-element y in A let $S(y)$ be the set of all elements in A whose p'-part is y. Then $A = \bigcup S(y)$ and $|A| = \sum_y |S(y)|$. By Lemma 7 $|S(y)| = p^{2c(y)}$ and so

$$|A| = \sum_y p^{2c(y)}, \tag{III}$$

where y ranges over all p'-elements in A.

By Lemma 8 $\Gamma(y)^2 \geq p^{2c(y)}$ for any p'-element y in A. Suppose that the result is false. Then by Lemma 8 $\Gamma(x)^2 > p^{2c(x)}$ and so by (III)

$$|A| \geq \sum_y \Gamma(y)^2 > \sum_y p^{2c(y)} = |A|.$$

This contradiction implies the result.

<u>Proof of the Theorem.</u> Let

$$n = \frac{1}{p(p+\varepsilon)} [p^2 + \Gamma(x_2)\, \Gamma(x_4)\, \Gamma(x_{10})].$$

If $p = 3$ then $\varepsilon = -1$, $\Gamma(x_2) = 3$, $\Gamma(x_{10}) = -1$ and $\Gamma(x_4) = 1$. Thus $n = 1$. Suppose that $p > 3$. Then

$$n = \frac{1}{p(p+\varepsilon)} [p^2 \pm p] = \frac{p \pm 1}{p + \varepsilon}.$$

As $p > 3$ and n is an integer it follows that $n = 1$. The Theorem now follows from Lemma 6.

CLASSICAL GROUPS AS GALOIS GROUPS

J.H. Walter
University of Illinois, Urbana, IL 61801

G. Belyĭ [3] has shown that the groups $GL(n,q)$, and for odd q, the groups $SO(2n+1,q)$, $U(n,q)$, and $GSp(2n,q)$, $q \neq 9$, are Galois groups $Gal(K/L_G)$ where L_G is a cyclotomic number field whose choice depends on the group being represented. Of course, L_G can be chosen to be the maximal abelian extension $\mathbb{Q}_{ab}$ of the rational number field $\mathbb{Q}$. As Belyĭ points out, the problem of showing that a given group is a Galois group is much more feasible when one considers extensions of $\mathbb{Q}_{ab}$ instead of $\mathbb{Q}$. This amounts to studying the subgroup $Gal(\mathbb{C}/\mathbb{Q}_{ab})$ of the absolute Galois group $Gal(\mathbb{C}/\mathbb{Q})$. Our principal result is the following. The notation for the classical groups is discussed at the end of this introduction.

<u>Theorem I</u>. Let $\bar{G}$ be a group satisfying one of the following.[1]

$$PGL(n,q) \supseteq \bar{G} \supseteq PSL(n,q), \quad n \geq 2; \tag{1a}$$

$$PGU(2n+1,q) \supseteq \bar{G} \supseteq PSU(2n+1,q), \quad n \geq 1; \tag{1b}$$

$$PGU(2n,q) \supseteq \bar{G} \supseteq PSU(2n,q), \quad n \geq 3; \tag{1c}$$

$$PGSp(2n,q) \supseteq \bar{G} \supseteq PSp(2n,q), \quad n \geq 2; \tag{1d}$$

$$PGO^-(2n,q) \trianglerighteq \bar{G} \supseteq P\Omega^-(2n,q), \quad n \geq 3; \tag{1e}$$

$$PGO^+(2n,q) \trianglerighteq \bar{G} \supseteq P\Omega^+(2n,q), \quad n \geq 4; \tag{1f}$$

$$PGO(2n+1,q) \supseteq \bar{G} \supseteq P\Omega(2n+1,q), \quad n \geq 3, \quad q \text{ odd}. \tag{1g}$$

Then there exists a cyclotomic extension $L_{\bar{G}}$ of the rational number field

---

[1] Up to isomorphism type, this list contains all projective classical groups with exactly one nonabelian composition factor. Taking advantage of the isomorphism between certain classes of classical groups, we have omitted some redundancies to avoid some awkward problems with minimal situations.

A list of the linear groups obtainable as Galois groups by these methods appears in Proposition 9.

$\mathbb{Q}$ and a Galois extension $K/L_{\bar{G}}$ such that $\bar{G}$ is isomorphic to $Gal(K/L_{\bar{G}})$.

Our argument proceeds by verifying certain group theoretical conditions which Belyǐ [3] has shown to be sufficient to establish the representation of these groups as Galois groups. Belyǐ's result is the following.

<u>Theorem II</u>. (Belyǐ [3]). Let G be a finite group acting irreducibly on a vector space V. Suppose that

$$G = \langle a,b \rangle$$

where a has an eigenspace on V of codimension 1. Then there exists a cyclotomic extension L_G of Q and a Galois extension K/L_G such that $Gal(K/L_G)$ is isomorphic to G provided $N_{GL(V)}(G) = GC_{GL(V)}(G)$ and $Z(G)$ is a direct factor of $N_{GL(V)}(\langle a \rangle)$.

This result is the combination of Theorems I and II of Belyǐ [3] together with Hilbert's irreducibility theorem. Theorem I of [3] asserts that a finite group G is a Galois group if $G = \langle a,b \rangle$ and $A_G(a^G, b^G, (ab)^G)$ is rigid as defined in Feit's article "Rigidity and Galois Groups". Theorem II of [3] shows that $A_G(a^G, b^G, (ab)^G)$ is rigid when G is irreducible and a has an eigenspace of codimension 1.

We choose a to be a transvection or homology in G and take for b a particular element whose minimal polynomial has an irreducible factor of highest possible degree among the elements of G. Set $H = \langle a,b \rangle$. The focus of our argument is to characterize the normal subgroup $K = \langle a^H \rangle$. This is a group generated by either transvections or homologies. There is a large number of possibilities for K. We use the characterizations of groups generated by transvections given (in complete form) by W. Kantor [6] and groups generated by homologies given by A. Wagner [10,11]. The important case of groups generated by reflections (homologies of order 2) was also done by V. Serezkin and A. Zalesskiǐ [7]. The main problem in this paper is to eliminate the extraneous cases. Following ideas introduced by E. Artin [1,2], we accomplish this by comparing the order of a certain Sylow subgroup of $\langle b \rangle$ with the order of K using Zsigmondy primes. In order to secure all the conditions of Theorem II, we must be certain to choose a and b so that $G = \langle a,b \rangle$ is as large as possible with respect to the inclusion in (1). This is the purpose of the analysis in the first section of this paper. Then having that G is a Galois group in this case, we use an argument based on Hurwitz's formula to realize all these groups as Galois groups.

In Theorem III at the end of this paper, we have listed those groups which are shown to be generated by a transvection or homology and another element.

We use for the most part a conventional notation for designating the classical groups. Let X be one of the symbols L, U, Sp, O^+, O^-, O. Then $X(V)$ denotes a classical group $X(n,q)$ acting on a vector space V of dimension n over a field $F = F_q$ of q elements with the exception of $U(n,q)$ where $F = F_{q^2}$. Of course, $GL(n,q)$ is used rather than $L(n,q)$. Then $GX(V) = N_{GL(V)}(X(V))$ is the group of similitudes. By $SX(V) \cong SX(n,q)$, we mean the subgroup of elements of determinant 1. By $O^+(V) = O^+(n,q)$ and by $O^-(V) \cong O^-(n,q)$, we mean respectively the first and second orthogonal groups and by $\Omega^+(V) \cong \Omega^+(n,q)$ and $\Omega^-(V) \cong \Omega^-(n,q)$, we mean their commutator subgroups. Of course, this distinction is made only when n is even; when n is odd, write $O(V) \cong O(n,q)$ and $\Omega(V) \cong \Omega(n,q)$. When the occasion arises, we embellish this notation by writing $O_i^\varepsilon(n,q)$ and $\Omega_i^\varepsilon(n,q)$, $\varepsilon = +$ or $-$, where, for $i = 1,2$, the orthogonal groups are defined by bilinear forms with square and nonsquare discriminant respectively. Following Wagner [11], we write $R\Omega^\varepsilon(n,q)$ for a subgroup of $O^\varepsilon(n,q)$ generated by a class of reflections. There are two isomorphic choices, and unlike Wagner [11], we do not distinguish between them. Finally when $X(V)$ or $X(n,q)$ designates any of these groups, $PX(V)$ or $PX(n,q)$ designates the corresponding projective group. $\Sigma L(n,q)$ designates an extension of $SL(n,q)$ by a field automorphism.

Our methods treat the entire class of groups in a uniform manner. For that reason, we have included the results Belyĭ obtained. The most interesting case is the orthogonal groups. For the interpretation of (1e), (1f), and (1g), we note the following sequences. First let q be odd, $\varepsilon = +$ or $-$. We have

$$\begin{aligned} &PGO_1^\varepsilon(2n,q) \triangleright PO_1^\varepsilon(2n,q) \triangleright PSO_1^\varepsilon(2n,q) \triangleright P\Omega_1^\varepsilon(2n,q), \\ &PGO_2^\varepsilon(2n,q) \triangleright PO_2^\varepsilon(2n,q) \triangleright PSO_2^\varepsilon(2n,q) = P\Omega_2^\varepsilon(2n,q), \end{aligned} \tag{2}$$

and $PGO_i^\varepsilon(2n,q)/P\Omega_i^\varepsilon(2n,q)$ is dihedral of order 8 or 4 according as $i = 1$ or 2.

$$PO_i^{\varepsilon}(2n,q) \triangleright PR\Omega_i^{\varepsilon}(2n,q) \triangleright P\Omega_i^{\varepsilon}(2n,q).$$

But $PR\Omega_1^{\varepsilon}(2n,q) \ntriangleleft PGO_1^{\varepsilon}(2n,q)$ while $PSO_1^{\varepsilon}(2n,q) \triangleleft PGO_1^{\varepsilon}(2n,q)$. Also

$$PGO(2n+1,q) = PO(2n+1,q) = PSO(2n+1,q) \triangleright P\Omega(2n+1,q).$$

When q is even

$$PGO^{\varepsilon}(2n,q) = PO^{\varepsilon}(2n,q) = PSO^{\varepsilon}(2n,q) \triangleright P\Omega^{\varepsilon}(2n,q),$$

and $PO(2n+1,q) \cong PSp(2n,q)$.

Designate by F_q a finite field of q elements. By F_q^+ and $F_q^{\times}$, we mean the additive and multiplicative groups of F_q. When $F_q \supset F_r$, F_q^+ may be regarded as an F_r-space, and we write F_q/F_r for this construction. Designate by tr_{F_q/F_r} and by N_{F_q/F_r} the trace and norm mappings associated with the extension F_q/F_r.

By $O_p(G)$ and $O_{\infty}(G)$, we mean the maximal normal p-subgroup and maximal normal solvable subgroup of a group G.

A homology is a transformation which fixes a hyperplane called its axis and leaves invariant a complementary subspace called its center. Homologies of order 2 are called reflections. A transvection t likewise leaves fixed a hyperplane called its axis. The 1-dimensional image of $t-1$ is called the center of t.

1 IRREDUCIBLE ELEMENTS

Let V be a vector space over F, and set $Z = Z(GL(V))$. For $G \subseteq GL(V)$, set $\bar{G} = GZ/Z$ and take $\bar{G}$ to satisfy one of the seven cases in (1). Thus $F = F_q$ except when G is a unitary group, in which case $F = F_{q^2}$. Let p be the prime dividing q. Let G_1 and G_o denote respectively the inverse images in $GL(V)$ of the first and last terms in (1x), x = a,b,c,d,e,f,g.

<u>Proposition 1</u>. In accordance with Table I below, there exists a homology or transvection $a \in G_1$ and an element $b \in G_1$ such that $\langle b\rangle$ acts irreducibly on a subspace V' and one of the following holds.

$$G_1 = G_o\langle a\rangle, \quad b \in G_o \tag{3}$$

$$G_1 = G_o\langle b\rangle, \quad a \in G_o \tag{4}$$

$$G_1 = G_o\langle a,b\rangle, \; G_1/G_o \text{ is dihedral of order 4 or 8.} \tag{5}$$

TABLE I

Group G_1	Type of a	Structure of G_1	Dim V/V'	$\lvert b\rvert$	$\lvert\langle b\rangle \cap Z(G_1)\rvert$
(a) $GL(n,q)$, $n\geq 2$	homology	(3)	0	$(q^n-1)/(q-1)$	$(q-1,n)$
	transvection	(4)	0	q^n-1	$q-1$
(b) $GU(2n+1,q)$, $n\geq 1$	homology	(3)	0	$(q^{2n+1}+1)/(q+1)$	$(q+1,2n+1)$
(c) $GU(2n,q)$, $n\geq 3$	homology	(3)	1	$(q^{2n-1}+1)/(q+1)$	1
(d) $GSp(2n,q)$, $n\geq 2$	transvection	(4)	0	$(q^n+1)(q-1)$	$q-1$
(e) $GO^-(2n,q)$, q odd $n \geq 3$	reflection	(5)	0	$(q^n+1)(q-1)$	$q-1$
$GO^-(2n,q)$ q even $n \geq 3$	transvection	(3)	0	$(q^n+1)(q-1)$	$q-1$
(f) $GO^+(2n,q)$, q odd $n \geq 4$	reflection	(5)	2	$(q^{n-1}+1)(q-1)$	$q-1$
$GO^+(2n,q)$, q even $n \geq 4$	transvection	(3)	2	$2(q^{n-1}+1)(q-1)$	$q-1$
(g) $GO(2n+1,q)$, q odd $n \geq 3$	reflection	(3)	1	q^n+1	1

$\lvert Z(G_1)\rvert = q-1$ except in cases (b) and (c) where $\lvert Z(G_1)\rvert = q^2-1$. The homologies in (b) and (c) have order $q+1 > 2$.

<u>Proof</u>. We treat the various cases in succession.

a) $G_1 = GL(V) \cong GL(n,q)$. Take $V = F_{q^n}/F_q$. Set $B_o = F_{q^n}^{\times}$, and set $B = B_o \cap SL(V)$. Because $\det b = N_{F_{q^n}/F_q}(b)$ for $b \in B$ and $N_{F_{q^n}/F_q}(F_{q^n}^{\times}) = F_q^{\times}$, it follows that $|B_o/B| = q-1$. So $|B| = (q^n-1)/(q-1)$. As $|SL(V) \cap Z(G_1)| = (q-1,n)$, $|B \cap Z(G_1)| = (q-1,n)$. Take $B = \langle b\rangle$, and let a generate the group A with a given axis and center. Then $\det A = F_q^{\times}$. So $G_1 = \langle a\rangle G_o$ and (3) holds. By taking a to be a transvection and replacing B by B_o, $G_1 = \langle b\rangle G_o$ and (4) holds.

b) $G_1 = GU(V) \cong GU(2n+1,q)$. Set $m = 2n+1$; then $m \geq 3$. Set $V = F_{q^{2m}}/F_{q^2}$. For $x,y \in V$, set

$$f(x,y) = tr_{F_{q^{2m}}/F_{q^2}}(xy^{q^m}) = \sum_{j=0}^{m-1} x^{q^{2j}} y^{q^{2j+m}}. \tag{6}$$

Because m is odd, (6) defines a hermitian form, which is clearly non-degenerate. Let $U(V)$ be the unitary group determined by f. Set $B_o = F_{q^{2m}}^{\times}$, $B_1 = B_o \cap GU(V)$, $B_2 = B_o \cap Z(G_1)$, $B_1^* = B_o \cap U(V)$ and $B = B_o \cap SU(V)$. For $\lambda \in B_1$, let $e(\lambda)$ be the multiplier of f determined by λ. Then $f(x\lambda,y\lambda) = e(\lambda)f(x,y)$ where $e(\lambda) \in F_{q^2}^{\times}$. As f is hermitian, $e(\lambda)^q = e(\lambda)$; so $e(\lambda) \in F_q^{\times}$. By (6),

$$tr_{F_{q^{2m}}/F_{q^2}}((\lambda^{q^m+1} - e(\lambda))xy^{q^m}) = 0$$

for all $x,y \in F_{q^{2m}} = V$. Hence $B_1 = \{\lambda \in F_{q^{2m}}^{\times} \mid \lambda^{q^m+1} \in F_q\}$. But $N_{F_{q^{2m}}/F_{q^m}}(B_o) = F_{q^m}^{\times}$. Hence we obtain an exact sequence

$$1 \longrightarrow B_1^* \longrightarrow B_1 \xrightarrow{\;e\;} F_q^{\times} \longrightarrow 1$$

and $|B_1^*| = |\ker N_{F_{q^{2m}}/F_{q^m}}| = q^m+1$. So $|B_1| = (q^m+1)(q-1)$. Let $\lambda \in B_1^*$. Then $\det \lambda^{q+1} = N_{F_{q^{2m}}/F_{q^2}}(\lambda)^{q+1} = N_{F_{q^{2m}}/F_q}(\lambda) = N_{F_{q^m}/F_q}(N_{F_{q^{2m}}/F_{q^m}}(\lambda)) = 1$. So take b_1^* to be a generator of B_1^*, and set $b = b_1^{*q+1}$. Then $B = \langle b\rangle$

and $|B| = (q^m+1)/(q+1)$.

Next take A to be the group of homologies with given center V_1 and axis $V_2 = V_1'$. Let a generate A. It follows that $|\det a| = |a| = q+1$ and that $U(V) = SU(V)A$. Since $e:\lambda \longmapsto e(\lambda)$ is a homomorphism with kernel $U(V)$ and $e(B_2) = e(B_1) = e(G_1)$, it now follows that $G_1 = U(V)B_2 = SU(V)AB_2 = G_o\langle a\rangle$. So (3) holds.

c) $G_1 = GU(V) \cong GU(2n,q)$. Take V to be a vector space of dimension $2n$ over $F = F_{q^2}$. Let V' be a hyperplane in V, and let V'' be a complement. Define a nondegenerate hermitian sesquilinear form f' on V' by (6), and extend f' to a nondegenerate hermitian sesquilinear form f on V by setting $f(x,x) = 1$ for some $x \in V''$ and $f(x,y) = f(y,x) = 0$ for $x \in V''$ and $y \in V'$. Because $\dim V' \geq 4$, we may define B' to be the subgroup of $SU(V')$ which is determined exactly as the subgroup B of $U(V)$ was determined in case (b). Let $B' = \langle b'\rangle$, and take $b \in SU(V)$ so that $b_{|V'} = b'$ and $b_{|V''} = 1_{|V''}$. Then $|b| = q^{2n-1}+1$. Set $B = \langle b\rangle$. It is clear that $B \cap Z(G_1) = 1$. But as in case (b), $e(Z(G_1)) = F_q^{\times}$. Then $G_1 = U(V)Z(G_1)$.

Choose A to be the group of homologies with a fixed center V_1 and axis $V_1^{\perp}$ and let a be a generator of A. Then $G_1 = SU(V)AZ(G_1) = G_o\langle a\rangle$. So again (3) holds, and we have verified Table I in this case.

d) $G_1 = GSp(V) \cong GSp(2n,q)$. Let $V = F_{q^{2n}}/F_q$, and define a nondegenerate alternating form g on V by taking for $x,y \in V$

$$f(x,y) = tr_{F_{q^{2n}}/F_q}(xy^{q^n} - yx^{q^n}). \tag{7}$$

Let $GSp(V)$ be the group of symplectic similitudes determined by f. Set $B_o = F_{q^{2n}}^{\times}$, and set $B = B_o \cap GSp(V)$. For $\lambda \in B$, let $e(\lambda)$ be the multiplier of f determined by λ. Then from (7), $tr_{F_{q^{2n}}/F_q}((e(\lambda)-\lambda^{q^{n+1}})\cdot(xy^{q^n}-yx^{q^n})) = 0$ for all $x,y \in V$. Hence $B = \{\lambda \in B_o \mid \lambda^{q^n+1} = e(\lambda) \in F_q\}$. Set $B_1 = B \cap Sp(V)$; then $|B_1| = q^n+1$, and B_1 acts irreducibly on V. On the other hand, $N_{F_{q^{2n}}/F_{q^n}}(F_{q^{2n}}^{\times}) = F_{q^n}^{\times}$. Hence $e(B) = F_q^{\times}$, and $GSp(V) = Sp(V)B$. Let b be a generator of B, and take a to be any transvection in $Sp(V)$. Then (4) holds in this case.

e) $G_1 = GO^-(V) \cong GO^-(2n,q)$. Take $V = F_{q^{2n}}/F_q$, and define a nondegenerate quadratic form g by setting for $x \in V$,

$$g(x) = tr_{F_{q^n}/F_q}(xx^{q^n}). \tag{8}$$

The associated bilinear form f has the value $f(x,y) = tr_{F_{q^n}/F_q}(xy^{q^n}+yx^{q^n})$ for $x,y \in V$. Set $B_o = F_{q^{2n}}^{\times}$ and $B = B_o \cap GO^{\varepsilon}(V)$, $\varepsilon = +$ or $-$. Let $B_1 = B \cap O^{\varepsilon}(V)$ and $B_2 = F_q^{\times}$. Let $e(\lambda)$ be the multiplier of q determined by $\lambda \varepsilon B$. Then from (8), it follows that $tr_{F_{q^n}/F_q}((e(\lambda)-\lambda^{q^n+1})xx^{q^n}) = 0$ for all $x \in V$. Hence $B = \{\lambda \in B_o |$ $\lambda^{q^n+1} = e(\lambda)\}$. Since $N_{F_{q^{2n}}/F_q}(F_{q^{2n}}^{\times}) = F_{q^n}^{\times}$, $e(B) = F_q^{\times}$. Then $GO^{\varepsilon}(V) = O^{\varepsilon}(V)B$. Also $B_1 = \ker N_{F_{q^{2n}}/F_q}$; so $|B_1| = q^n+1$ and $B_1 \subseteq SO^{\varepsilon}(V)$. But now using the formulas for the orders of the orthogonal groups, it follows that $O^{\varepsilon}(V) = O^-(V)$ (this argument is given in detail in the footnote to (13)). Also $|B:B_1| = q-1$. So $|B| = (q^n+1)(q-1)$. On the other hand, $|B_1 \cap B_2| = (q-1,2)$ so $|B_1B_2| = |B|/(q-1,2)$.

Consider now that q is odd. Then $|B:B_1B_2| = 2$. There exists a 2-element $u \in B \backslash B_1B_2$ such that $u \in F_{q^2}^{\times}$ as q^2-1 divides $|B|$. When n is even, $e(u) = u^2$ and $|e(u)| = |q-1|_2$, so $|u| = 2|q-1|_2$. When n is odd, $|q^n+1|_2 = |q+1|_2$ and also $e(u) = u^{q+1}$. So $|u| = |q+1|_2|e(u)|_2 = |q^2-1|_2 = 2|q-1|_2$ or $2|q+1|_2$ according as $q \equiv 1 \pmod 4$ or $q \equiv 3 \pmod 4$.

Let b be a generator for B, and take a to be a reflection in $O^-(V)$ with center $V_a = x_aF_q$. Because $e(u) \notin F_q^2$, $g(x_a)F_q^2 \neq g(ux_a)F_q^2$. Hence a and uau^{-1} have distinct spinorial norms. When n is odd and $q \equiv 3 \pmod 4$, $GO^-(V) = GO_1^-(V)$, and by (2) $GO^-(V)/\Omega^-(V)Z(G_1)$ is dihedral of order 8. Then $SO^-(V) = \langle a,uau^{-1}\rangle\Omega(V)$. So $G_1 = \langle a,b\rangle\Omega(V) = \langle a,b\rangle G_o$ and (5) holds. When n is even or n is odd and $q \equiv 1 \pmod 4$, $GO^-(V) = GO_2^-(V)$ by (2). Then $GO^-(V)/\Omega^-(V)$ is dihedral of order 4, and $GO^-(V) = \langle a,u\rangle\Omega^-(V)Z(G_1) = \langle a,b\rangle\Omega^-(V) = \langle a,b\rangle G_o$. So again (5) holds.

Consider next that q is even. Then $B = B_1 \times B_2$, and $e(B_2) = (F_q^{\times})^2 = F_q^{\times}$. Take b to be a generator of B. Then $GO^-(V) = O^-(V)B_2 = O^-(V)Z(G_1)$. Take a to be a transvection in $O^-(V)$. Then $O^-(V) = \langle a\rangle\Omega(V)$. Hence (3) holds in this case.

f) $G_1 = GO^+(V) \cong GO^+(2n,q)$. Let V be a vector space of dimension $2n$ over F_q and let V' and V'' be complementary subspaces of dimensions $2n-2$ and 2 respectively. Identify V' with $F_{q^{2n-2}}/F_q$ and define a quadratic form g' on V' using (8). Then V' is the orthogonal sum of $n-1$ elliptic planes and also when $n-1$ is even the orthogonal sum of $n-1$ hyperbolic planes. On V'' define a quadratic form g'' so that V'' is either an elliptic or hyperbolic plane according as n is even or odd. Define a quadratic form $g = g' \perp g''$ on V by stipulating $V' \perp V''$, $g_{|V'} = g'$, and $g_{|V''} = g''$. Then V is an orthogonal sum of an even number of elliptic planes or of an odd number of hyperbolic planes. Hence g has Witt index n, and we may take $GO^+(V)$ to be the group of similitudes determined by g. The groups $GO^+(V')$ and $GO^+(V'')$ determined by g' and g'' are now taken as subgroups of $GL(V)$ fixing the given complementary subspace.

Let $B' = F_{q^{2n-2}}^{\times} \cap GO(V')$, $B_1' = B' \cap O^-(V)$, and $B_2' = F_q^{\times}$, all acting by multiplication on V' as in case (e). Then $B' = B_1'B_2'\langle u'\rangle$ where u' is a 2-element of $F_{q^2}^{\times}$ acting on V' which is chosen exactly as u was chosen in case (e) with $u' = 1$ when q is even. By (e) when q is odd, $|u'| = 2|q-1|_2$ when $n-1$ is even and $|u'| = 2|q-1|_2$ or $2|q+1|_2$ when $n-1$ is odd according as $q \equiv 1 \pmod 4$ or $q \equiv 3 \pmod 4$. Let $e'(u')$ be the multiplier of g' associated with u'. Then $e'(u')$ is a nonsquare in F_q; in fact, $|e'(u')| = |q-1|_2$ since $t \longmapsto e'(t)$ defines a homomorphism of $GO^+(V')$.

Let $B_1' = \langle b_1'\rangle$. Then $B_1' \subseteq O^-(V-) \subseteq O^+(V)$. Also $|B_1'| = q^{n-1}+1$. Take $b_1 = (b_1')^{|q+1|_2}$ when n is even and $q \equiv 3 \pmod 4$, and in all other cases take $b_1 = b_1'^2$. Set $B_1 = \langle b_1\rangle$. Then $|B_1| = (q^{n-1}+1)/|q+1|_2$ or $(q^{n-1}+1)/(q-1,2)$, respectively. In any event, $|B_1|$ is odd. Likewise in $GL(V'')$, let B_2'' designate the group of scalar transformations, and choose generators b_2' and b_2'' of B_2' and B_2'', respectively, so that $b_2'b_2''$ acts as a scalar transformation of V. Set $b_2 = (b_2'b_2'')^{|q-1|_2}$, and $B_2 = \langle b_2\rangle$. Then $|B_2| = (q-1)/|q-1|_2$, and $|B_2|$ is odd. As $(q^{n-1}+1,q-1) = 2$, $(|B_1|,|B_2|) = 1$ and B_1B_2 is cyclic. Then $|B_1B_2| = (q^{n-1}+1)(q-1)$ when q is even, and $|B_1B_2| = (q^{n-1}+1)(q+1)/2|q+1|_2$ or $(q^{n-1}+1)(q-1)/2|q+1|_2$ when q is odd, the latter case occurring only when n is even and $q \equiv 3 \pmod 4$. In any event, $|B_1B_2|$ is odd.

We will now define B to be a cyclic group of the form $B_1B_2\langle u\rangle$ where $u = u'u''$ by choosing an appropriate element u'' of

GL(V"). Indeed, take first the case that n is odd. Then V" is a hyperbolic plane and contains a hyperbolic basis x_1, x_2. Thus $f(x_1, x_2) = 1$ where f is the bilinear form associated with g, and $g(x_1) = g(x_2) = 0$. Let $e'(u') = \alpha$, and define u" by setting $u''x_1 = x_2$ and $u''x_1 = x_2\alpha$. Hence

$$|u''| = 2|\alpha| = 2|e'(u')| = 2|q-1|_2 = |u'|. \tag{9}$$

Hence both u'^2 and u''^2 act as scalar transformations on V' and V" respectively. Let $e''(t'')$ denote the multiplier of g" associated with an element t" of $GO^+(V'')$, and let e(t) denote the multiplier of g associated with an element t of $GO^+(V)$. Now clearly $e''(u'') = \alpha$. So $e(u) = \alpha$. Now because $e'(u'^2) = e''(u''^2) = \alpha^2$, $u'^2 = \alpha^2 1_{V'}$, and $u''^2 = \alpha^2 1_{V''}$. Hence $u^2 = \alpha^2 1_V$. Clearly $[B_1B_2, u] = 1$. So with choice of u, set $B = B_1B_2\langle u\rangle$. Then B is cyclic. Using (9), it follows that $|B| = 2(q^{n-1}+1)(q-1)$ when q is even, and $|B| = (q^{n-1}+1)(q-1)$ when q is odd. Furthermore, $|B_2\langle u\rangle| = 2(q-1)$. For $\lambda 1_V \in B_2$, $e(\lambda 1_V) = \lambda^2$; so $|e(B_2)| = |B_2| = (q-1)/|q-1|_2$ as $|B_2|$ is odd. On the other hand, $|e(u)| = |\alpha| = |q-1|_2$. So $e(B_2\langle u\rangle) = F_q^{\times}$. Hence $GO^+(V) = O^+(V)B$ in this case. Furthermore, by the definition of u", it follows that u acts irreducibly on V". Hence B acts irreducibly on both V' and V".

Next consider the case that n is even. Then V' is an elliptic plane. Then we may treat $GO^-(V'')$ just as in case (e). In particular, we may identify V" with F_{q^2}/F_q. Then there exists a 2-element u" of $F_{q^2}^{\times}$ such that $|u''| = 2|q-1|_2$ or $2|q+1|_2$ according as $q \equiv 1 \pmod 4$ or $q \equiv 3 \pmod 4$. Then $|u''| = |u'| = |u|$. With this choice of u, $[B_1B_2, u] = 1$ and $(|B_1B_2|, |u|) = 1$. So B is cyclic. In fact, $|B| = |B_1B_2||u| = 2(q^{n-1}+1)(q-1)$ or $(q^{n-1}+1)(q-1)$ according as q is even or odd. Taking u' as an element of $F_{q^2}^{\times}$ acting on V' and u" as an element of $F_{q^2}^{\times}$ acting on V", we have that $e'(u') = u'^{q+1}$ and $e''(u'') = u''^{q+1}$. Without changing $|u''|$, we may replace u" by an element u''^k, k odd, so that $e'(u') = e''(u'')$. Then u is an element of $GO^+(V)$, and the multiplier e(u) of g associated with u satisfies $e(u) = e'(u')$ and $|e(u)| = |q-1|_2$. Then as in the case when n is odd, $e(B_2\langle u\rangle) = F_q^{\times}$. Hence again $GO^+(V) = O^+(V)B$. Furthermore, $u'' \notin B_2''$; hence u" acts irreducibly on V", and B acts irreducibly on both V' and V".

Let b be a generator of B. Choose a to be a transvection or reflection according as q is even or odd. When q is even, $O^+(V) = \Omega(V)\langle a\rangle$; and, since $e(B_2) = F_q^{\,x}$, $GO^+(V) = O^+(V)Z(GO^+(V))$. Hence (3) holds in this case. So consider q to be odd. Then exactly as in case (e), when $GO^+(V) = GO_1^+(V)$, $O^+(V) = \langle a,uau^{-1}\rangle\Omega(V)$ and $G_1 = \langle a,b\rangle G_o$ with G_1/G_o dihedral of order 8. When $GO^+(V) = GO_2^+(V)$, $G_1 = \langle a,u\rangle\Omega(V)Z(G_1) = \langle a,b\rangle G_o$. So (5) again holds.

g) $G_1 = GO(V) \cong GO(2n+1,q)$. Let V be a vector space of dimension $2n+1$ over F_q. When q is even, $V^\perp$ has dimension 1 and $GO(V)$ acts irreducibly on $V/V^\perp$ as $GSp(V/V^\perp)$. So we may consider q to be odd. Let V' be a hyperplane of V and let V'' be a complement to V'. On V' define a quadratic form g' just as in case (e), and extend g' to a quadratic form g on V by stipulating that $V' \perp V''$ and that $g(x'') = 1$ for some nonzero vector $x'' \in V''$. Define the subgroup B' of $GO^-(V')$ exactly as the subgroup B_1 of $O^-(V)$ was defined in case (e). Let b' be a generator of B', and let b be the element of $O(V)$ defined so that $b_{|V'} = b'$ and $b_{|V''} = 1_{V''}$. Set $B = \langle b\rangle$. Then $|B| = q^n+1$.

For $v \in GO(V)$, let $e(v)$ be the multiplier of g determined by v. Let f be the symmetric bilinear form obtained from g. Calculating determinants from the relation $f(vx,vy) = e(v)f(x,y)$, we obtain $e(v)^{2n+1} = (\det v)^2$. Hence $e(v) \in F_q^{\,2}$. But $Z(GO(V)) \cong F_q^{\,x}$; hence $e(Z(GO(V))) = e(GO(V)) = (F_q^{\,x})^2$. Thus $GO(V) = SO(V)Z(GO(V))$ since $O(V) = SO(V)\langle -1_V\rangle$.

Finally choose a_1 to be a reflection in $O(V)$ which is not contained in $\Omega(V)$, and set $a = -a_1$. Then $SO(V) = \Omega(V)\langle a\rangle$. So $GO(V) = \langle a\rangle\,\Omega(V)Z(GO(V))$, and (3) holds. This proves Proposition 1.

2 THE STRUCTURE OF <a,b>

In the previous section, we have chosen b to satisfy Proposition 1 and a to be a homology of order $q+1$, a reflection, or a transvection. In cases (c), (f) and (g) where $V \neq V'$, we must further specify the choice of a. In this case, $\langle a\rangle$ is determined by specifying the center V_1 of a since its axis will be $V_1^\perp$. We will require in cases (c), (f) and (g) that

$$V_1 \not\subseteq V' \quad\text{and}\quad V_1 \not\subseteq V''. \tag{10}$$

With this choice of a, set $H_1 = \langle a,b\rangle$ and $K_1 = \langle a^{H_1}\rangle$. Then $K_1 \triangleleft H_1$ and $H_1 = K_1B$ where $B = \langle b\rangle$.

By a <u>primitive element</u> of B, we mean an element b_p which acts on V' as a primitive element of the field F_{q^m} where V' has been given as F'/F with $F'/F = F_{q^m}/F_q$ or in cases (b) and (c) as $F_{q^{2m}}/F_{q^2}$ and $B_o = F'^{\times}$ as in Section 1. Then $F[b_p] = F'$. It follows from Table I that q^m-1 divides $|B|$ or m is even and $q^{m/2}+1$ divides $|B|$. This implies that B always contains primitive elements. Note that b_p is primitive if and only if $|b_p|$ does not divide q^s-1 for $s < m$.

<u>Lemma 2</u>. Let b_p be a primitive element of B, and set $B_p = \langle b_p\rangle$. Choose a to satisfy (10) and set $H_p = B_pK_1$. Then H_p acts irreducibly on V and K_1 acts completely reducibly on V.

<u>Proof</u>. Because b_p acts irreducibly on V', it suffices to consider only the cases (c), (f), and (g). When a is a homology, its invariant subspaces are either contained in V_1 or $V_1^{\perp}$ or they contain V_1 or $V_1^{\perp}$. When a is a transvection, they are either contained in $V_1^{\perp}$ or they contain V_1. But now $\dim V_1 = 1$ and $\dim V' \geq 3$. Then as $V'' = V'^{\perp}$, it follows that $V_1^{\perp} \not\subseteq V' \not\subseteq V_1$ and $V_1^{\perp} \not\subseteq V'' \not\subseteq V_1$. On the other hand, (10) implies that $V_1 \not\subseteq V' \not\subseteq V_1^{\perp}$ and $V_1 \not\subseteq V'' \not\subseteq V_1^{\perp}$. So a leaves invariant neither V' nor V'', and, since $\dim V' \neq \dim V''$, a does not permute these subspaces. Hence H_p acts irreducibly on V, which implies that K_1 acts completely reducibly on V.

<u>Lemma 3</u>. Let b_p be a primitive element of B. Then $C_{G_1}(b_p) = BG(V'')$ where $G(V'') = \{g \in G_1 | g|_{V'} \in Z(G_1)_{|V'}\}$.

<u>Proof</u>. Since $C_{G_1}(b_p)$ leaves invariant V', it suffices to show that $C_{G(V')}(b_p) = B'Z'$ where $B' = B_{|V'}$ and $Z' = Z(G_1)_{|V'}$. In the proof of Proposition 1, we identified V' with F'/F where $F'/F = F_{q^m}/F_q$ or $F_{q^{2m}}/F_{q^2}$ and $m = \dim V'$, and we showed that $B_o' \cap G_1 = B'Z'$ where B_o' is the group $F'^{\times}$ acting by multiplication on V. Then F' is a maximal subfield of the central simple algebra $\mathrm{Hom}_F(V',V')$. Since $F[b_p] = F'$, $C_{\mathrm{Hom}_F(V',V')}(b_p) = F'$. So $C_{G(V')}(b_p) = B'Z'$.

Let ℓ,m be integers with $\ell,m > 1$. A <u>Zsigmondy prime</u> is defined to be a prime which divides ℓ^m-1 but not ℓ^k-1 for $k < m$. A result of Zsigmondy ([1], [5; IX, 8]) shows that for all pairs (ℓ,m) Zsigmondy primes exist except in the cases $(\ell,m) = (2,6)$ or $(2^k-1,2)$, $k > 1$. When $\ell = p^j$, the Zsigmondy primes determined by (p,jm) are also primes for (ℓ,m). When $q = p^j$ and $(p,jm) \neq (2,6)$ or $(2^k-1,2)$ define a <u>Zsigmondy number</u> $s(q,m)$ to be the Zsigmondy prime s for which $|p^{jm}-1|_s$ is a maximum; otherwise, set $s(2,6) = 9$ and $s(2^k-1,2) = 2^k$. Then still $s(q,m)$ does not divide p^r-1 for $r < jm$. The <u>degree</u> of $s(q,m)$ is defined to be m. A crucial fact about Zsigmondy primes $s = s(q,m)$ is that $s \equiv 1 \pmod m$ so that $s \geq m+1$. Associate to each classical group G over F_q the Zsigmondy number $s_q(G)$ of maximum degree among the numbers $s(q,m)$ dividing $|G|$. For the convenience of the reader, we list the orders of the classical groups G_o of (1).

$$|SL(n,q)| = q^{n(n+1)/2}(q^n-1)(q^{n-1}-1)\dots(q^2-1), \tag{11a}$$

$$|SU(2n+1,q)| = q^{(n+1)(2n+1)}(q^{2n+1}+1)(q^{2n}-1)\dots(q^3+1)(q^2-1), \tag{11b}$$

$$|SU(2n,q)| = q^{n(2n+1)}(q^{2n}-1)(q^{2n-1}+1)\dots(q^3+1)(q^2-1), \tag{11c}$$

$$|Sp(2n,q)| = q^{n^2}(q^{2n}-1)(q^{2n-2}-1)\dots(q^2-1), \tag{11d}$$

$$|\Omega^-(2n,q)| = \frac{1}{(2,q-1)}\, q^{n(n-1)}(q^n+1)(q^{2n-2}-1)\dots(q^4-1)(q^2-1), \tag{11e}$$

$$|\Omega^+(2n,q)| = \frac{1}{(2,q-1)}\, q^{n(n-1)}(q^n-1)(q^{2n-2}-1)\dots(q^4-1)(q^2-1), \tag{11f}$$

$$|\Omega(2n+1,q)| = \frac{1}{(2,q-1)}\, q^{n^2}(q^{2n}-1)(q^{2n-2}-1)\dots(q^2-1). \tag{11g}$$

Since $|G_1/G_o|$ divides 8 or q^2-1, $s_q(G)$ is defined for $G_1 \supseteq G \supseteq G_o$. Likewise when X is a group such that $|X|$ divides q^m-1 and $s(q,m)$ divides $|X|$, set $s_q(X) = s(q,m)$. Then by Proposition 1, $s_q(G_1) = s_q(B)$. Let B_s be the Sylow s-subgroup of B where s divides $s_q(B)$. Then B_s is a Sylow s-subgroup of G_1. In all cases, $(s_q(B),|Z(G_1)|) = 1$ as $|Z(G_1)| = q-1$ or q^2-1, the latter case occuring only in cases (b) and (c).

It is clear that an element b_s of B of order $s_q(B)$ is a primitive element of B. Recall that an irreducible linear group L acting on a vector space V is imprimitive if it permutes a set of proper irreducible direct summands of V belonging to some normal subgroup; otherwise it is said to be a primitive linear group. Thus all monomial groups are imprimitive.

Lemma 4. Let b_s be an element of B of order $s = s_q(B)$. Then $\langle b_s \rangle$ is a primitive linear group on V' except in the cases $(q,m) = (2,6)$ or $(4,3)$, in which cases

$$V' = W_1 \oplus W_2 \oplus W_3 \tag{12}$$

where $\dim W_i = 2$ or 1 according as $q = 2$ or 4; and except in the case $(q,m) = (3,2)$.

Proof. Suppose that $V' = W_1 \oplus W_2 \oplus \ldots \oplus W_t$, $t > 1$ where each W_i is an irreducible $\langle c_s \rangle$-subspace for some $c_s \in \langle b_s \rangle$. Since b_s is a primitive element of B, $|c_s| < |b_s|$. In particular, if s is a prime, $c_s = 1$. In that case b_s permutes the set $\{W_i | i = 1,2,\ldots,t\}$ cyclically. Hence $t = s$. Also as $c_s = 1$, $\dim W_i = 1$. Hence $\dim V' = s$. But this implies that $s = s(p,s)$ for some prime p, which contradicts $(s,s(p,s)) = 1$. When $s = s(p,2) = 2^k$ with $p = 2^k - 1$, the minimum polynomial $f(x)$ of b_s has the form $x^2 - a$ where $a \in F_p$ inasmuch as $t = 2$ in this case. Then $2^k = |b_s| = 2|a|$. As $(p-1, 2^{k-1}) = 2$, $k = 2$, whence $s = 4$ and $q = p = 3$. Finally consider that $s = s(2,6) = s(4,3) = 9$. In this case, the minimal polynomial $f(x)$ of b_s has the form $f(x) = \phi_9(x) = \phi_3(x^3) = x^6 + x^3 + 1$ over F_2 and $f(x) = x^3 - \alpha$, $\alpha \in F_4$, over F_4. This implies (12).

In order to apply the results of W. Kantor [6] and A. Wagner [10,11], we provide the following lemma.

Lemma 5. Choose a to be a reflection or transvection in G_1 so that (10) holds. Then either K_1 acts irreducibly on V, and, when a is a transvection, $O_p(K_1) = 1$ and K_1 is generated by a conjugacy class of transvections, or (12) holds with each W_i of dimension 2 and K_1-invariant and with $V = V'$.

Proof. Let L_1 be a minimal normal subgroup of K_1 containing an element $t_1 \in a^{H_1}$. Let $K_1^* = K_1 N_B(L_1)$. Then $K_1^* \triangleleft H_1 = K_1 B$. Now H_1 acts by conjugation on the set $L_1^{H_1}$, and K_1^* is the stabilizer of each element of this set. Clearly $L_1^{H_1} = L_1^B$, and $B/N_B(L_1)$ acts cyclically on $L_1^B = \{L_1, L_2, \ldots, L_k\}$. Thus $L_i = L_1^{b_i}$ where b_i, $i = 1,2,\ldots,k$, form a transversal to $N_B(L_1)$ in B. By the minimality of L_1, $L_1 = \langle t_1^{K_1}\rangle$ since $\langle t_1^{K_1}\rangle \triangleleft K_1$. Then $L_i = \langle t_i^{K_1^*}\rangle$ where $t_i = t_1^{b_i}$, $i = 1,2,\ldots,k$. Thus all elements of a^{H_1} are contained in some L_i, $i = 1,2,\ldots,k$.

Let W_i be an irreducible L_i-subspace of V. For any $t \in a^{H_1}$, let V_t be its axis and U_t its center. Then $V_t \supseteq W_i$ implies that t is contained in the kernel of the representation of L_i on W_i. Consequently when $t \in t_i^{K_1^*}$, $U_t \subseteq W_i$ since W_i is L_i-invariant. The irreducibility of W_i implies that $W_i = \langle U_t | t \in t_i^{K_1^*}\rangle$. Thus to each L_i there corresponds a unique irreducible L_i-subspace W_i, $i = 1,2,\ldots,k$. Because $L_i \triangleleft K_1^*$, each W_i is K_1^*-invariant. In particular, if $W_j \neq W_i$, then $U_t \nsubseteq W_i$ for $t \in t_j^{K_1^*}$; so $V_t \supseteq W_i$ and the subgroup L_j is in the kernel of the rperesentation of K_1 on W_i. Thus the representations afforded by distinct subspaces W_i are inequivalent.

Let $B(W_1) = \langle b \in B | bW_1 = W_1\rangle$. Then $K_1^* \subseteq B(W_1)K_1 \subset H_1$, and $B(W_1)K_1 \triangleleft H_1$. Hence $B(W_1)K_1$ is the stabilizer of each W_i in the action of H_1 on $\{W_i | i = 1,2,\ldots,k\}$. As K_1 is completely reducible, $V = \sum_{i=1}^k W_i$. Let b_{i_j}, $j = 1,2,\ldots,\ell$, form a transversal of $B(W_1)$ in B, and set $W_{i_j} = b_{i_j}W_1$. By the inequivalence of the subspaces W_{i_j}, $V = W_{i_1} \oplus W_{i_2} \oplus \cdots \oplus W_{i_\ell}$. When $b_s \in B(W_1)$, then $V' = [V, b_s] \subset W_1$. As $\dim V' < \frac{1}{2}\dim V$, $W_1 = V$ and $\ell = 1$ in this case.

Consider next that $b_s \notin B(W_1)$. Take first the case that b_s acts imprimitively on V'. Then (12) holds. So $\dim V \leq 3 \dim W_1 + 2 \leq 8$ by Lemma 4. Hence $\ell < 9$. But $|b_s| = 9$ in this case. So $b_s^3 \in B(W_1)$ and $\ell = 3$. This forces $V = V'$ and (12) to hold. Hence we may take b_s to have order $s_q(B) = s_q(G_1)$. It is a basic property of Zsigmondy primes of degree m that $s_q(G_1) \equiv 1 \pmod m$. Then $s_q(G_1) > m$. But these degrees can be determined from (11) as follows[1]:

$$s_q(SL(n,q)) = s(q,n), \quad n \geq 2; \tag{13a}$$

---

[1]At this point in the proof, one may distinguish $O^+(2n,q)$ from $O^-(2n,q)$ by their Zsigmondy numbers as mentioned in case (e) of Section 1. The point to be made is that because $s_q(\Omega^+(2n,q)) < s_q(\Omega^-(2n,q))$, $|b_s|$ does not divide $|\Omega^+(2n,q)|$.

$$s_q(SU(2n+1,q)) = s(q,4n+2)),\ n \geq 1; \tag{13b}$$

$$s_q(SU(2n,q)) = s(q,4n-2),\ n \geq 2; \tag{13c}$$

$$s_q(Sp(2n,q)) = s(q,2n),\ n \geq 2; \tag{13d}$$

$$s_q(\Omega^-(2n,q)) = s(q,2n),\ n \geq 3; \tag{13e}$$

$$s_q(\Omega^+(2n,q)) = s(q,2n-2),\ n \geq 4; \tag{13f}$$

$$s_q(\Omega(2n+1,q)) = s(q,2n),\ n \geq 2;\ q \text{ odd}. \tag{13g}$$

From this it follows that $s_q(G_1) > m > \frac{1}{2}\dim V$. But $|b_s| = s_q(G_1)$ is a prime; so $\langle b_s\rangle$ permutes $\{W_{i_j} \mid j = 1,2,\ldots,\ell\}$ cyclically and thus $\ell = s_q(G_1) = \dim V/\dim W_1$. Consequently $\dim W_1 = 1$, and $m = \dim V - 1$ since $s_q(G_1) \equiv 1 \pmod m$. But by (13), either $m \geq \dim V$ or $m = \dim V-2$. Thus we conclude that $\ell = 1$ and $V = W_1$ or that (12) holds with $V = V'$. In the former case, V is an irreducible subspace for each L_i, $i = 1,2,\ldots,k$. As $K_1 = L_1L_2\ldots L_k$, V is an irreducible K_1-space.

Because K_1 is an irreducible group, and p divides q, $O_p(K_1) = 1$. In order to show that K_1 is generated by a conjugacy class of transvections, we argue that $K_1 = L_1$. In any event, the characterization of Kantor [6; Theorem II] is immediately applicable to the group L_1. If $k > 1$, the transvections in L_2 induce automorphisms on L_1. However, the automorphism groups of the groups on Kantor's list are well-known. The only groups with outer automorphism groups of orders divisible by p are the classical groups, and these automorphisms are obtained from field or graph automorphisms. In other words, $|N_{K_1}(L_1)/L_1C_{K_1}(L_1)|_p = 1$ since K_1 is a linear group containing L_1. As L_2 is generated by elements of order p, $L_2 \subseteq L_1C_{K_1}(L_1)$. Let $M(V) = \mathrm{Hom}_F(V,V)$ and $C_1 = C_{M(V)}(K_1)$. By famous results of Schur and Wedderburn, $C_1 = F_{q^s}$ for some $s \geq 1$. Thus $|C_1{}^x|_p = 1$. Hence $L_2 \subseteq L_1$. Arguing with L_1 and L_2 interchanged, we conclude that $L_1 = L_2$. This contradiction shows $k = 1$. The minimality of L_1 now shows that K_1 is generated by a conjugacy class of transvections.

Lemma 6. Let B_s be a Sylow subgroup of B containing b_s, and assume that K_1 acts irreducibly on V under the conditions of Lemma 5. Then $B_s \subseteq K_1$.

Proof. By Lemma 3, $[K_1,b_s] \neq 1$. Therefore b_s induces a nontrivial automorphism of K_1. By the results of Kantor [6; Theorem II] and Wagner [10,11], K_1 is either one of the groups given in (1) or K_1 is a symmetric group S_m acting on a subgroup of a homocyclic group of rank $m-1$ and order dividing $(q-1)^{m-1}$ or $K_1 \cong 3 \cdot A_6$. The classical groups arising in this manner are either defined over a subfield of F or over F_2, F_3 or F_4. The homocyclic group (when it is nontrivial) is obtained from a torus in G_1. Now $B \subseteq N_{G_1}(K_1)$, and no element of $N_G(K_1)$ induces a field, graph or twisted graph automorphism on K_1 when K_1 is a classical group. Hence $N_{G_1}(K_1)/K_1C_{G_1}(K_1)$ has a structure given by a factor group G_1/G_o of one of the groups of Proposition 1. In all other cases, the outer automorphism group of K_1 has order at most 2. Thus $N_{G_1}(K_1)/K_1C_{G_1}(K_1)$ has order dividing $q-1$, $q+1$, or 8.

On the other hand, it follows from (13) that the numbers $s_q(G_1)$ are odd and relatively prime to $q(q^2-1)$ except in case (a) when $n = 2$, where $s_q(G_1)$ is still relatively prime to $q(q-1)$. Consequently $B_s \subseteq K_1C_{G_1}(K_1)$. Now $C_{G_1}(K_1) \subseteq F^{*x}$ where F^* is an extension of the field F of definition of G_1. But certainly F^* is not a maximal subfield of the central simple algebra $\mathrm{Hom}_F(V,V)$. Hence $[F^*:F]$ is a proper divisor of $\dim V$. Let r denote the degree of $s_q(G_o)$. Then $r > \frac{1}{2} \dim V \geq [F^*:F]$. Then $(s_q(G_1), |F^*|-1) = 1$. Hence $|B_s|$ is relatively prime to $|C_{G_1}(K_1)|$. Thus $B_s \subseteq K_1$.

Proposition 7. Let b be chosen to satisfy Proposition 1. Then in accordance with Table I, a reflection or transvection a can be chosen in G_1 so that

$$K_1 \supseteq G_o. \tag{14}$$

Proof. Choose a so that (10) is satisfied in cases (c), (f), and (g). Furthermore when $F = F_2$, $\dim V = 6$ and V has the decomposition (12), choose a so that $a W_1 \neq W_1$. Then by Lemmas 5 and 6, K_1 acts primitively on V and $B_s \subseteq K_1$. Hence

$$|B_s| \text{ divides } |K_1|. \tag{15}$$

Also K_1 is given by Kantor [6; Theorem II] and Wagner [10], [11; III, 3 and Appendix]. We will systematically deal with these possibilities.

i) Consider first the case that $\dim V = 2$. This occurs only when $G_1 \cong GL(2,q)$. Then K_1 is an irreducible subgroup of $SL(V)$ generated by transvections. Thus by [6], $K_1 \cong SL(2,r)$ where $r = 5$ or r divides q. When r divides q, (15) implies that $s_q(G_o) = s(q,2) = s(r,2)$. Then $r = q$, and (14) is obtained. This leaves the case $r = 5$ and $K_1 \cong SL(2,5)$. Then $|a_1|$ divides 120. When $|a_1| = 2$ or 5, r again divides q. So we may take $|a_1| = 3$. As $|K_1| \equiv 0 \pmod 5$, $q = 3^{2k}$, $k \geq 1$. Then $s_q(K_1) = s(3^{2k},2) = s(3,4k)$. Since $|K_1| = 8 \cdot 3 \cdot 5$, $s_q(K_1) = 5$ and $k = 1$. Thus $G_o \cong SL(2,9)$. As $PSL(2,9) \cong A_6$ and as $|xy| = 4$ when $x = (126)$ and $y = (12345)$, it follows that a_1 can be chosen to obtain (14). Thus in the remaining cases, $(s_q(G_1), q(q^2-1)) = 1$. Thus $(s_q(G_1), |Z(G_1)|) = 1$ and $s_q(G_1) > 3$.

ii) Let $d = \dim V$. From the aforementioned results of Kantor and Wagner, K_1 can be a symmetric group S_m with $m = d$, $d+1$, or $d+2$ or K_1 can have the form $E \rtimes S_d$ where E is obtained from a homocyclic group of rank $m-1$ and order dividing $(q-1)^{m-1}$ by taking the elements of determinant 1 ([12; III, 3]). In all these cases, no nontrivial outer automorphisms are induced by elements of B. Thus $B \subseteq K_1 Z(G_1)$. In all these cases, a is either a reflection or transvection; so G_1 is never a unitary group. Therefore, $F = F_q$. Since the noncentral elements of B do not leave invariant any hyperplane in V', they do not centralize a^{H_1}. Hence $C_B(K_1) = B \cap Z(G_1)$. Let $t = |B/B \cap Z(G_1)|$. Then $t = 2(q^{(d-2)/2}+1)$, $q^{(d-2)/2}+1$, or $(q^d-1)/(q-1)$ by virtue of Table I.

As b_s is primitive in B, the minimal polynomial $f(x)$ for b_s in its action on V' is irreducible of degree equal to $\dim V'$. But $f(x)$ divides $(x^s-1)/(x-1)$ where $s = s_q(B) = |b_s|$. Therefore $\dim V' \leq s-1$. On the other hand, b_s is represented as a cyclic subgroup of S_m of prime power order. Hence $s \leq m$. Let $v'' = \dim V''$. Then we have obtained the following relation:

$$d-1 \leq d-v'' + 1 \leq s \leq m \leq d + k \leq d + 2 \tag{16}$$

where $0 \leq k \leq 2$ and $0 \leq v'' \leq 2$. Then $m-3 \leq s \leq m$. But then a maximal cyclic subgroup of S_m which contains $\langle b_s \rangle$ has order at most js where $1 \leq j \leq 3$. When $O_\infty(K_1) \neq 1$, $m = d$, and (16) implies that $v'' = 1$. Then G_1 is given by case (c) or (g). Consequently $(t,q-1) = 1$. Then

$\langle b \rangle \cap O_\infty(K_1) = 1$. So as $\langle b \rangle \subseteq C_{K_1}(b_s)$,

$$t = js. \tag{17}$$

By combining (16) and (17), we will obtain contradictions. Take first the case that $v'' = 2$. Then K_1 satisfies (1e) and is a subgroup of $O^+(2n,q)$. By (16) or (17) and Table I, we obtain

$$2n-1 \leq s = t/j = \frac{(q-1,2)}{j}(q^{n-1}+1) \leq 2n+2. \tag{18}$$

The middle term increases exponentially with n. It is easy to check that there is only one solution, namely $(q,n) = (2,4)$ since $n \geq 4$. In this case, $s = 9$ and $t = 18$. Hence by (16), $9 \leq m \leq 10$. But neither S_9 nor S_{10} contains an element of order 18.

Take next the case $v' = 1$. Then G_1 satisfies (1c) or (1g) and K_1 is a subgroup of $GU(2n,q)$ or $GO(2n+1,q)$. From (16), $m-2 \leq s \leq m$. Now from (17), $s = t$ or $t/2$. However, (16) rules out the possibility that $s = \frac{1}{2}(2^5+1)$ or $\frac{1}{2}(3^3+1)$ corresponding to the choices $(q,n) = (2,3)$ when $G_1 \cong GU(2n,q)$ and $(3,3)$ when $G_1 \cong GO(2n+1)$.

Finally consider that $v' = 0$. Then from (16)

$$d+1 \leq s \leq m \leq d+2. \tag{19}$$

As $m-s \leq 1$, (17) implies that $s = t$. In case (a), this implies that $s = (q^d-1)/(q-1)$ where $d \geq 3$, and this contradicts (19). Otherwise $s = t = q^n+1$ when $d = 2n$ or $q^{2n+1}+1$ when $d = 2n+1$. As s is odd, q is even. Then a solution occurs when $(q,d) = (2,4)$, in which case $G_1 \cong GSp(4,2)$ and $m = 6$. But it is well-known that $GSp(4,2) \cong S_6$. So $K_1 = G_1$ and (14) holds in this case. In all other cases, (16) is contradicted.

iii) In the cases we next consider, K_1 is defined as a subgroup of G_1 where G_1 itself is defined over F_q where $q = p^k$ for some k. The values of k are determined by congruences on p listed in the table, and the groups presented in the table are those having this property given by [6; Theorem II], [11], [12; III, Appendix]. Now in all cases $s_q(G_1)$ divides $|K_1|$ by (15). But $s_q(G_1) = s(p^k,m) = s(p,km) \geq km+1$. Hence it is sufficient to consider only the smallest possible values for k in ruling out these cases. Table II provides this information. The groups G_1 listed in this table are only those allowed by (1), and subject to this restriction they are the smallest classical group defined over $F_p k$ containing K_1. When there is more than one possibility for K_1 as a subgroup of a given group only the maximal such example need be considered. For this reason, the group $O^+(4,3)$ does not appear in Table II although it is listed in [11]. Only cases where K_1 is not a normal subgroup of G_1 are listed.

TABLE II

K_1	G_1	$\|K_1\|$	$s_q(G_1)$
U(3,2)	GL(3,p), $p \equiv 1 \pmod 3$	$2^3\ 3^4$	s(p,3)
	GU(3,p), $p \equiv -1 \pmod 3$		$s(p^2,3) = s(p,6)$
U(4,2)	GL(4,p), $p \equiv 1 \pmod 3$	$2^6 \cdot 3^5 \cdot 5$	s(p,4)
	GU(4,p), $p \equiv -1 \pmod 3$		$s(p^2,4) = s(p,8)$
$3 \cdot A_6$	GL(3,4)	$2^3 \cdot 3^2 \cdot 5$	s(4,3) = 9
$(Z_4 \circ Q_8 \circ Q_8)A_6$	GL(4,p), $p \equiv 1 \pmod 4$	$2^8 \cdot 3^2 \cdot 5$	s(p,4)
	GU(4,p), $p \equiv 3 \pmod 4$		$s(p^2,4) = s(p,8)$
$RO^+(4,5)$	GL(4,p), $p \equiv \pm 1 \pmod 5$	$2^6 \cdot 3^7 \cdot 5^2$	s(p,4)
	GU(4,p), $p \equiv \pm 2 \pmod 5$		$s(p^2,4) = s(p,8)$
ΣL(3,4)	GU(4,3)	$2^7 \cdot 3^4 \cdot 5 \cdot 7$	s(9,4) = 41
PΩ(5,3)	GO(5,p), $p \equiv 1 \pmod 3$	$2^6 \cdot 3^4 \cdot 5$	s(p,4)
	GU(5,p), $p \equiv -1 \pmod 3$		$s(p^2,4) = s(p,8)$
PO(5,3)	$GO_1(6,p)$, $p > 2$	$2^6 \cdot 3^4 \cdot 5$	s(p,4), s(p,6)
$3 \cdot R\Omega^-(6,3)$	GU(6,p), $p \equiv -1 \pmod 3$	$2^9 \cdot 3^7 \cdot 5 \cdot 7$	$s(p^2,6) = s(p,12)$
	$GO^-(6,p)$, $p \equiv 1 \pmod 3$		s(p,6)
Sp(6,2)	GO(7,p), $p>2$	$2^9 \cdot 3^4 \cdot 5 \cdot 7$	s(p,6)
$2 \cdot O^+(8,2)$	$GO_1(8,p)$, $p>2$	$2^{14} \cdot 3^5 \cdot 5^2 \cdot 7$	s(p,6), s(p,8)

$GO_1(n,p)$ denotes the orthogonal group determined by a form of discriminant 1. The two choices for $s(p,m)$ for this group are made according as it has type $GO^+(n,p)$ or $GO^-(n,p)$. The notation $n\cdot X$ denotes a central extension of X by Z_n. The groups $3\cdot A_6$ and $3\cdot R\Omega^-(6,3)$ are generated by transvections over a field of characteristic 2. The groups $U(3,2)$ and $U(4,2)$ are generated by homologies of order 3. The remaining groups are generated by reflections.

W. Feit has defined a large Zsigmondy prime s to be a Zsigmondy prime for which $|p^m-1|_s > m+1$. (Here is it not necessary to have p prime.) When $p \geq 3$ and $m \geq 3$, Feit [4] has refined the argument of Artin [1] to show that large Zsigmondy primes exist except in the case $(p,m) = (3,4)$ or $(3,6)$.

We have defined $S(q,m)$ so that it is a large Zsigmondy prime whenever $q \geq 3$, $m \geq 3$, and $(q,m) \neq (3,4)$ or $(3,6)$. Then because of (15), we may conclude that $|B_s| > m+1$ when $s = s(p,m)$ except when $(p,m) = (3,4)$ or $(3,6)$, and $|B_s| \geq m+1$ in all cases. Using these conditions together with the conditions on p in the second column only four possibilities for (K_1,G_1) remain, namely, $(3\cdot A_6, GL(3,4))$, $(PO(5,3), GO^+(6,3))$, $(Sp(6,2), GO(7,3))$, and $(O^+(8,2), GO(8,3))$. The case $(PO(5,3), GO^+(6,3))$ does not occur as we only consider $GO^+(2n,q)$ when $n \geq 4$.

In the first case, $|B| = 63$. Hence B contains an element b_p of order 7, which is primitive. By Lemma 3, b_p induces a nontrivial automorphism on K_1. But A_6 has no elements of order 7. In the last two cases, we see from Table I that $|B/Z(G_1)| = 3^3+1 = 28$. Also $|Z(G_1)| = 2$. Thus K_1 admits the faithful action of a cyclic group of order 28. But in $Sp(6,2)$, an element of order 7 is self-centralizing by virtue of its action on the 6-dimensional space over F_2 on which $Sp(6,2)$ is defined. To investigate $2\cdot O^+(8,2)$, we represent it as the group E_8/Z where E_8 denotes the Euclidean reflection group and Z represents its center, which has order 2. The centralizer of a generating reflection is well-known to be the Euclidean reflection group of type A_7, and so the centralizer has the shape $Z_2 \times S_8$. In this group the centralizer of an element of order 7 does not contain an element of order 4. This contradiction eliminates the possibilities in Table II.

iv) There now remain the cases where K_1 is an irreducible classical group acting on V. The irreducibility of K_1 implies that V may be regarded as F^*-vector space where F^* is the extension field of F

which is the centralizer of K_1 in $\mathrm{Hom}_F(V,V)$. But as a subgroup of $GL(V)$, F^* is represented completely reducibly with irreducible subspaces of dimension $[F^*:F]$. Such transformations never commute with reflections or transvections as they leave no 1-dimensional subspace invariant. When $[F^*:F] > 1$, then K_1 is defined over the same field F that G_1 is defined. On the basis of the group order formulas (11), this implies that

$$s_q(G_1) = s_q(K_1) \tag{20}$$

where $q = |F|$.

From the values of $s_q(G_1)$ given in (13), there are only four anomalous cases for K_1 on the basis of (20) using the additional condition that both G_1 and K_1 are defined on the same space. Three of these cases arise when $G_1 = GL(m,q)$:

α) $K_1 = SU(m,q)$, $m = 2n+1 \geq 3$, $|F| = q^2$, $s_q(K_1) = s(q,2m)$,

β) $K_1 = Sp(m,q)$, $m = 2n \geq 4$, $|F| = q$, $s_q(K_1) = s(q,m)$,

γ) $K_1 = O^-(m,q)$, $m = 2n \geq 4$, $|F| = q$, $s_2(K_1) = s(q,m)$.

Set $N_1 = N_{G_1}(K_1)$. Then $N_1 \cong GU(m,q)$, $GSp(m,q)$, or $GO^-(m,q)$ in cases (α), (β), (γ), respectively. Then $N_1 \supseteq H_1 = BK_1$. Hence $C_{N_1}(B_s) = C_{G_1}(B_s) = B$. On the other hand, $V = V'$ in all these cases. So by Lemma 3 and Table I, we obtain the contradiction, $|C_{N_1}(B_s)| = (q^m+1)(q-1) < q^{2m}-1 = |B|$.

There remains the case $G_1 \cong GSp(2n,q)$ and $K_1 \cong O^-(2n,q)$. Here q is even as G_1 and hence also K_1 are generated by transvections (cf. Table I). Then $GSp(2n,q) = Sp(2n,q) \times Z_{q-1}$. All inclusions of $O^-(2n,q)$ are known to be conjugate in $GL(V)$ according to [6][1]. Thus we may take K_1 so that it is defined with respect to a quadratic form g_1 given by (8). Without loss of generality we may suppose that G_1 is defined relative to the alternating form (7), which is the polarization of (8). Thus B can be taken to be the group defined by the analysis of case (1c).

We next argue that

[1]The argument can be obtained by studying the relationship between the transvections of K_1 and the underlying geometry.

$$N_{GL(V)}(B_s) = N_{K_1}(B_o)B_o. \tag{21}$$

Indeed, by Lemma 3, $C_{GL(V)}(B_s) = B_o$. Hence B_o is a self-centralizing normal subgroup of $N_{GL(V)}(B_s)$ and $N_{GL(V)}(B_s) \supseteq N_{GL(V)}(B_o)$. Since B_s is a Sylow subgroup of B_σ, $N_{GL(V)}(B_s) = N_{GL(V)}(B_o)$. But $B_o \subseteq M(V) = \mathrm{Hom}_{F_q}(V,V)$ as the multiplicative group of a maximal subfield $E = F_{q^m}$. Thus the elements of $N_{GL(V)}(B_o)$ induce field automorphisms of E, and we obtain an epimorphism

$$1 \longrightarrow B_o \longrightarrow N_{GL(V)}(B_o) \xrightarrow{\theta} \mathrm{Gal}(E/F_q) \longrightarrow 1 \tag{22}$$

Because E is finite, $H^2(\mathrm{Gal}(E/F_q),E) = 1$, and (22) splits. Indeed, for $\sigma \in \mathrm{Gal}(E/F_q)$, let d_σ be the element of $GL(V)$ which satisfies

$$d_\sigma(x(\lambda)) = x(\lambda^\sigma) \tag{23}$$

where $x(\lambda)$ is the element of $V = F_{q^m}/F_q$ satisfying $x(\lambda)1 = \lambda$. Let $D = \{d_\sigma | \sigma \in \mathrm{Gal}(E/F_q)\}$. Then

$$N_{GL(V)}(B_o) = B_oD. \tag{24}$$

But it is now clear that each d_σ leaves invariant the form g_1 defined by (8). Hence $D \subseteq K_1$. Because $N_{GL(V)}(B_s) = N_{GL(V)}(B_o)$, (23) now implies (21).

Now we are in a position to conclude the argument. We cannot rule out the existence of K_1 in this case. Consequently, choose a transvection $a_2 \in G_1 \setminus K_1$, and suppose that $\langle a_2,b\rangle \neq G_1$. Set $H_2 = \langle a_2,b\rangle$ and $K_2 = \langle a_2^{H_2}\rangle$. Then the preceding analysis applies to H_2 and K_2 as well as to H_1 and K_1 providing that a_2 is chosen so that a_2 does not leave invariant the subspaces W_i, $i = 1,2,3$, in (12) when $G_1 \cong Sp(6,2)$. In this case, b_s has order 9, of course, and the subspaces W_1, W_2, W_3 may be taken to be mutually orthogonal elliptic planes, each of which admitsthe action of b_s^3. Because all elements of order 9 are conjugate in K_1, this is the only manner in which such an element acts. Then if a transvection leaves the subspaces W_i invariant, it must fix two of them and act nontrivially on the third, say W_1. But because W_1

is elliptic $O(W_1) \cong GL(2,2) \cong S_3$. Hence a transvection of G_1 leaving invariant W_i, $i = 1,2,3$, belongs to K_1. Since $a_2 \notin K_1$, it now follows that $K_2 = O^-(2n,q)$.

Let t_1 be an element of $GL(V)$ such that $K_2^{t_1} = K_1$, and let $t_2 \in K_1$ be chosen so that $B_s^{t_1 t_2} = B_s$. Set $t = t_1 t_2$. Then $t \in N_{GL(V)}(B_s)$. For the purpose of showing that $t \in K_1$, we now argue that $t \in G_1$. Let g_1' and f_1' be the forms $g_1': x \longmapsto g_1(t^{-1}x)$ and $f_1': (x,y) \longmapsto f_1(t^{-1}x, t^{-1}y)$. Then f_1' is obtained from g_1' by polarization. Because $K_1 = K_2^t$, it follows that $g_1' = \chi g_1^{\sigma}$ where χ is a linear character of K_1 and σ an automorphism of F. Then $\chi^2 = 1$. Hence by polarization $f_1' = f_1^{\sigma}$. Because f_1^{σ} determines the same symplectic group as does f_1, it follows that $G_1^t = G_1$. But $N_{GL(V)}(G_1) = G_1$. So $t \in G_1$. Then $t \in N_{G_1}(B_s)$. But $N_{G_1}(B_s) \cap N_{K_1}(B_s)B_o = N_{K_1}(B_s)(B_o \cap G_1) = N_K(B_s)B = N_{K_1}(B)$. Hence by (21), $t \in N_{K_1}(B_s)$. Then $K_2 = K_1^t = K_1$. This contradicts the choice of a_2. This concludes the proof of the proposition.

3 PROOFS OF THEOREMS I AND III

Proposition 8. Theorem I holds with $\bar{G} = \bar{G}_1$.

Proof. Let $G^* = \langle a,b \rangle$ where a and b are chosen in accordance with Proposition 7. Then Propositions 1 and 7 imply

$$G_1 = G^* Z(G_1). \tag{25}$$

It suffices to show that G^* can be represented as a Galois group in accordance with Theorem II. Because of (25), $G_1 = N_{GL(V)}(G^*) = GC_{GL(V)}(G^*)$. Consequently, it remains to show that $Z(G^*)$ is a direct factor of $N_{G^*}(\langle a \rangle)$. Let $N_a = N_{G_1}(\langle a \rangle)$. Because $Z(G^*) \subseteq Z(G_1)$, it suffices to show that $Z(G_1)$ is a direct factor of N_a. Let V_1 be the center of A. Then it is a center for all the elements of $\langle a \rangle$. Hence N_a leaves V_1 invariant. Let ρ be the restriction homomorphism $\rho: N_a \longrightarrow GL(V_1)$. But $\rho(Z(G_1)) = GL(V_1))$. By factoring this restriction homomorphism, we obtain a split exact sequence

---

[1]Let ϕ denote an isomorphism of K_1 onto K_2 and denote by ϕ_t the isomorphism induced by the conjugation $k \mapsto k^t$. Then $\phi\phi_t$ is an automorphism of K_1, and the characterization of the automorphisms of $O^-(2n,q)$ gives this result.

$$1 \longrightarrow M_a \longrightarrow N_a \longrightarrow Z(G_1) \longrightarrow 1 \tag{26}$$

Since $Z(G_1)$ is central, it is a direct factor. This proves the proposition.

Proposition 9. Theorem I holds for the groups $\bar{G}$ listed in (1). Furthermore, the following linear groups G can be represented as $Gal(K/L_{G*})$ where L_{G*} is the cyclotomic number field determined by $G*$ in Proposition 8.

$$GL(n,q) \supseteq G \supseteq SL(n,q), \quad n \geq 2 \tag{27a}$$

$$U(n,q) \supseteq G \supseteq SU(n,q), \quad n \geq 3,\ n \neq 4 \tag{27b}$$

$$GSp(2n,q),\ Sp(2n,q), \quad n \geq 2 \tag{27c}$$

$$GO^-(2n,q),\ O^-(2n,q),\ R\Omega_2^-(2n,q),\ SO_2^-(2n,q),\ \Omega_2^-(2n,q),\ n \geq 3 \tag{27d}$$

$$GO^+(2n,q),\ O^+(2n,q),\ R\Omega_2^+(2n,q),\ SO_2^+(2n,q),\ \Omega_2^+(2n,q),\ n \geq 4 \tag{27e}$$

$$SO(2n+1,q),\ \Omega(2n+1,q), \quad n \geq 3,\ q \text{ odd}. \tag{27f}$$

Proof. Denote images of linear groups in projective groups by bars, and $G* = \langle a,b \rangle$. Let $Z* = G* \cap Z(G_1)$. In order to prove Theorem I, take $G \subseteq G*$ so that $\bar{G}$ satisfies (1x), x = a,b,c,d,e,f,g. By (3), (4), and (5), $\bar{G}_1/\bar{G}$ is either dihedral or cyclic and thus the same is true for $G*/GZ*$. When $G*/GZ*$ is dihedral, it follows from Theorem B of Feit's "Rigidity and Galois Groups" that G as well as $G*$ is represented as $Gal(K/L_{G*})$ where L_{G*} is the cyclotomic number field determined by $G*$. But when $G*/GZ*$ is cyclic of order m, the same is true. Indeed, the index of ramification of the cusps in this case will be $e_1 = e_2 = m$ and $e_3 = 1$. Using Hurwitz's formula with these values, it follows that G is the covering group of an algebraic curve of genus 0, which is what is required.

To obtain that the groups G listed in (27) are representable as $Gal(K/L_{G*})$, it suffices to show that $G \triangleleft G*$ and $G*/G$ is dihedral or cyclic or that $G \cap Z* = 1$. In (27a) and (27b), $G* \cong GL(n,q)$ and $U(n,q)$ respectively; so $G*/G$ is cyclic. In (27c), $G* = GSp(V)$; so take

$G = Sp(V)$. Then $G^* = BG$; so G^*/G is cyclic. In the remaining cases, we may assume that $G \cap Z^* \neq 1$. This eliminates (27f) and also gives the result in (27d) and (27e) when $G_1 \cong GO_2^{\pm}(2n,q)$ since the G^*/GZ^* is a four subgroup in this case from which the four subgroups in (27d) and in (27e) are obtained. But when $G^* = GO_1^{\pm}(V)$, $O_1^{\pm}(V)$ is the only normal complement to B. There are no other normal subgroups with a cyclic or dihedral factor group since $|B| > 2$. This proves the proposition.

By slightly altering our choices for a,b, we obtain the following results.

Theorem III. The following classical groups G are generated by two elements a,b, one of which is a homology or transvection.

$$GL(n,q) \supseteq G \supseteq SL(n,q), \quad n \geq 2 \tag{28a}$$

$$U(n,q) \supseteq G \supseteq SU(n,q), \quad n \geq 3, \; n \neq 4 \tag{28b}$$

$$GSp(2n,q), \; Sp(2n,q), \quad n \geq 2 \tag{28c}$$

$$GO^-(2n,q), \; R\Omega^-(2n,q), \quad n \geq 3 \tag{28d}$$

$$GO^+(2n,q), \; R\Omega^+(2n,q), \quad n \geq 4 \tag{28e}$$

$$O(2n+1,q), \; \Omega(2n+1,q), \quad n \geq 3, \; q \text{ odd} \tag{28f}$$

Proof. The group $G^* = \langle a,b \rangle$ discussed in the previous proposition is the first listed group in each of the above cases. In cases (28a) and (28b), the remaining groups except for $SL(n,q)$ and $SU(n,q)$. But these last two groups can be obtained using a transvection rather than a homology. This requires the use of [6] rather than [10] for the characterization of K_1. Table I remains the same and the argument given in the paper can be repeated verbatim with this change.

To obtain $Sp(V) \cong Sp(2n,q)$, set $B_1 = B \cap Sp(V)$ and let $B_1 = \langle b_1 \rangle$. Then $Sp(V) = \langle a,b_1 \rangle$ by the argument of Proposition 7. The important point is that $B_s \subseteq B_1$. A similar replacement obtains $R\Omega^-(2n,q)$, $R\Omega^+(2n,q)$ and $\Omega(2n+1,q)$.

4 RATIONALITY

It has been shown by Belyĭ [3] and also by J. Thompson [8] that the cyclotomic field L_G mentioned in Theorem II is the field $K_{a,b,ab} =$

$Q[\chi_u(a), \chi_u(b), \chi_u(ab), u = 1,2,\ldots,k]$ where $\{\chi_u | u = 1,2,\ldots,k\}$ is the set of irreducible characters of G. Belyĭ's result provides rigidity without specifying the element ab. This is a distinct advantage, but the price paid for it is that it is not possible to calculate L_G in general without this knowledge. In [9], Thompson meets this problem directly. In particular, he shows that $L_G = \mathbb{Q}$ when $G = O^-(\ell-1,2)$ and $O^+(\ell-1,2)$ where χ is a prime for which 2 is a primitive root.

However, the characters of the groups Sp(4,3), Sp(6,2) and $O^+(8,2)$ are known to be rational. (These are the Euclidean reflection groups modulo centers.) Thus $L_G = \mathbb{Q}$ in these cases, a result which is known. However, the same is true for $\Omega^+(8,2)$, which was not previously known to be a Galois group over $\mathbb{Q}$. (This follows from the use of Hurwitz's formula as explained above.)

REFERENCES

1. Artin, E. (1955). The orders of the linear groups. Comm. Pure and Applied Math. 8, 355-366.
2. Artin, E. (1955). The orders of the classical simple groups. Comm. Pure and Applied Math. 8, 455-472.
3. Belyĭ, G.V. (1980). On Galois extensions of a maximal cyclotomic field. Math. USSR Izveztia. Amer. Math. Soc. Trans. 14, 247-256.
4. Feit, W. Extensions of cuspidal characters of $GL_m(q)$. To appear.
5. Huppert, B. & Blackburn, N. (1982). Finite Groups II. Springer Verlag.
6. Kantor, W. (1979). Subgroups of classical groups generated by long root elements. Trans. Amer. Math. Soc. 248, 347-379.
7. Serezkin, V.N. & Zaleskii, A.E. (1981). Finite linear groups generated by reflections. Math USSR Izvetzia. Amer. Math. Soc. Trans. 17, 477-503.
8. Thompson, J.G. Some Finite groups which appear as Gal(L/K) where $K \subseteq Q(\mu_m)$. To appear in J. Alg.
9. Thompson, J.G. Primitive roots and rigidity. These Proceedings.
10. Wagner, A. (1978). Collineation groups generated by homologies of order greater than 2. Geom. Dedicata. 7, 387-398.
11. Wagner, A. Determination of the finite primitive reflection groups over an arbitrary field of characteristic not 2. Geom. Dedicata. I, 9 (1980), 239-253. II, 10 (1981), 183-189. III, 10 (1981), 475-523.

SOME REMARKS ON THE PRINCIPAL IDEAL THEOREM

R. Foote
Department of Mathematics, University of Vermont
Burlington, VT 05401

Following the completion of the classification of finite simple groups, a number of natural avenues of research lay open to group theorists whose efforts were directed by the classification program. Although I intend to participate in the revision scheme led by Gorenstein and Lyons I also plan to channel my research into those areas of algebraic number theory where finite groups play a non-trivial role. At this early stage it is not clear to me whether some, or indeed any, of the wealth of techniques developed during the classification will be applicable to the arithmetic problems I am interested in; nonetheless, I believe that the perspectives that group theorists have achieved through their intimacy with both the local intricacies and the global structural generalities occuring in finite groups will prove to be a valuable asset. Ultimately it is the interplay between number theory and group theory which should lead to the most fruitful veins of research. This note, however, which sketches my initial explorations, is chiefly group theoretical.

The Principal Ideal Theorem (P.I.T.) asserts: for any finite group G,

the transfer homomorphism $V:G \to G'/G''$ is the trivial map.

This was proved originally by P. Furtwangler in [5] and established via different methods by a number of other mathematicians; a more detailed account of the history and (sketches of) the techniques of these proofs may be found in [1, pp. 122-133]. The Principal Ideal Theorem was formulated to verify the (last to be proven) class field theory conjecture of Hilbert: if K is a number field, $K^{(1)}$ its Hilbert class field, then

every fractional ideal of K becomes principal when extended to $K^{(1)}$.

In [7, Chapter V.13] the reduction E. Artin achieved in 1927 of Hilbert's conjecture to the group theoretic statement of the P.I.T. is given; the link comes from G being the Galois group of $K^{(2)}$ over K ($K^{(2)}$ is the Hilbert class field of $K^{(1)}$), whence $G' = \mathrm{Gal}(K^{(2)}/K^{(1)})$, $G/G' = \mathrm{Gal}(K^{(1)}/K)$, $G'' = 1$, and the transfer $G/G' \to G'$ corresponds to extension of ideal classes.

My explorations of class field theory beginning from the P.I.T. viewed chiefly from its group theoretic formulation have (initially) two main goals: to acquire a better understanding of why this transfer must be trivial, and to discover conditions under which the transfer $G \to H/H'$ is trivial, for other subgroups H of G. First, some focusing is in order: since G'' plays no role in these questions we may factor it out, so G' is henceforth abelian. Furthermore, most of these questions about transfer reduce to the same questions for G a p-group, p a prime, so G is also assumed to be a p-group. Fix notation: $\bar{G} = G/G' = \langle\bar{x}_1\rangle\times\langle\bar{x}_2\rangle\times\ldots\times\langle\bar{x}_n\rangle$, $|\bar{x}_i| = e_i$, x_i a preimage of $\bar{x}_i$, so G' is a $\mathbb{Z}\bar{G}$-module spanned by $\{[x_i,x_j] \mid 1 \le i < j \le n\}$. The usual formula for the transfer [6, Theorem 7.3.3] gives that if $\bar{H} = \langle\bar{x}_2,\bar{x}_3,\ldots,\bar{x}_n\rangle$,

$$V(x_1) = \prod_{h\in\bar{H}} (x_1^{e_1})^h .$$

It is not clear from the proofs of the P.I.T. occuring in the literature whether the vanishing of the transfer is a purely "combinatorial" phenomenon (i.e. the above expression for $V(x_1)$ can be reduced to the identity by formal manipulation of the relations which define G as a metabelian group), or whether it can be deduced from the $\mathbb{Z}\bar{G}$-module structure of G'. Not unlike such results as the Cayley-Hamilton Theorem, the truth appears to lie somewhere between these two reasons. To back this assessment up two special cases of the P.I.T. are discussed here: a complete ("combinatorial") proof is given when $n = 2$, and a ("structural") proof is outlined when G' is elementary abelian. Both of these arguments appear to be dissimilar to proofs in the literature.

<u>Lemma 1</u>. For $x,y \in G$, $a \in \mathbb{Z}^+$,

i) $[x,y^a] = [x,y]^{1+y+y^2+\ldots+y^{a-1}}$, and

ii) $[x^a,y] = [x,y]^{1+x+x^2+\ldots+x^{a-1}}$.

Proof. These known formulas can be checked directly by induction on a (remember $G'' = 1$).

Theorem 1. If $n = 2$, V is the trivial homomorphism.

Proof. Since G is a p-group, $G = \langle x_1, x_2 \rangle$, $G' = \langle [x_1,x_2]^\alpha \mid \alpha \in \mathbf{Z}\bar{G} \rangle$, and, by symmetry, it suffices to prove $V(x_1) = 1$. Let $x_i^{e_i} = [x_1,x_2]^{\alpha_i}$, $\alpha_i \in \mathbf{Z}\bar{G}$, $i = 1,2$. Thus

$$
\begin{aligned}
V(x_1) &= (x_1^{e_1})^{1+x_2+\ldots+x_2^{e_2-1}} \\
&= [x_1,x_2]^{\alpha_1(1+x_2+\ldots+x_2^{e_2-1})} \\
&= [x_1,x_2]^{(1+x_2+\ldots+x_2^{e_2-1})\alpha_1} && \text{since } \bar{G} \text{ is abelian,} \\
&= [x_1,x_2^{e_2}]^{\alpha_1} && \text{by Lemma 1,} \\
&= [x_1,[x_1,x_2]^{\alpha_2}]^{\alpha_1} \\
&= [x_1,[x_1,x_2]^{\alpha_1}]^{\alpha_2} && \text{since } \bar{G} \text{ is abelian,} \\
&= [x_1,x_1^{e_1}]^{\alpha_2} = 1.
\end{aligned}
$$

Theorem 2. If G' is elementary abelian, V is the trivial homomorphism.

Sketch of Proof. Let G be a minimal counterexample so wlog $V(x_1) \neq 1$. By minimality $|Z(G) \cap G'| = p$. Let $b = x_1^{e_1}$, $B = \langle b^{\bar{H}} \rangle$ so B is a cyclic $F_p\bar{H}$-module which is not annihilated by the trace element of $\bar{H}$; thus B is a free $F_p\bar{H}$-module of dimension $r = |\bar{H}|$ and $B \leq C_{G'}(x_1)$. Since free $F_p\bar{H}$-modules are injective (see [3, Theorem 62.3]) and $C_{G'}(x_1)$ admits $\bar{H}$, $C_{G'}(x_1) = B \times D$, for some $F_p\bar{H}$-module D; however, $Z(G) \cap G'$ being cyclic forces $D = 1$ and $C_{G'}(x_1) = B$.

Next let $1 = h_1, h_2, \ldots, h_r$ be representatives of $\bar{H}$, let $a_i = [x_1,h_i]$, $b_i = [b,h_i]$ so $B_0 = \langle b_2, b_3, \ldots, b_r \rangle$ is isomorphic as an $F_p\bar{H}$-module to the augmentation ideal of $F_p\bar{H}$ and has dimension $r-1$. Let $\gamma = \bar{1}+\bar{x}_1+\bar{x}_1^2+\ldots+\bar{x}_1^{e_1-1}$ and let C_i be the $F_p\langle\bar{x}_1\rangle$-module generated by a_i, $2 \leq i \leq r$. By Lemma 1 $a_i^{\gamma} = b_i$ $(\neq 1)$ so, as argued above, C_i is a free $F_p\langle\bar{x}_1\rangle$-module; furthermore, since the unique one dimensional $F_p\langle\bar{x}_1\rangle$-submodules $\langle b_i\rangle$ of C_i span an $r-1$ dimensional space, $A = \langle C_2, C_3, \ldots, C_r\rangle = C_2 \times C_3 \times \ldots \times C_r$. Since $\dim C_{G'}(x_1) = r$ and A is injective $G' = A \times A_0$, where A_0 is a non-trivial cyclic $F_p\langle\bar{x}_1\rangle$-module; in particular,

$$r+1 \leq \dim_{F_p} G' \leq r + p^{e_1}.$$

Since A may not admit $\bar{H}$ some further refinement of the module structure of G' is necessary. In the end, a contradiction comes from the facts that $\tau = \sum_{h \in \bar{H}} h$ annihilates an $F_p\bar{H}$-submodule A^* of G' of the same dimension as A (A^* is a direct sum of $F_p\bar{H}$ augmentation ideals) and τ also annihilates all commutators $[x_i, x_j]$ with $2 \leq i,j \leq n$ (this can be deduced from Lemma 1); the details involve a few pages of argument of an elementary nature (which show $b \in \langle A^*, [x_i, x_j], 2 \leq i,j \leq n\rangle$).

Unfortunately, neither of the proofs of Theorems 1 or 2 seem to extend to yield the general case, although the possibility of recovering the full P.I.T. by these methods has not been ruled out. Their strength may lie in generalizations of the P.I.T. even for only a restricted class of groups.

E. Artin's argument relating the extension of ideal classes to transfer shows more generally that if $G = \mathrm{Gal}(K^{(2)}/K)$, $G' \leq A \leq G$ with $A' = 1$, and L is the fixed field of A, then every ideal of K becomes principal when extended to L if and only if the transfer $V_A: G \to A$ is the trivial homomorphism. In practice, arithmetic information about L (which is a subfield of $K^{(1)}$) is much more easily computed than properties of the fixed field of $V_A(G)$ (which is an extension of $K^{(1)}$ within $K^{(2)}$), so any group theoretic theorems restricting $V_A(G)$ in terms of G/A and A/G' would be in the right direction.

One open result which appears accessible by the above methods is: if G is an arbitrary p-group, $G' \leq A \leq G$ with $A' = 1$ and $V_A: G \to A$ the transfer homomorphism, then $|V_A(G)| \leq |A:G'|$. Indeed, a slight modification to the end of the proof of Theorem 2 extends it to give this conjecture when G' is elementary provided $A \leq \phi(G)$. Furthermore, this conjecture should at least be easier to investigate in

2-generator groups where, as demonstrated by the proof of Theorem 1, the commutator calculations are relatively simple. Viable conjectures on the relationship between the isomorphism types of G/A and A/G' to $V_A(G)$ are more difficult to formulate. Indeed, it may be demonstrable that (other than restrictions such as those described above) no non-trivial relationship need hold (negative results of this nature are established in [2]); this possibility must also be studied.

An additional avenue to explore is the applicability of the techniques of the proof of Theorem 2 to Demuskin groups. A profinite p-group P is a Demuskin group if

i) $\dim H^1(P) < \infty$,

ii) $\dim H^2(P) = 1$, and

iii) the cup product $H^1(P)\times H^1(P) \to H^2(P)$ is a non-degenerate bilinear form.

Such groups occur, in particular, as Galois groups of maximal p-extensions of k, where k is a finite extension of $\mathbb{Q}_p$ which contains all p^{th} roots of 1 (see [8, p. II-30]). A Demuskin group P has the property that for each $x \in P$ there is an open normal subgroup U of finite index in P such that the transfer $P \to U/\phi(U)$ is trivial at x; conversely, if P has cohomological dimension 2, this "vanishing" property forces P to be a Demuskin group (see [8, pp. I-85, I-86]). In fact, in the case when P is the Galois group described above, the transfer $P \to \phi(P)/\phi(\phi(P))$ is identically trivial. Interesting families of pro p-groups with certain "transfer vanishing" properties thus arise in natural contexts, so it would be instructive to see what structural restrictions these properties impose on larger classes of pro p-groups, say, those of finite cohomological dimension. In particular, new characterizations of Demuskin groups such as an alternate proof of the Main Theorem of [4] may be discovered. The methods used to prove Theorem 2 seem to apply to these problems since, in a minimal counterexample framework where the transfer of a group into a normal elementary abelian section is non-trivial, some of the module structure of this section can be pinned down.

Finally, I reiterate that in all the possible research problems I described the number theoretic context should not be discarded. In the event that abstract group theoretic generalizations of the P.I.T. are hard to come by, returning G to its role as the Galois group of $K^{(2)}$ over K may give enough arithmetic information (or coprime action, if K is Galois over some subfield) to make significant progress.

REFERENCES

1. Chandler, B. & Magnus, W. (1982). The History of Combinatorial Group Theory: A Case Study in the History of Ideas. New York: Springer-Verlag.
2. Chang, S.-M. & Foote, R. (1980). Capitulation in class field extensions of type (p,p). Canadian J. Math. 32, 1229-1243.
3. Curtis, C. & Reiner, I. (1966). Representations of Finite Groups and Associative Algebras. New York: Wiley.
4. Dummit, D. & Labute, J. (1983). On a new characterization of Demuskin groups. Inventiones Math. 73, 413-418.
5. Furtwangler, P. (1930). Beweis der Hauptidealsatzes fur Klassenkörper algebraischer Zahlkörper. Abh. Math. Sem. Univ. Hamburg, 7, 14-36.
6. Gorenstein, D. (1968). Finite Groups. New York: Harper and Row.
7. Janusz, J. (1973). Algebraic Number Fields. New York: Academic Press.
8. Serre, J.-P. (1965). Cohomologie Galoisienne. Springer Lecture Notes #5.

MODULAR FORMS AND THE THEORY OF THOMPSON SERIES

G. Mason*†
Department of Mathematics, University of California, Santa Cruz
California 95064

1 INTRODUCTION

This paper is a brief survey of some results obtained recently by the author concerning some connections between finite groups and modular forms -- a subject currently languishing under the rubric of "moonshine." This is a somewhat unfortunate title, as there appears to be emerging a smattering of *theory*, as opposed to mere observation and conjecture, which may prove useful in understanding the origin of such connections.

One of the stated goals of the Rutgers Conference and of these Proceedings is to present problems and potential areas of research for today's group-theorist, and with regard to this one may simply refer to the original papers of Thompson [8], [9] and Conway-Norton [1] and ask for an explanation. It was already made clear during the Santa Cruz Conference of 1979 that here was a field ripe for cultivation -- I refer in particular to the last few lines of Ogg's article in [5], and indeed there has been a great deal of interest in "Monstrous Moonshine." Real progress, however, has been slow, and we seem as far from an understanding today as we were at the advent of Moonshine some five or more years ago.

To the extent that it was practiced by the original moonlighters, "Moonshine" meant connections between the Monster and modular functions arising from certain function fields of genus zero. In this paper we wish to raise the question: how general are these phenomena? This entails replacing the Monster by an arbitrary group G and replacing modular functions by modular forms of various levels and (integral) weights. The notion of *Thompson Series for* G embodies the connections between G and the modular forms and they are the main object of study.

---

*Much of the research reported on here was performed during the author's tenure at the Max Planck Institut für Mathematik, Bonn.

†Research also supported by a grant from the National Science Foundation.

One may think of a Thompson Series as an "equivariant modular form," and the question arises to what extent does the theory of modular forms carry over to this more general situation? The main result of this paper is that, under suitable circumstances, certain Thompson series admit a theory of <u>Hecke operators</u>.

The paper is organized as follows: in Section 2 we reproduce some of the standard facts concerning modular forms, through Hecke operators. In Section 3 we give the definition of Thompson series together with some important examples, and state a result (Theorem 1) which can be regarded as an existence proof of moonshine for an arbitrary finite group. In Section 4 we discuss Hecke operators for Thompson series. Proofs are invariably omitted: we hope to publish a fuller account of these matters elsewhere.

2 MODULAR FORMS

First we have the upper half plane

$$\mathfrak{H} = \{z \in \mathbb{C} \mid \operatorname{im} z > 0\}.$$

There is a natural left action of $GL_2^+(\mathbb{R})$ (group of 2×2 real matrices with positive determinant) on $\mathfrak{H}$ given by

$$z \longmapsto \frac{az+b}{cz+d} \quad \text{for} \quad \begin{pmatrix} a & b \\ c & d \end{pmatrix} \in GL_2^+(\mathbb{R}).$$

Now let k be a non-negative integer. Then there is a right action of $GL_2^+(\mathbb{R})$ on the space of functions on $\mathfrak{H}$ called the "kth stroke operator." Namely, if $\alpha = \begin{pmatrix} a & b \\ c & d \end{pmatrix} \in GL_2^+(\mathbb{R})$ and if f is a function on $\mathfrak{H}$ we write

$$f \longmapsto f|_k\alpha$$

where

$$f|_k\alpha = (\det \alpha)^{k/2}(cz+d)^{-k} f(\alpha z).$$

Certain discrete subgroups of $GL_2^+(\mathbb{R})$ play a central role. We set

$$\Gamma = SL_2(Z)$$

and for each positive integer N set

$$\Gamma(N) = \{\alpha \in \Gamma | \alpha \equiv I(\text{mod } N)\}$$

$$\Gamma_0(N) = \left\{\begin{pmatrix} a & b \\ c & d \end{pmatrix} \in \Gamma | c \equiv 0(\text{mod } N)\right\}$$

$$\Gamma_1(N) = \left\{\begin{pmatrix} a & b \\ c & d \end{pmatrix} \in \Gamma_0(N) | a \equiv d \equiv 1(\text{mod } N)\right\}.$$

(Here I is the identity matrix.) We note that there is a diagram

$$\begin{array}{ccccccccc} 1 & \longrightarrow & \Gamma(N) & \hookrightarrow & \Gamma & \xrightarrow{\nu} & SL_2(Z/NZ) & \longrightarrow & 1 \\ & & \uparrow{\scriptstyle id} & & \uparrow & & \uparrow & & \\ 1 & \longrightarrow & \Gamma(N) & \hookrightarrow & \Gamma_0(N) & \xrightarrow[\nu]{} & B & \longrightarrow & 1 \end{array}$$

where each row is short exact, ν is reduction modulo N, and B is a Borel subgroup (= upper triangular matrices) of $SL_2(Z/NZ)$.

We note also that as a group of transformations of $\mathfrak{h}$, $GL_2^+(\mathbb{R})$ is not faithful but has as kernel the group of scalar matrices. In general we shall fail to distinguish between elements of $GL_2^+(\mathbb{R})$ as either matrices or transformations of $\mathfrak{h}$.

Generally we have to deal with subgroups G of $GL_2^+(\mathbb{R})$ which are commensurable with Γ, that is $\Gamma \cap G$ has finite index in both G and Γ. In this situation we set

$$Y_G = G\backslash\mathfrak{h},$$

the orbit space of G on $\mathfrak{h}$. Then Y_G is a Riemann surface of finite volume which can be completed to a compact Riemann surface, which we denote by X_G, by adjoining <u>cusps</u>. We will not go into details concerning the analytic structure, but in all cases which concern us $i\infty$ will be a cusp and the corresponding local variable is $q_0 = e^{2\pi iz/c}$ where c is the smallest positive real number such that $z \to z+c$ is a transformation in G. In particular, we set

$$X_N = X_{\Gamma(N)},\ X_0(N) = X_{\Gamma_0(N)},\ X_1(N) = X_{\Gamma_1(N)},$$

and note that* $q = e^{2\pi iz}$ is the appropriate local variable at $i\infty$ on $X_1(N)$ and $X_0(N)$.

A <u>meromorphic modular form for</u> G <u>of weight</u> k is a function f on $\mathfrak{h}$ such that the following conditions hold:

i) $f|_k\alpha = f$ for all $\alpha \in G$.

ii) f is meromorphic in $\mathfrak{h}$.

iii) f is meromorphic at the cusps of X_G.

We shall dispense with a precise discussion of what (iii) actually means, only noting that if $i\infty$ is a cusp then f will have a Fourier expansion at $i\infty$ of the shape

$$f(z) = \sum_{n \geq M} a(n)q_o^n, \text{ some } M \in \mathbb{Z}.$$

We obtain a <u>holomorphic modular form for</u> G by replacing "meromorphic" by "holomorphic" in the preceding.

If $k = 0$, condition (i) means exactly that f is G-invariant, that is

$$f(\alpha z) = f(z) \quad \text{for} \quad \alpha \in G.$$

We denote by $F(G)$ the field of all such G-invariant meromorphic functions on X_G, setting

$$F_0(N) = F(\Gamma_0(N)).$$

For $k > 0$ we denote by

$$M(G,k)$$

the space of holomorphic modular forms for G of weight k. It is a finite-dimensional vector space over $\mathbb{C}$. Of special importance here is the case $G = \Gamma(N)$. As $\Gamma(N) \trianglelefteq \Gamma$ there is a natural action of Γ on $M(\Gamma(N),k)$, and we are particularly interested in the semi-invariants of $\Gamma_0(N)$. If f is such a semi-invariant then $\mathbb{C}f$ affords a linear character of $\Gamma_0(N)/\Gamma(N)$, in particular f is invariant under $\Gamma_1(N)$ and

---

*q has this meaning throughout the present paper.

the representation of $\Gamma_0(N)/\Gamma_1(N) \cong (\mathbf{Z}/N\mathbf{Z})^*$ on $\mathbf{C}f$ is given by a linear character of $(\mathbf{Z}/N\mathbf{Z})^*$, i.e., a Dirichlet character

$$\varepsilon:(\mathbf{Z}/N\mathbf{Z})^* \to \mathbf{C}$$

For a given N and ε we set

$$M(N,k,\varepsilon) = \{f \mid \text{(i) } f \in M(\Gamma(N),k) \quad \text{(ii) } f|_k\alpha = \varepsilon(d)f \text{ for all } \alpha = \begin{pmatrix} a & b \\ c & d \end{pmatrix} \in \Gamma_0(N)\}.$$

So $M(N,k,\varepsilon)$ is the space of all semi-invariants of $\Gamma_0(N)$ which transform according to the Dirichlet character ε. Note that each f in $M(N,k,\varepsilon)$ lives on $X_1(N)$, hence has a Fourier expansion at $i\infty$ of the form

$$f(z) = \sum_{n\geq 0} a(n)q^n.$$

We need one more function space. Namely, we know that each form f in $M(N,k,\varepsilon)$ is holomorphic and hence has a Fourier expansion with no principal part at each cusp. If in fact each of these Fourier expansion has no constant, so that f has expansion

$$f(\tau) = \sum_{n>0} a(n)\tau^n$$

in the local variable τ at the cusp in question, we call f a <u>cusp-form</u>. The subspace of all such cusp-forms is denoted by

$$S(N,k,\varepsilon).$$

<u>Examples</u>. We present some illustrative examples with a view to later applications. It will not necessarily be obvious that the examples we present have the desired properties -- we refer the reader to textbooks, e.g., [7], for proofs.

1. θ-<u>functions</u>. Let L be an even lattice of dimension $2r$, i.e., $L \cong \mathbf{Z}^{2r}$ and L carries a positive definite integral quadratic form δ taking only even values. We set

$$\theta_L(z) = \sum_{n\geq 0} a(n)q^n$$

where

$$a(n) = |\{x \in L | \delta(x) = 2n\}|.$$

Then $\theta_L(z) \in M(N,r,\varepsilon)$ where

$$\varepsilon(d) = (\frac{\Delta}{d}) \quad \text{(Kronecker symbol)}$$

and Δ = discr. L. Moreover N is essentially the exponent of L^*/L. For example if L is the E_8 root-lattice or the Leech lattice then L is self-dual, so $N = \Delta = 1$ and $\theta_L(z)$ belongs to $M(1,4,id)$ or $M(1,12,id)$ respectively. We have

$$\theta_{E_8}(z) = 1 + 240 \sum_{n\geq 1} \sigma_3(n)q^n,$$

$$\theta_{Leech}(z) = 1 + 196,560q^2 + \ldots,$$

where

$$\sigma_3(n) = \sum_{\substack{d|n \\ d>0}} d^3.$$

Actually it is not necessary to insist that L be an even lattice, though the theory is a little cleaner in this case.

Suppose now that L is not only even but unimodular $(\Delta = 1)$ and that L admits a group of isometries G of determinant 1. We may then also consider the fixed sublattice L^g for $g \in G$ and its θ-function. We then have

$$\theta_{L^g}(z) \in M(N_g,r_g,\varepsilon_g)$$

where N_g = order of g, $r_g = 1/2 \dim L^g$, ε_g = Kronecker symbol. We refer the reader to Lemma 2 of [8].

2. η-functions. We define the Dedekind η-function by the product

$$\eta(z) = q^{1/24} \prod_{n=1}^{\infty} (1-q^n).$$

Now let S be a finite set of cardinality divisible by 24, and denote by $A(S)$ the alternating group on the set S. Thus an element π of $A(S)$ is an even permutation of S and may be written as a product of disjoint cycles. Schematically we may write

$$\pi \longleftrightarrow 1^{j_1} 2^{j_2} \ldots$$

if π has j_i cycles of length i, $i \geq 1$, and in this way we associate π to a partition of $|S|$ since of course

$$|S| = \sum_{i \geq 1} i j_i.$$

We set

$$\eta_\pi(z) = \eta(z)^{j_1} \eta(2z)^{j_2} \ldots$$

the corresponding product of η-functions. As $24 \mid |S|$ we see that

$$\eta_\pi(z) = q^{|S|/24} + \text{higher terms}.$$

Proposition 1. $\eta_\pi(z) \in S(N_\pi, k_\pi, \varepsilon_\pi)$ where

k_π = 1/2 number of cycles of $\pi = 1/2 \Sigma j_i$

$N_\pi = o(\pi)h$ for some integer h (depending on π) dividing g.c.d. $(24, o(\pi))$.

ε_π = explicitly given real-valued Dirichlet character (mod N_π), $\varepsilon_\pi : (\mathbf{Z}/N_\pi\mathbf{Z})^* \to \pm 1$.

For example if $|S| = 24$ and π = identity then

$$\eta_{id}(z) = \Delta(z) = q \prod_{n=1}^{\infty} (1-q^n)^{24} = q - 24q^2 + 252q^3 - \ldots,$$

the famous Klein's modular form, or discriminant. We have

$\Delta(z) \in S(1,12,\mathrm{id})$.

We complete this section with a brief survey of <u>Hecke operators</u> -- needless to say we only scratch the surface of this important topic. For more complete information see, for example, Chapters II and VII of Lang's book [2]. We fix a space $S = S(N,k,\varepsilon)$ of cusp-forms, assuming that $S \neq 0$ so that there are non-zero forms $f \in S$, say

$$f = \sum_{n\geq 1} a(n)q^n.$$

For each positive integer d there are operators

$$U_d : f \longmapsto \sum_{d|n} a(n)q^{n/d}$$

$$V_d : f \longmapsto \sum a(n)q^{dn}$$

and for each positive integer m we define the mth Hecke operator on S to be

$$T(m) = \sum_{d|m} \varepsilon(d)d^{k-1}V_d \circ U_{m/d}.$$

For example if $m = p$ is prime then

$$T(p) = U_p + \varepsilon(p)p^{k-1}V_p.$$

It can be shown that the $T(m)$, $m \in \mathbb{N}$, generate a commutative algebra H of $\mathbb{C}$-linear operators on S, the <u>Hecke algebra</u>. This algebra can also be described in terms of the double cosets of $\Gamma_0(N)$ in $GL_2(Z)$. Furthermore S carries a natural non-degenerate form with respect to which the elements of H are unitary, so that S has a basis of <u>eigenforms</u>, i.e., forms which are simultaneous eigenvalues for the operators in H. The arithmetic significance of this is the following: to each $f \in S$ as above we may associate its Dirichlet series (or Mellin transform), namely

$$D_f = \sum_{n\geq 1} \frac{a(n)}{n^s}.$$

Here $s \in C$ is such that $\mathrm{Re}(s)$ is big enough to ensure the convergence of D_f. Hecke proved the following beautiful result: f is a (suitably normalized) eigenform if, and only if, D_f has an Euler product, in which case we have

$$\sum_{n\geq 1} \frac{a(n)}{n^s} = \prod_p \left(\frac{1-a(p)}{p^s} + \frac{\varepsilon(p)p^{k-1}}{p^{2s}} \right)^{-1},$$

the product running over all rational primes p. In particular a(n) is a multiplicative function.

For example, we have already remarked that $\Delta(z)$ lies in S(1,12,id), and in fact this space is 1-dimensional. Thus $\Delta(z)$ is necessarily an eigenform for $\mathcal{H}$, and we obtain that if

$$\Delta(z) = \sum_{n\geq 1} \tau(n)q^n$$

then

$$\sum_{n\geq 1} \frac{\tau(n)}{n^s} = \prod_p \left(1 - \frac{\tau(p)}{p^s} + \frac{p^{11}}{p^{2s}} \right)^{-1}.$$

This is a non-trivial result -- it was conjectured by Ramanujan and first proved by Mordell. We shall see an "equivariant" analogue in Section 4.

3 THOMPSON SERIES

Fix a finite group G, with R the ring of $\mathbb{Z}$-valued generalized characters of G. In its most general form, a Thompson series for G is a formal q-expansion

$$\Gamma_G = \sum_{n\geq M} \gamma_n q^n$$

satisfying the following:

a) Each $\gamma_n \in R$.

b) For each $g \in G$, the q-expansion

$$\Gamma_g = \sum_{n\geq M} \gamma_n(g)q^n$$

is a meromorphic modular form.

We retain this notation throughout. In this paper we focus attention on two kinds of Thompson series; we say that Γ_G is a holomorphic Thompson series if (a) and (b)' hold, and is a meromorphic Thompson series if (a) and (b)" holds. Here,

b)' For each $g \in G$,

$$\Gamma_g \in M(N_g, k_g, \varepsilon_g).$$

b)" For each $g \in G$,

$$\Gamma_g \in F_0(N_g).$$

This nomenclature is hardly ideal, but will suffice for present purposes. Following [1], we call γ_n the nth <u>Head</u> <u>character</u> of Γ (or G).

An important remark: as each $\gamma_n \in R$ we have $\gamma_n(g) \in \mathbb{Z}$ for $g \in G$, so we are dealing exclusively with modular forms with rational integral coefficients. So if $\langle g\rangle = \langle h\rangle$ then $\Gamma_g = \Gamma_h$.

<u>Examples</u>.

1. If $G = \langle 1\rangle$ consists only of the identity then a Thompson series for G is simply any modular form with rational integral coefficients.

2. Suppose that $G \cong Z_p$ has prime order p. Then G has just two rational conjugacy classes with representatives 1, g say, and the condition that $\gamma_n \in R$ is exactly that $\gamma_n(1) \equiv \gamma_n(g) \pmod p$. So Γ is a Thompson series for Z_p when

$$\Gamma_1 \equiv \Gamma_p \pmod p,$$

that is the corresponding forms are congruent (mod p).

The subject of congruence of modular forms is currently a very active subject in number theory -- see for example Ribet's talk at Warsaw [6]. One may take the view that the theory of Thompson series in general is an attempt at a non-abelian version of this.

3. (<u>Monstrous</u> <u>Moonshine</u>). Our whole subject arose from the following example: let F_1 be the Fischer-Griess Monster. Then there is a meromorphic Thompson series

$$M_{F_1} = M = 1_{F_1} q^{-1} + \sum_{n\geq 1} \mu_n q^n$$

with the following properties:

a) Each μ_n is a character of F_1.

b) Let $g \in F_1$. Then there is an integer N_g such that not only do we have

$$M_g \in F_0(N_g)$$

but also there is a group H_g such that

$$\Gamma_0(N_g) \trianglelefteq H_g \leq GL_2^+(\mathbb{R}),$$

such that H_g has genus zero (that is, X_{H_g} is topologically a sphere), and such that M_g is a normalized generator of the field of meromorphic functions on X_{H_g} (that is, each such meromorphic function is a rational function of M_g).

c) If g, N_g are as in (b) then

$$N_g = o(g)h$$

where h is a divisor of g.c.d. $(24, o(g))$ which depends on g. In particular, if g is the identity element of F_1 it follows that

d) $M_1 = j-744$, where

$$j(z) = q^{-1} + 744 + 196884q + \ldots$$

is the modular function.

As references for this result we refer the reader to [1], [9] and Fong's article in [5].

4. θ-functions. Let L be an even unimodular lattice with quadratic form δ as in Section 2, and let G be a group if isometries of L of determinant 1 (not all of these conditions are strictly necessary for what we wish to say). Let

$$L_n = \{x \in L \mid \delta(x) = 2n\}$$

and notice that L_n admits G as a group of permutations. If λ_n is the permutation character of G afforded by L_n we set

$$\theta_{L,G} = \theta = \sum_{n \geq 0} \lambda_n q^n.$$

It is not hard to see that if $g \in G$ then

$$\theta_g = \theta_{L^g} = \theta\text{-function of } L^g,$$

and the assertion that θ is a holomorphic Thompson series follows from the discussion of Example 1 in Section 2. We refer the reader to [8] for further discussion.

5. ETA-<u>functions</u>. We use the notation of Example 2 of the last section.

<u>Proposition 2</u>. There is a holomorphic Thompson series $\Omega_{A(S)} = \Omega$ for $A(S)$ such that for each $\pi \in A(S)$ we have

$$\Omega_\pi = \eta_\pi(z).$$

Recall that $\eta_\pi(z)$ is actually a cusp-form. This example is discussed at length in [3].

6. As a final example we state a theorem which bears a strong resemblance to Example 3 and which holds* for any finite group.

<u>Theorem 1</u>. G is a finite group. Then there is a positive integer d and a meromorphic Thompson series for G

$$M_G = M = 1_G q^{-d} + \sum_{n>-d} \mu_n q^n$$

with the following properties:

a) If $g \in G$ then $M_g \in F_0(N_g)$ where N_g satisfies the same conditions as (c) of Example 3.

b) There is a monic polynomial $p(x) \in \mathbf{Z}[x]$ of degree d such that

$$M_1 = p(j).$$

We should add that there is nothing unique about the Thompson series we have attached to G in this theorem. The difference between Monstrous Moonshine and the general case represented by Theorem 1 is that

---

*In theorem 1 one could, for example, take $p(x) = x$ and each μ_n to be multiplies of the trivial character. The point is that there are (in general many) ways to pick $p(x)$ so that the μ_n are non-trivial.

for F_1 the μ_n are characters (as opposed to mere generalized characters), the function M_1 corresponding to the identity element is j up to a constant rather than a polynomial in j, and finally the genus zero condition necessarily has no analogue in general. As Conway has impressed upon the author several times, the genus zero condition is the true source of Monstrous Moonshine, and Theorem 1 is an affirmation of this assertion.

4 HECKE OPERATORS ON THOMPSON SERIES

Once again we adopt the notation of Example 2 in Section 2, and for each $\pi \in A(S)$ we let N_π, k_π, ε_π be as in Proposition 1.

Theorem 2. Let d be an odd integer greater than 1. Then the function

$$\pi \to \varepsilon_\pi(d)d^{k_\pi-1}$$

is a generalized character of $A(S)$.

We remark that the theorem in general is false for even integers. We will not say anything about the proof of the theorem, though it is not without interest, but explain its bearing on the theory of Thompson series. We denote by ψ_d the generalized character of $A(S)$ given by Theorem 2 for odd d, so that

$$\psi_d(\pi) = \varepsilon_\pi(d)d^{k_\pi-1}, \quad \pi \in A(S).$$

Let Ω be the Thompson series of Proposition 2,

$$\Omega = \sum_{n\geq n_0} \alpha_n q^n$$

with α_n the corresponding head characters. We can then define for each integer d

$$U_d:\Omega \to \sum_{d|n} \alpha_n q^{n/d},$$

$$V_d:\Omega \to \sum \alpha_n q^{dn},$$

and for each odd integer m:

$$T(m)\Omega = (\sum_{d|m} \psi_d V_d \circ U_{m/d})\Omega.$$

We assert that $T(m)\Omega$ is again a Thompson series for $A(S)$. Indeed its coefficients are certainly generalized characters of $A(S)$ in view of the fact that each ψ_d is. Furthermore if $\pi \in A(S)$ then $\Omega_\pi = \eta_\pi$ is a cusp-form and we have

$$[T(m)\Omega]_\pi = T(m)\eta_\pi$$

where $T(m)$ on the right side of this equation is the usual Hecke operator acting on the cusp-form η_π. Thus our generalized Hecke operator $T(m)$ preserves Thompson series.

The value of this is the following: starting with the Thompson series Ω, we can construct many new Thompson series by applying the operators $T(m)$. It is in general difficult to construct interesting Thompson series, and this construction provides a lot, all potentially interesting! To get a better idea of this, first note that the cusp-form η_1 corresponding to the identity element is given by

$$\eta_1 = \Delta^{|S|/24}(z),$$

a form lying in $\mathcal{S} = \mathcal{S}(1,|S|/12,\mathrm{id})$. Now although the Hecke algebra $\mathcal{H}$ can be diagonalized on $\mathcal{S}$ as a C-space, it seems quite likely that it acts irreducibly on the Z-module of forms in $\mathcal{S}$ all of whose coefficients lie in Z (such forms form a lattice in $\mathcal{S}$). This being the case, we see that we can generate (via the $T(m)$'s) Thompson series such that the form corresponding to the identity element ranges over a lattice in $\mathcal{S}$.

Concerning the prime 2 we have the following result.

<u>Theorem 3</u>. Let S be a finite set of cardinality divisible by 24 and let G be one of the following groups:

i) $|S| = 24$ and $G \cong M_{24}$.

ii) $|S| - 1$ is a prime p and $G \cong L_2(p)$.

In each case consider the usual permutation representation of G on the set S. Then in previous notation, the function

$$\psi_2 : \pi \longmapsto \varepsilon_\pi(2)2^{k_\pi - 1}, \quad \pi \in G,$$

is a character of G.

So in these cases at least, we can define Hecke operators $T(m)$ for <u>all</u> $m \in \mathbb{N}$, applied to the basic Thompson series Ω. We finish with two illustrative examples.

<u>Example 1</u>. $\underline{G} \cong L_2(47)$: Here we are taking $|S| = 48$ and considering the permutation representation of G on 48 points. Let Ω be the corresponding Thompson series. Thus, for example,

$$\Omega_1 = \eta(z)^{48} = \Delta(z)^2 = q^2 - 48q^3 + 1080q^4 - 15040q^5 + \dots$$

$$\Omega_{47} = \eta(z)\eta(47z) = q^2 - q^3 - q^4 + q^7 + \dots$$

where the subscripts 1, 47 refer to the elements of G of that order.

We set $T(2)$ to be the Hecke generator whose existence is guaranteed by Theorem 3, with

$$\Lambda = T(2)\Omega = \sum_{n \geq 1} \lambda_n q^n$$

the "new" Thompson series. Thus

$$\Lambda_1 = T(2)\Delta^2 = q + 1080q^2 + 143{,}820q^3 + \dots$$

$$\Lambda_{47} = q - q^2 + q^4 + \dots$$

We set

$$M = \Lambda/\Omega = \sum_{n \geq 1} \mu_n q^n,$$

that is, the Thompson series obtained by formerly inverting Ω and multiplying by Λ. Since, for each $g \in G$, the forms Λ_g and Ω_g both have the same weight, level and character then the Thompson series

$$M_g = \Lambda_g/\Omega_g = \sum_{n \geq 1} \mu_n(g) q^n$$

is a modular function of the appropriate level.

A closer inspection of the poles of M_g, i.e., the zeros of Ω_g show that in fact M_g is a hauptmodul and in fact one can verify that, up to a constant, M_g coincides with one of the modular functions of Conway-Norton [1] associated to an element of the Monster of the same order as g. For example

$$\Lambda_1/\Omega_1 = q^{-1} + 1128 + 196884q + \ldots = j + 384.$$

So in effect we have constructed the head characters of F_1 restricted to an $L_2(47)$ subgroup. It is not known, however, if $L_2(47)$ is a subgroup of F_1!

Example 2. $G \cong M_{24}$: We take $|S| = 24$. Again let $\Omega = \Omega_{M_{24}}$ be the corresponding Thompson series.

Theorem 4. $\Omega_{M_{24}}$ is an eigenform for the Hecke operators $T(m)$. In fact if

$$\Omega_{M_{24}} = \sum_{n\geq 1} \alpha_n q^n$$

then

$$T(m)\Omega_{M_{24}} = \alpha_m \Omega_{M_{24}},$$

that is the mth head character of M_{24} is the eigenvalue for $T(m)$.

This result is discussed at length in [4], where it was proved without using Hecke operators. In fact part of the motivation for this paper was to be able to make statements like that of Theorem 4. The assertion of Theorem 4 involves a number of combinatorial identities involving the head characters of M_{24}. Also, since $\Omega_1 = \Delta$, Klein's modular form, $\Omega_{M_{24}}$ may be regarded as the equivariant analogue of Δ promised in Section 2.

Finally, we can form the Dirichlet series

$$D_{M_{24}} = \sum_{n\geq 1} \frac{\alpha_n}{n^s}$$

and deduce from the foregoing, just as in the original case, that there is an Euler product

$$D_{M_{24}} = \prod_p \left(1 - \frac{\alpha_p}{p^s} + \frac{\psi_p}{p^{2s}}\right)^{-1}.$$

In particular, the head characters for M_{24} are multiplicative!

REFERENCES

1. Conway, J. & Norton, S. (1979). Monstrous Moonshine. Bull. Lond. Math. Soc. 11, 308-339.
2. Lang, S. (1976). Introduction to Modular Forms. New York: Springer-Verlag.
3. Mason, G. Frame shapes and rational characters of finite groups. To appear in J. Alg.
4. Mason, G. M_{24} and certain automorphic forms. To appear in Proceedings of the Montreal Conference on finite groups.
5. Proceedings of the Santa Cruz Conference on finite groups. Ed. by B. Cooperstein and G. Mason. A.M.S., Symposium, vol. 37, 1980.
6. Ribet, K. (1983). Congruences between modular forms. I.C.M. Proceedings Warsaw.
7. Schoeneberg, B. (1974). Elliptic Modular Functions. New York: Springer-Verlag.
8. Thompson, J. (1979). Finite groups and modular functions. Bull. Lond. Math. Soc. 11, 347-351.
9. Thompson, J. (1979). Some numerology between the Fischer-Griess monster and the elliptic modular function. Bull. London Math. Soc. 11, 352-353.

AN APPLICATION OF ULTRAPRODUCTS TO FINITE GROUPS

R.H. Gilman*
Mathematics Department, Stevens Institute of Technology
Hoboken, NJ 07030

We would like to investigate the use of new techniques in finite group theory, namely techniques from logic. Since logic does not have much to say about finite structures, this combination may seem implausible; but it is not hopeless.

Consider the following conjecture (whose origin we do not know).

<u>Conjecture</u>. If G is a finite simple group, then every $g \in G$ is a commutator.

Even though all finite simple groups are known, it does not seem easy to check the conjecture directly. For any finite group G and element $g \in G'$ (where G' is the commutator subgroup of G), define $L(g)$ to be the minimum length of g as a product of commutators. From [3] and [4] we have the following results: First, the conjecture above can be checked from the character table of G; second, if G' has order $|G'| \leq [(n+2)!n!]/2$, then $L(g) \leq n$ for all $g \in G'$. Finally there is no finite upper bound on $L(g)$ as g ranges over all elements of all finite groups.

If G has an element g with $L(g) = n$, then certainly G contains h with $L(h) = m$ for all m with $1 \leq m \leq n$. Let π be a set of primes; we will consider how $L(g)$ behaves as g ranges over the π-elements of G. We are not so much interested in a particular result as we are in seeing how ultraproducts might be useful in finite group theory.

We begin with some lemmas. Let N be the natural numbers, G a finite group, and $\langle g \rangle$ the cyclic subgroup generated by $g \in G$.

<u>Lemma 1</u>. If $g,h \in G$ and $n \in N$, then $(gh)^n = g^n h^n k$ for some k with $L(k) \leq n$.

*I would like to thank the Rutgers Mathematics Department for its support during the special year on group theory.

Lemma 2. If $g \in G$ and $h \in \langle g\rangle$, then $L(h) \leq 3L(g)$; if $g \in G$ and $\langle h\rangle = \langle g\rangle$, then $L(h) = L(g)$.

Lemma 1 is proved by induction on n, and Lemma 2 follows from the fact that every $h \in \langle g\rangle$ is a product of at most 3 generators of $\langle g\rangle$. The finiteness of G is important in the proof of Lemma 2.

As above let π be a set of prime divisors of $|G|$. Let π' be the complementary set of primes. By a π-number we mean an element $n \in N$ whose prime divisors lie in π. A π-element of G is one whose order is a π-number, and likewise for a π-group. Finally let $e(G)$ be the exponent of G and let $e(G,\pi)$ be the largest factor of $e(G)$ which is a π-number.

Now we will use ultraproducts to investigate $L(g)$. Let $\{G_i | i \in N\}$ be an infinite sequence of finite groups, and let G^* be the ultraproduct (over a non-principal ultrafilter) of the G_i's. G^* is a certain quotient of the direct product $\prod_{i=1}^{\infty} G_i$, and we let P denote the projection onto G^*. Each $g \in G^*$ is $P(\{g_i\})$ for some sequence $\{g_i\}$ with $g_i \in G_i$. Let N^* be the corresponding ultrapower of N, and use P again to denote the projection from $\prod_{i=1}^{\infty} N$ to N^*. N^* is a nonstandard model of N, and we make N^* into an extension of N by identifying $n \in N$ with $P(\{n_i\})$ where $n_i = n$ for all i. N is the set of standard members of N^*.

The action of N on each G_i gives an action of N^* on G^*. If $n = P(\{n_i\})$ and $g = P(\{g_i\})$, then $g^n = P(\{g_i^{n_i}\})$. Since each G_i is finite, we can choose n_i so that $g^{n_i} = 1$ whence $g^n = 1$ too. The least n with this property is $|g|$, the order of g. If $\{\pi_i\}$ is a sequence of sets of primes, then $\pi = P(\prod_{i=1}^{\infty} \pi_i)$ is a set of pseudo-primes (that is, each $n \in \pi$ can only be factored trivially). Because $g_i \in G_i$ can be written as a product of a π-element and a π'-element, the same is true for $g \in G^*$. For the complete story on transferring results from $\{G_i\}$ and N up to G^* and N^* and back down, see [1, Chapter 5] or [2, Chapter 4] or [5, Chapter 6]. Introductions to nonstandard analysis are given in [6] and [7].

Proposition 1. There exists an integer $n_0 \in N$ with the following property: If G is a finite group, π a set of primes and g a

π-element of G', then one of the following occurs:

i) $L(g) \leq n_0 e(G,\pi')$

ii) For some π-element h, $L(h) < L(g) \leq n_0 L(h)$.

Here is a sketch of the proof of Proposition 2: Suppose the proposition is false and for each $j \in N$ pick G_j, π_j, and a π_j-element g_j such that

$$L(g_j) > je(G_j,\pi_j')$$

and there is no π_j-element $h \in G_j$ with

$$L(h) < L(g_j) \leq jL(h).$$

Define N^*, G^*, π as above and let $g = P(\{g_j\})$. Let $e = e(G^*,\pi')$. Our conditions imply that for all standard integers, j

$$L(g) > je \tag{1}$$

and there is no π-element $h \in G^*$ with

$$L(h) < L(g) \leq jL(h). \tag{2}$$

In particular $L(g)$ is not a standard integer. Because $L(h) = L(h^{-1})$ and $L(hk) \leq L(h) + L(k)$ for h, k in $G^{*\prime}$, the following sets are subgroups of $G^{*\prime}$:

$$J = \{h | L(h) < jL(g) \text{ for some } j \in N\}$$

$$K = \{h | L(h) < L(g)/j \text{ for all } j \in N\}.$$

Since J and K are subgroups, they are closed under raising elements to finite powers. By Lemma 2, J and K are closed under raising elements to arbitrary powers. In particular if $h \in J$, then $h = pq$ where p is a power of h which is a π-element and q is a power of h which is a π'-element, and p and q both lie in J.

As K contains all commutators in G^*, K is normal and G^*/K is abelian. Let $\bar{J} = J/K$ and use bars to denote images in $\bar{J}$. Define

$$S = \{h | L(h) < L(g)/3\}.$$

$\bar{S}$ is a set of generators for $\bar{J}$. Suppose $h \in S$ and $h = pq$ as above. By Lemma 2 $L(p) < L(g)$, so by (2) above $p \in K$. By Lemma 1 and (1) h^e is in K for any $h \in S$. Likewise $\alpha: \bar{J} \to \bar{J}$, $\alpha(\bar{h}) = \bar{h}^e$ is well defined and is an endomorphism of $\bar{J}$. As α maps $\bar{S}$ to $\bar{1}$, α is trivial on $\bar{J}$. But since e is relatively prime to the order of g, $\alpha(\bar{g}) \neq \bar{1}$, and this contradiction establishes Proposition 1.

Proposition 1 says something about the values of $L(g)$ on the set of π-elements. Investigating $\bar{J}$ further might reveal more information. Our construction of $\bar{J}$ follows [8], and we have not used the fact that $\bar{J}$ is a complete arcwise connected normed group with norm induced by L.

By a general result of logic any statement about finite groups proved using ultraproducts has a standard proof by deductions from the axioms for N and the definition of finite group. Even when the standard proof is at hand, ultraproducts and other techniques from logic are still valuable. They offer a new language for discussing finite groups, and recasting problems in a new language can give new insights.

REFERENCES

1. Bell, J.L. & Slomson, A.B. (1971). Models and Ultraproducts, North-Holland, Amsterdam.
2. Chang, C.C. & Keisler, H.J. (1977). Model Theory, 2nd ed. North-Holland, Amsterdam.
3. Gallagher, P.X. (1962). Group characters and commutators. Math. Z. 79, 122-126.
4. Gallagher, P.X. (1965). The generation of the lower central series. Canad. J. Math. 17, 405-410.
5. Gratzer, G. (1979). Universal Algebra, 2nd ed., New York: Springer Verlag.
6. Machover, M. & Hirschfeld, J.H. (1969). Lectures on Non-Standard Analysis. Lecture Notes in Mathematics 94. Berlin: Springer Verlag.
7. Stroyan, K.D. & Luxemburg, W.A.J. (1976). Introduction to the Theory of Infinitesimals. New York: Academic Press.
8. van den Dries, L. & Wilkie, A.J. On Gromov's theorem concerning groups of polynomial growth. Preprint.

FIXED-POINT-FREE AUTOMORPHISM GROUPS

D. Parrott
Department of Mathematics, University of Adelaide
Adelaide, South Australia

Let A be a group of automorphisms of the finite group G and define $C_G(A) = \{x \in G | x^a = x \ \forall a \in A\}$. If $C_G(A) = 1$ then A is a _fixed-point-free_ group of automorphism of G. Throughout this note we assume that

A _is a fixed-point-free group of automorphism of_ G and consider the general problem:

1) _For given_ A, _determine the structure of_ G.

At the turn of the century, Frobenius conjectured that if $A \cong Z_p$ (p prime) then G is nilpotent. Two special cases of this conjecture appeared in Burnside's book of 1911:

If $|A| = 2$ _then_ G _is abelian_, and _if_ $|A| = 3$ _then_ G _is nilpotent of class at most_ 2.

In 1959 J.G. Thompson proved the most important result on fixed-point-free automorphism groups:

If $|A| = p$, p _prime_, _then_ G _is soluble_.

As it had been known for many years that if $|A| = p$, p prime, and G is soluble then G must be nilpotent, Thompson's theorem completed the proof of Frobenius' conjecture. As in the proof of Frobenius' conjecture, (1) has usually been considered as two separate questions:

1a) _For a given group_ A _show that_ G _is soluble_;

1b) _For given_ A _and soluble_ G, _determine the structure of_ G -- _in particular, determine the nilpotent height of_ G.

Since Thompson's result there have been a number of solutions to (1a) for particular choices of A. We list some examples:

$A \cong Z_4$ (Gorenstein-Herstein, see [2]);

$A \cong Z_2 \times Z_2$ (Glauberman, see [2]);

A elementary abelian (Martineau [4]);

$|A| = rst$, with $(|A|, |G|) = 1$ (Rowley [6]).

In all these cases A is abelian. Recently, B. Dolman and the author have considered this problem for non-abelian A, namely $A \cong S_3$, the symmetric group on 3 letters. In his Ph.D. thesis [1], B. Dolman proved that if $A \cong S_3$ and $(|A|, |G|) = 1$ then G is soluble; while in [5] the author showed (1a) holds for $A \cong S_3$ and G of even order coprime to 3.

If one considers a minimal counterexample G to either of these two results it is easily seen that G is a non-abelian simple group. Thus the theorems are a consequence of the classification of simple groups, or more precisely the classification of simple groups of order coprime to 3. Of course both theorems are proved directly, using the existence of the automorphism group A.

The two key results in the proofs of these theorems are Glauberman's Z(J)-theorem (see [2]) and the following result of E. Schult [7] (which is the solution to (1b) for $A \cong S_3$):

_If_ $A \cong S_3$ _and_ G _is soluble then_ G' _is nilpotent_.

In [1] B. Dolman uses these two results to analyze the structure of a maximal A-invariant subgroup of G (note that if G is a minimal counterexample, any proper A-_invariant_ subgroup of G is soluble). This information is used to study the centralizer $C_G(a)$ of an involution $a \in A$. As $C_G(a)$ is not A-invariant, it is difficult to determine the structure of $C_G(a)$ -- for example $C_G(a)$ may be non-soluble. (It is here that Glauberman's Z(J)-theorem is particularly useful.) He is eventually able to derive a contradiction using the structure of the maximal A-invariant subgroups, $C_G(a)$ and $C_G(x)$ for various p-elements x of G.

There are a number of ways these results may be extended. For example, consider $A = D_{2p}$ (the dihedral group of order $2p$, p prime) and $(|G|, p) = 1$. Unfortunately the structure of a soluble group G which admits $A \cong D_{2p}$ (fixed-point-free) is not as simple (if $p > 3$) as for $A \cong S_3$. In fact Schult [7] shows that the only groups A which force a soluble group G to have G' nilpotent are $A = Z_2 \times Z_2$, S_3 or Z_p (p prime).

More generally one could consider A non-abelian of order pq (p,q distinct primes). In the same paper Schult shows for such A

and soluble G that G has nilpotent length at most 2. Thus the main problem here would be that Glauberman's theorem could not be applied to arbitrary subgroups of G (unless of course $|G|$ is assumed to be odd or of order coprime to 3). Perhaps the most likely cases which could be considered under the method outlined above would be $A \cong D_{2p}$ or perhaps $A \cong A_4$ (the alternating group on 4 letters) and $(|A|, |G|) = 1$.

If $A \cong S_3$, the assumption that $(|G|, 3) = 1$ could be dropped. However in this case, the most basic result in the study of fixed-point-free automorphism groups does not hold; namely the existence of a unique A-invariant Sylow p-subgroup of G for and prime $p||G|$. The same difficulty occurs if $A \cong D_{2p}$ and $p||G|$ or $A \cong A_4$ and $|G|$ is even.

Finally, the successful solution of some of these problems could then be used to consider $A \cong A_5$. The structure of a soluble group G with $A \cong A_5$ has been investigated in [3], and the nilpotent length is at most 5.

REFERENCES

1. Dolman, B. (1983). Groups admitting a fixed-point-free group of automorphisms isomorphic to S_3. Ph.D. Thesis, University of Adelaide.
2. Gorenstein, D. (1968). Finite Groups. Harper and Row, New York.
3. Kurzweil, H. (1983). Die Gruppe A_5 als fix punktfreie Automorphismengruppe. Illinois J. Math. 27, 67-76.
4. Martineau, R.P. (1972). Elementary abelian fixed-point-free automorphism groups. Quart. J. Math. Oxford (2) 23, 205-212.
5. Parrott, D. Finite groups which admit a fixed-point-free automorphism group isomorphic to S_3. To appear.
6. Rowley, P. (1981). Solvability of groups admitting a fixed-point-free automorphism of order rst, I. Pacific J. Math. 95 (2), 12-46.
7. Schult, E. (1966). Nilpotence of the commutator subgroup in groups admitting fixed-point-free operator groups. Pacific J. Math. 17, 323-347.